# ADVANCED PROGRAMMING AND PROBLEM SOLVING WITH PASCAL

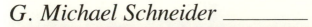

G. *Michael Schneider* ———— *Steven C. Bruell*

# ADVANCED PROGRAMMING AND PROBLEM SOLVING WITH PASCAL

**John Wiley & Sons**    *New York*    *Chichester*    *Brisbane*    *Toronto*

*Library of Congress Cataloging in Publication Data:*

Schneider, G. Michael.
   Advanced programming and problem solving with
PASCAL.

   Includes bibliographies and index.
   1. PASCAL (Computer program language)
2. Structured programming.  I. Bruell, Steven C.
II. Title.

QA76.73.P2S35     001.64'24     81-1344
ISBN 0-471-07876-X           AACR2

Printed in the United States of America

10 9 8 7 6 5 4 3 2 1

To *Ruthann, Benjamin, Rebecca and Sandy*

# PREFACE

This textbook is designed for use in a second course in computer programming at the undergraduate level. It follows closely the outline of the course entitled CS2, Computer Programming II, as described in ACM Curriculum '78. (*Communications of the ACM,* March 1979.) The only background required is a one-semester, college-level programming course that introduces some fundamental concepts of computer programming and any high-level programming language. Anything similar to the course CS1, Computer Programming I, as described in Curriculum '78, would be appropriate.

Our primary goal is to prepare students to address and manage *all* aspects of the program development process and to be comfortable and productive in the environment of large, "real-world" programming projects. Of necessity, the first course in computer programming usually discusses at length the syntax of a particular language. It rarely does justice to the many other important concerns that are a fundamental part of programming. In this text we introduce new programmers to a wide range of additional problems concerning the specification, design, coding, implementation, testing, and maintenance of software. If the first course is viewed as teaching students how to *code,* this book should be viewed as teaching students how to *program* and how to program *well*.

The five parts of this text are organized in the following manner.

1. Part I introduces, as a second language, the block-structured language Pascal developed by Niklaus Wirth and in widespread use throughout the world. This language is used throughout the remainder of the text to teach the principles of programming. Students who were introduced to Pascal in their first course may omit Part I or use it as a language review.

2. Part II introduces the principles of programming style, expression, and elegance. Every program, subprogram, and program fragment used in the text will follow the style guidelines developed here. Students will see numerous examples of what we feel is well-written code and will begin to understand the importance of good programming style in reading, understanding, and working with computer programs.

3. Part III introduces important program design concepts such as structured coding, modularization, stepwise refinement, programming teams, and the top-down design of both algorithms and data structures. The exercises in this part describe several large, team-oriented programming projects designed to give students practice in using these newly learned design techniques.

4. Part IV introduces additional advanced concepts in the area of data structures—pointers and linked structures, files, stacks, queues, strings, and com-

plex multilevel data structures. We also present basic algorithms for manipulating these data structures (e.g., additions, deletions, searching, sorting).

5.  Finally, Part V introduces and studies problems arising during the implementation phase of programming—efficiency, analysis of algorithms, debugging techniques, testing, formal verification, documentation standards, and program maintenance.

Too often, instructors of second programming courses have been forced to select course readings from different and unrelated sources, each of which addressed only one of the preceding topics. The course material might typically come from (1) a language reference manual, (2) a book on programming style, (3) a book on structured programming and top-down design, and (4) tutorials or journal articles on testing methods, verification techniques, and the analysis of algorithms. This approach can be expensive as well as inconvenient to the student, who must try to organize and assimilate unrelated readings.

We have tried to solve this problem by collecting together and integrating into a single text the set of topics that ACM has described as appropriate to a second programming course. This text offers a unified view of the important but eclectic set of subjects that, taken together, are called advanced programming.

One other pedagogical tool that we have utilized and that we feel is very important is the use of large, complex case studies to illustrate programming principles. Frequently, textbooks will illustrate the need for some advanced design tool, programming technique, or stylistic aid using "toy" programs that contain as little as 25 to 50 lines of code. Students then become confused because the need for such a high level of sophistication is not at all apparent from this simplified program. We have tried to illustrate many of the key points in the book using programs whose complexity requires and necessitates such development tools. Students will then be able to see the importance of these programming techniques and the chaos that can result from failing to utilize them. For example, the case study at the end of Chapter 6 presents the complete specification, design, and top-down implementation of a discrete event simulation program containing about 750 lines of code. At the end of Chapter 9 we have designed and coded a text editor that illustrates the use of some interesting data structures. That program contains 43 modules and 1450 lines of code.

These (and other) programs demonstrate the complexity of large software projects and also provide the opportunity for interesting, nontraditional programming problems other than the usual "write a program." Exercises at the end of some chapters request students to read and review these case studies and then modify and/or extend them to meet new specifications. These maintenance projects can also be the basis for team-oriented assignments to give students additional experience in the problems of effectively working and communicating with other programmers.

We thank several people who assisted us in the preparation of this book. Steve Collins coded and tested the text editor in Chapter 9. Jon Spear debugged and tested all the other programs; in addition, he made many helpful suggestions on program structure and organization and gently chided us when the code did not rigidly adhere to our own stylistic guidelines. Terrie Christian assisted in the typing of the final manuscript. We specially thank Sandra Whelan, who is truly an "artist with a blue pencil"; Sandy edited and reedited the manuscript until it met her high standards.

*G. Michael Schneider*
*Steven C. Bruell*

# CONTENTS

*Part I*

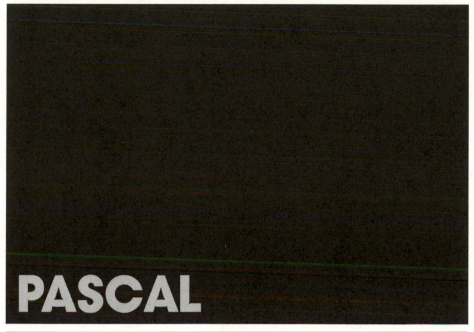

PASCAL

# Chapter 1

# THE PASCAL LANGUAGE

## 1.1 INTRODUCTION

This book assumes that readers know how to code in some high-level programming language. Its purpose is to teach how to program and to teach how to program and code *well*.

It is important to distinguish between coding and programming. *Computer programming* is the entire series of steps involved in solving a problem on a computer. While this, of course, includes *coding*—the process of writing statements in a particular computer language—it also involves many other steps. We must first select the method to solve the problem (called the *algorithm*) and then organize and outline the proposed solution. After coding the solution, we must locate and remove all errors, test the program for correctness, and finish writing whatever documentation is necessary to insure that the program can be easily understood, used and, if necessary, changed. Figure 1.1 summarizes the steps in the overall programming process. (Note that these steps are not sequential; most will overlap.)

In addition to introducing programming operations, we will teach how to program and code well. Too often, basic programming courses stress correctness as their single objective. Users begin to think that they have satisfactorily completed a task as soon as the program produces a correct answer. However, while correctness is of primary importance, it is not the only basis on which to judge a program. We will be introducing many criteria to evaluate the ''beauty'' and ''elegance'' of a computer program as well as the reasons why these characteristics are important.

To introduce and teach these two basic objectives—how to program and how to program well—we will be using the language called Pascal.[1] Pascal was developed

---

[1]Pascal is *not* an acronym and is therefore not written in capital letters. It was named after the French mathematician and religious fanatic, Blaise Pascal (1623-1662).

1. *Defining the Problem.* Developing a clear, unambiguous set of specifications describing exactly what the problem is, what input will be provided, and what results are desired.
2. *Outlining and Structuring the Solution.* Developing an outline of the overall solution, including descriptions of the various tasks and subtasks and their interrelationships.
3. *Selecting Solution Methods.* Selecting the best algorithms and data structures to perform each task.
4. *Coding.* Selecting a programming language and translating algorithms into that language.
5. *Debugging.* Locating and correcting all errors in the program.
6. *Testing and Verifying.* Guaranteeing to some level of satisfaction that the program is working properly and will give correct results for all cases.
7. *Documenting.* Producing written documents for the end-user and the programmers who may have to work with the program in the future.
8. *Maintaining the Program.* Keeping the working program current. This involves correcting mistakes discovered after the program has been running as well as modifying the program in response to changes in problem specifications.

---

*Figure 1.1.* The steps involved in computer programming.

by Professor Niklaus Wirth at the Eidgenossische Technische Hochschule (ETH) in Zurich, Switzerland. Work on the language began around 1970. By the latter part of the decade it had become extremely popular as an introductory programming language in colleges and universities and as a production programming language in industry and government.

One might reasonably ask why one should expend the effort to learn yet another language when our initial assumption was that readers were already familiar with at least one high-level programming language—probably FORTRAN, BASIC, or COBOL. However, Pascal has three features that make it far superior for introducing advanced concepts in programming.

1. *Richness of Control Structures.* Pascal offers a wide range of iterative and conditional control statements: **repeat-until, while-do, if-then-else, for-do, case-end, begin-end, goto.** This facilitates the writing of well-structured, readable programs, which will be of critical importance throughout this text.
2. *Richness of Data Structures.* Aside from the simple array, Pascal provides record, set, file, pointer, subrange, and user-defined data types. These data structures may themselves be recursively combined to form extremely complex (and interesting!) data structures.
3. *The Support of Some Interesting Programming Concepts.* Pascal directly supports certain concepts that are interesting to discuss in advanced programming classes, such as recursion and call-by-reference/call-by-value parame-

ters. This allows students to use programming ideas presented in the classroom and text.

Pascal is not the only language that provides these facilities. However, other languages that do (e.g., PL/1 and ALGOL-68) are *so* rich in programming features that they place enormous demands on machine resources to support them and human resources to learn them. Pascal, because of its modest size and data-typing facilities, meets the goals described by Edsger Dijkstra when accepting the 1972 ACM Turing Award (the highest award in Computer Science).

> A lesson we should have learned from the past is that the development of richer or more powerful programming languages was a mistake in the sense that these baroque monstrosities, these conglomerations of idiosyncrasies, are really unmanageable, both mechanically and mentally. I see a great future for very systematic but very modest programming languages.

## 1.2  BASIC PASCAL CONCEPTS

The remainder of this chapter will introduce the Pascal programming language. Because of limited space, it cannot replace a complete language reference manual. We will provide a sufficiently detailed description so that readers will be able to understand the programs and concepts introduced in the following chapters. However, for more specific information and additional detail about the language, readers should refer to the bibliography at the end of this chapter.

### 1.2.1  Scalar Data Types

A scalar data type is a set of ordered constants. That is, every pair of constants, $c_1$ and $c_2$, will satisfy one of these three conditions: $c_1 < c_2$, $c_1 = c_2$, or $c_1 > c_2$. There are four predefined scalar types in Pascal. (Figure 1.6 summarizes the relationships among the data types of Pascal.)

**1.2.1.1  Integers.**   The *integers* are the signed or unsigned whole numbers up to some implementation-defined limit, a constant called maxint. (For example, on the Control Data CYBER/74, maxint is $2^{48} - 1$.) Some examples of valid Pascal integers are:

*1487*      (but not 1,487)
*− 1*
*+300*
*maxint*
*0*

**1.2.1.2  Reals.**  The *reals* are the set of implementation–defined decimal values. There are two ways to represent real constants. In *decimal notation,* a real number consists of a sign (optional), a whole number, a decimal point, and a whole number. There must be at least one digit on either side of the decimal point.

| *Valid*[2] | *Invalid* | |
|---|---|---|
| 3.141592728 | 100 | (a valid integer, but not a valid real) |
| − 800.0 | 15. | (no digit to the right of the decimal point) |
| + 13.75 | .0215 | (no digit to the left of the decimal point) |
| − 0.00123 | | |
| 1234567.89000 | | |

A shorthand method for writing reals is called *scientific notation.* The number is written as a value (called the *mantissa*) times 10 to the appropriate power (called the *characteristic*). The mantissa may or may not have a decimal point but, in either case, the resulting numeral is of type real. The characteristic may *only* be of type integer.

| *Valid*[3] | | *Invalid* | |
|---|---|---|---|
| 15.31E4 | (153,100) | + 15E.5 | (only whole number characteristics) |
| + 7.424E − 2 | (0.07424) | − 163.E + 5 | (if the mantissa includes a decimal point, there must be digits on both sides) |
| − 9.0E + 12 | (−9,000,000,000,000) | | |
| 6E − 5 | (.00006) | | |

**1.2.1.3  Boolean.**  The elements of the scalar data type boolean are the two constants true and false. In Pascal, all scalar types must be ordered. The ordering of the boolean data type is:

*false < true*

**1.2.1.4  Character.**  The elements of the scalar data type character (abbreviated char in Pascal) are all the individual characters that can be represented on a

---

[2]These numbers are valid only if they do not violate the machine–dependent precision and range constraints of your particular computer.
[3]Ibid.

specific machine. Unfortunately, this will differ on different machines, and no universal standard exists. The mapping of the specific characters of a character set onto the integers 0, 1, 2, . . . , N is called the *collating sequence*. The collating sequence for some well-known code sets is given in Appendix C.

To indicate an element of type char, we surround it with apostrophes (single quotation marks). To indicate the apostrophe character, we simply write it twice.

> 'A'
> 'Z'
> '>'
> '1'
> '''' (the single apostrophe character)

Remember that elements of the scalar data type char are always single characters. Constructs such as 'this book' are elements of a more complex data type called a *string*, which will be discussed in Chapter 7.

The syntax of the four basic scalar data types is summarized in Figure 1.2. To

Integer:

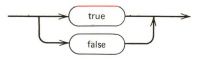

Real:

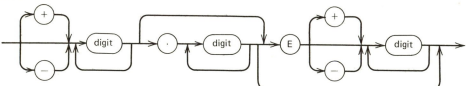

Boolean:

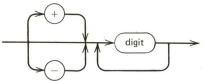

Character:

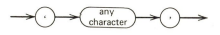

**Figure 1.2.** Syntax of the standard scalar data types.

introduce the syntax of statements in Pascal, we will be using a technique called a *syntax chart*. Each chart is composed of two specific classes of boxes.

1. *Rounded Boxes (Circles, Ellipses).* These boxes contain *terminal symbols* that will not be further defined. These objects correspond to things such as constants (true, . . .), operators (+, −, . . .), grouping marks (,[,],), or concepts that the reader is assumed to understand (e.g., digit).

2. *Rectangles.* These boxes contain nonterminal symbols that are further defined in some other named syntax chart.

These boxes are interconnected by directional lines. An object is said to be *syntactically valid* if you can generate it by proceeding through its syntax chart from beginning to end. For example, given:

Nonsense statement:

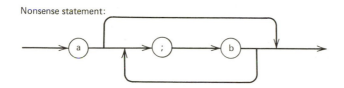

the following are syntactically legal "nonsense statements": a, a;b, a;b;b and a;b;b;b. However, b;a, a;, and a;b;c are all invalid.

## 1.2.2 Operators

The operators that apply to the scalar data types are as follows.

| | |
|---|---|
| + | Addition. |
| − | Subtraction (or unary negation). |
| * | Multiplication. |
| **div** | Integer division (divide and truncate). |
| **mod** | Modulo. A **mod** B is the integer remainder after dividing B into A. 9 **mod** 5 = 4. |
| / | Real division (divide and retain the fractional part). |
| > | Greater than. |
| >= | Greater than or equal to. |
| < | Less than. |
| <= | Less than or equal to. |
| = | Equal. |
| <> | Not equal. |
| **and** | Logical conjunction (A **and** B is true if and only if both A and B are true). |

or       Logical disjunction (A **or** B is true if and only if either A or B or both are true).

not      Logical negation (**not** A is true if and only if A is false).

Figure 1.3 summarizes the Pascal operators, the data types to which they apply, and the data type of the result. The last column indicates whether or not, for a particular binary operator, it is acceptable to mix an integer and real value and, if so, what the data type of the result will be. All other *mixed modes* (i.e., operands of incompatible types) are considered illegal in an operation.

### 1.2.3   Standard Functions

Pascal automatically provides a number of *standard functions* as part of the language, regardless of the particular implementation on which you are running. A particular installation may, however, choose to expand on this minimal set and provide additional standard functions not listed in Figure 1.4. The syntax for using any of these standard functions is:

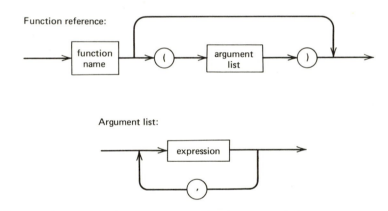

The set of available function names, the data type of the argument, and the data type of the result are summarized in Figure 1.4. There are two special types of functions in Pascal. Functions that return a value of a different type from that of the original argument (e.g., trunc, ord, or chr) are usually called *transfer functions*. Functions that return a boolean value (e.g., odd) are called *predicates*.

*Identifiers* or *names* are used throughout Pascal to name specific objects. These named objects can be a number of different things (e.g., programs, subprograms, new data types, variables, constants). However, in all cases the rules for forming an identifier are the same.

1.   The first character must be an uppercase or lowercase alphabetic character: 'A' .. 'Z', 'a' .. 'z'.

|  | Operand Type | | | | | |
|---|---|---|---|---|---|---|
| Operator | Integer | Real | Boolean | Char | Mixed Int-Real | |
| + | I | R | — | — | R | |
| − (unary or binary) | I | R | — | — | R | |
| * | I | R | — | — | R | |
| **div** | I | — | — | — | — | Data |
| **mod** | I | — | — | — | — | type of |
| / | R | R | — | — | R | the |
| > >= < <br> <= = < > } | B | B | B | B | B | result |
| **and** | — | — | B | — | — | |
| **or** | — | — | B | — | — | |
| **not** | — | — | B | — | — | |

(— = not allowed)

*Figure 1.3.* The basic Pascal operators.

2. All succeeding characters must be alphabetic or numeric. No special characters or punctuation marks are allowed.
3. No embedded blanks are allowed.
4. Identifiers may be as long as desired. However, some compilers may insist that the name be unique within the first N characters where N is a system-dependent parameter (usually 8 or 10). The syntax of an identifier is:

Identifier:

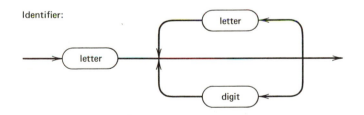

There are three specific classes of identifiers in Pascal. *Reserved identifiers,* sometimes called *reserved words,* are names reserved for a very specific and predefined purpose; they cannot be used for any other reason. For example, **and, div, mod, not,** and **or** are reserved identifiers and cannot be used, for example, as the name of a user variable. The list of all 35 reserved words appears in Appendix B.1. In this text, all reserved words within programs will be set in **bold type** to denote their status.

*Standard identifiers* are also used for a predefined purpose but may be redefined by the user. False, true, maxint, and all the standard function names (abs, arctan,

| Function Name | Argument Type | | | | Meaning |
|---|---|---|---|---|---|
| | Integer | Real | Boolean | Character | |
| abs | I | R | — | — | Absolute value |
| sqr | I | R | — | — | The square of the argument |
| trunc | — | I | — | — | Truncated integer value of a real trunc (5.83) = 5 |
| round | — | I | — | — | Rounded integer value round (5.83) = 6 |
| sin | R | R | — | — | Sine (radians) |
| cos | R | R | — | — | Cosine |
| arctan | R | R | — | — | Arctangent |
| ln | R | R | — | — | Natural logarithm (base e) |
| exp | R | R | — | — | Exponential function ($e^x$) |
| sqrt | R | R | — | — | Square root |
| ord | I | — | I | I | The *ordinal* function gives the ordinal position of a character in the internal collating sequence. In ASCII: ord('L') = 76 |
| chr | C | — | — | — | The *character* function gives the character corresponding to a specific integer. In ASCII: chr(35) = '#' |
| pred | I | — | B | C | The *predecessor* function. The constant immediately preceding the argument in the data type. In ASCII: pred('&') = '%' pred(8) = 7 |
| succ | I | — | B | C | The *successor* function. The constant immediately following the argument in the data type. succ(false) = true succ(−10) = −9 |
| odd | B | — | — | — | True if argument is odd, false otherwise. |

Data type of the result

**Figure 1.4.** Standard Pascal functions.

etc.) are examples of standard identifiers. The complete list of 39 standard identifiers is given in Appendix B.2. One important point to remember is that if the user redefines a standard identifier, the original meaning of that identifier is lost. If we choose to name a variable sqrt, we are no longer able to invoke the automatically provided square root routine in the Pascal library.

Finally, *user identifiers,* or *user names,* are any identifiers that do not fall into either of the previous two categories. (This implies that in this text we will never declare a user identifier with the same name as a standard identifier.) As we said earlier, they will be used to name a wide range of objects in Pascal.

### 1.2.4  A Pascal Program

A Pascal program will always be organized like the model in Figure 1.5. It is composed of three very distinct sections: the program heading, the declaration section, and the statement section. Figure 1.5 also points out some other key aspects of a Pascal program.

1. Pascal programs are totally free format, with no restrictions on columns or spacing. The selection of a format is based totally on legibility and readability. More than one statement may appear on a line. However, certain formats will obviously be more readable than others. Chapter 2 will develop stylistic guidelines for writing Pascal programs.

2. A comment is indicated by enclosing it within the brace characters: { }. Since many character sets do not include braces, almost all versions of Pascal will also accept the character pairs (* *) as comment delimiters. Comments may

```
program name (file₁, file₂ . . . ,fileᵣ); {program heading}
label declaration; {beginning of declaration section}
const declaration;
type declaration;
var declaration;
procedure declaration; {procedures and functions may appear in any order}
function declaration; {end of declaration section}
begin {beginning of the statement section}
     statement;
     statement;

        .
        .
        .

     statement
end. {end of the statement section}
```

*Figure 1.5.* The organization of a Pascal program.

be as long as desired, contain any character sequence (except } or *)), and appear wherever a blank space could properly appear.

3.    Semicolons are used as statement *separators,* not terminators. That is, they are used to separate statements from each other. If $s_1$, $s_2$, and $s_3$ represent any valid Pascal statements then

$s_1; \ s_2; \ s_3$

shows a valid use of semicolons. However, the semicolons indicated next by the arrows are unnecessary because they are not being used to separate two statements but a statement and a reserved word.

**begin** *; $s_1$ ; $s_2$ ; $s_3$ ;* **end**    .
    ↑            ↑

To make sense of this sequence, Pascal will insert *null statements* into the program as needed.

**begin** *null ; $s_1$ ; $s_2$ ; $s_3$ ; null* **end**

4.    Finally, Pascal programs terminate with a '.'.

## 1.3    THE PROGRAM HEADING

The *program heading,* as shown in Figure 1.5, provides a name for the entire program. The name can be any valid user identifier. The heading also lists the specific external data files to be used by the program to communicate with the outside world. Two standard files, input and output, are provided automatically by Pascal and correspond to the standard input and output devices—usually a keyboard and CRT or printer. We will limit our initial attention to these two files and will use the standard heading:

**program** *name (input,output);*

Chapter 9 will discuss further the file data type and files other than the two named here. In general, the syntax of the program heading is:

Program heading:

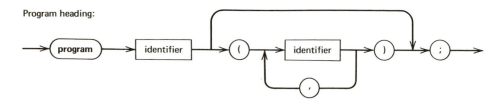

## 1.4   THE DECLARATION SECTION

The *declaration section* of a Pascal program consists of up to six specific types of declaration. All of them are optional. If any are present, however, each one must appear only once, in the order indicated in Figure 1.5 (except procedures and functions, which may appear in any order). It is important to remember that the declaration section simply contains static descriptions of various classes of data and not program statements to be executed by the computer. Those executable statements appear in the succeeding section of the program.

The remainder of this section will describe the syntax of the six available declarations. Examples of their proper use will appear throughout the book.

### 1.4.1   The Label Declaration

A *label* identifies an executable statement within a program. In Pascal, a label is a nonnegative, unsigned integer constant from one to four digits in length. To attach a statement label to any valid Pascal statement, precede that statement with the desired label followed by a colon. For example:

$$50: \quad s_1;$$
$$s_2;$$
$$999: \quad s_3$$

All labels so used must always be declared as follows.

Label declaration:

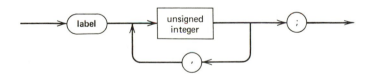

Referring to the previous example, our program would require a **label** declaration of the following form to be valid.

**label** *50,999;*

We cannot stress too early that labels are *rarely*, if ever, needed in a properly written Pascal program. If used, one or two are usually sufficient. The use of numerous statement labels is an early warning sign of a poorly organized program. This point will be repeatedly emphasized and discussed in detail in Chapter 4.

### 1.4.2  The Const Declaration

The **const** declaration assigns a name to a scalar constant. To improve readability, we may then use the name throughout the program in place of the actual constant.

Const declaration:

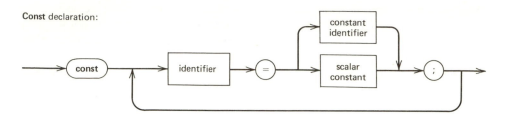

For example:

**const**

| | | |
|---|---|---|
| endofline | = ' '; | { end of line character } |
| hours | = 40; | { average hours worked per week } |
| minimumwage | = 3.10; | { the Federal minimum wage } |
| pi | = 3.14159; | { the value $\pi$ } |
| skyisblue | = true; | |
| upperbound | = maxint; | |
| lowerbound | = −upperbound; | |

We may now refer to these constants by their designated identifiers, as long as we remember that these names still represent constants and, therefore, may not be changed or modified in any way.

### 1.4.3  The Type Declaration

One of the most important concepts in Pascal and, indeed, in any programming language, is the *data type*. A data type is a set of data objects that "belong together" in the sense that all elements of one data type are treated uniformly and can be operated on by some uniform set of operators in a meaningful way. A good example of a data type is the set of integers. They are all formed in a uniform way and can be meaningfully combined using the standard integer operators: +, −, *, **div**, and **mod.** In addition, there are certain rules for determining whether an individual item belongs to a specific data type.

A data type can be thought of as a "template" for creating data elements of a particular form. Once we have created the data type, we can then create objects (e.g., variables, constants) of that type. If the data type is conceptualized as a "cookie cutter," the elements of the data type are the identically shaped "cookies" it produces. We have already introduced the four standard scalar data types: integer, real, boolean, and char. By standard, we mean that they exist automatically as part of the language, and the user need not define them further. (One other automatic type, text, will be introduced later.) All other data types in Pascal must be explicitly declared by the user through the **type** declaration. The general form of this declaration is:

Data type declaration:

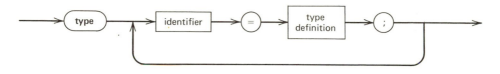

where "identifier" is used to provide an explicit name for this new data type and "type definition" contains the actual description of how to form this new data type. There are seven specific formats for the type definition mentioned here. They correspond to the seven additional classes of data types, beyond the standard ones, that are supported by Pascal. The following sections will describe briefly the syntactic rules for forming these new data types. Later chapters will present examples of the proper use of these data types in the context of developing complete programs. The first two sections describe two additional classes of scalar data types: the *user-defined* and the *subrange* types. The following sections describe the rules for forming the five classes of structured data types available within the language: the **array, record, set, file,** and pointer.

### 1.4.3.1  The User-Defined Scalar Data Type.
To create a *user-defined scalar data type*, the user specifically enumerates all constants of the new data type.

User–defined scalar type definition:

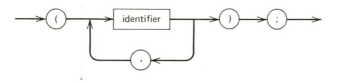

For example:

```
type
      majors   = (chemistry,physics,math,compsci);
      colors   = (red,yellow,blue,orange,green,purple);
```

The identifiers chemistry, physics, red, yellow, and so forth, represent the constants of these new data types, called majors and colors, and they are ordered in increasing left-to-right sequence (i.e., orange > red, green < purple). We can subsequently create variables of type majors or colors and assign values such as red or purple to these variables. In addition, we can apply the standard functions succ, pred, and ord to either data type.

> *succ(math)*     = *compsci*
> *succ(blue)*      = *orange*
> *pred(yellow)*    = *red*
> *ord(red)*        = *0 { the enumeration begins at 0 }*
> *ord(physics)*     = *1*

**1.4.3.2  The Subrange Data Type.**  A scalar *subrange data type* is composed of a contiguous portion, or subrange, of any of the other scalar data types presented so far.

Subrange type definition:

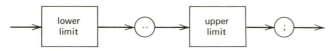

where lower limit, upper limit are:

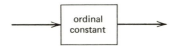

Lower limit and upper limit must be elements of the same type called the *base type*—either integer, boolean, character, or user-defined—and upper limit >= lower limit. The new data type incorporates all elements of the base type between lower limit and upper limit inclusive.

> **type**
>     *examscores*    = *0..100;*
>     *primary*       = *red. .blue;*     {*this will include red, yellow, blue*}
>     *letters*        = *'a'. .'z';*      {*Be careful. You may get more than the 26 characters. See Appendix A.2 for example.*}

The four standard scalar data types mentioned in Section 1.2.1—integer, real, boolean, character—plus the two classes of type just described represent the six classes of scalar data types available in Pascal. The set of five scalar data types not including real—integer, boolean, character, user-defined, and subrange—are usually referred to as the *ordinal data types*. They share one important characteristic: all elements (except the first and last) have a unique successor or predecessor. Thus, succ(x) or pred(x), where x is an element of an ordinal type, is a well-defined operation.

The scalar data types introduced so far could also properly be called *simple* data types because scalar variables are limited to assuming individual values of constants of that data type. No higher-level relationships or structures are possible. *Structured data types* are higher-level data types built up (ultimately) from collections of simple scalar data types, and they contain or imply some additional relationships among the various elements of that type. These structured types are described in the next four sections.

**1.4.3.3    The Array Data Type.**    An *array* is a collection of identically typed objects all referred to by the same name. The individual objects within the array are referenced by using an *array index* or *subscript*. An array reference in Pascal is written as arrayname[subscript]. A subscript can be an expression of arbitrary complexity but must evaluate to an element of an ordinal data type.

In Pascal an array is created by the following declaration.

Array type definition:

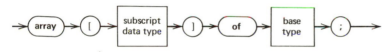

"Subscript data type" is the ordinal data type of the objects to be used in the subscript field. This declaration also determines the size of the array. The "base type" declaration describes the data type of the objects contained within the array.

**type**
      *example1* = **array** [0. .100] **of** *integer;*

This is a data type describing a 101-element integer array. If Z is a variable of type example1, the first element is referenced as Z[0], the second as Z[1], . . . , and the one hundred and first as Z[100]. (We will see how to create this variable Z in Section 1.4.4.)

**type**
      *example2* = **array** [*colors*] **of** *char;*

Using the declaration colors from Section 1.4.3.1, the preceding declaration would create an array of six elements, each containing a character value. If Z were declared to be of type example2, then Z[red] would refer to the first object, Z[yellow] would refer to the second, and Z[purple] would refer to the sixth and last.

The base type of an array can be anything, including another array. (This would create a multidimensional array.)

> **type**
>     *twod* = **array** [*1. .50*] **of array** [− *5. . + 5*] **of** *boolean;*

However, this form is clumsy and cumbersome. This shorthand notation is more convenient.

> **type**
>     *twod* = **array** [*1. .50, − 5. . + 5*] **of** *boolean;*

In general, for an arbitrary number of dimensions, the following format is used.

Generalized **array** type definition:

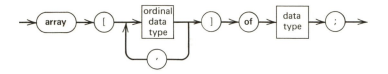

To reference that array, the following notation is used.

Array reference:

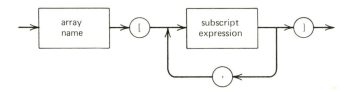

Other than the physical limitation of the amount of memory available, there are no restrictions on the number of dimensions permitted. Arrays will be discussed further in Chapter 7.

**1.4.3.4  The Record Data Type.**   The primary limitation of arrays is that all elements must be of the same type. However, Pascal provides a structured data type, the *record,* in which the individual elements (*fields*), need not be of identical data types. Records are created through the following declaration.

Record type definition:

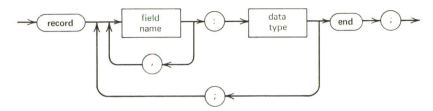

A record is composed of an arbitrary number of fields, each of which may be of any data type, including other structured types.

```
type
      studentrecord      =
          record
              name       : array [1. .20] of char;
              address    : array [1. .20] of char;
              year       : 1. .4;
              field      : majors; { see Section 1.4.3.1 }
              gpa        : real
          end; { of record studentrecord }
```

Studentrecord is a structured data type containing five distinct fields: two 20-element character arrays, a subrange, a user-defined field, and a real field. Access to the individual fields of the record is through a *path name:*

*record-variable.fieldname*

which selects out the specific "fieldname" of the "record-variable." For example, if mike were a variable declared to be of type studentrecord, then his year in school would be accessed through mike.year and his grade-point average would be accessed as mike.gpa. His name would be in the 20-element array indicated by:

*mike.name[1], . . . , mike.name[20]*

The data type of a field within a record can be another record, as shown in the example on the following page.

```
type
    mailrecord =
        record
            name                    :
                record
                    last            :  array [1. .20] of char;
                    first           :  array [1. .20] of char;
                    middleinitial   :  char
                end; { of name record }
            address                 :
                record
                    street          :  array [1. .20] of char;
                    city            :  array [1. .20] of char;
                    state           :  array [1. .20] of char;
                    zipcode         :  0. .99999
                end; { of address record }
            mailclass               :  integer;
            postage                 :  real
        end; { of record mailrecord }
```

If parcel were declared to be a variable of type mailrecord (Section 1.4.4 will show how to create the variable parcel), the following would be examples of path names to various fields within the record:

| | |
|---|---|
| *parcel.name.last[1]* | —first character of the last name |
| *parcel.address.zipcode* | —the zip code |
| *parcel.postage* | —the amount of postage |

Carrying this example one step further, we could imagine we had 100 parcels to be mailed with the information needed on each parcel contained in the data type mailrecord. Since each piece of information is identically organized, this would naturally suggest the array structure, with each element of the array as a single mailrecord.

```
type
    dailymail = array [1. .100] of mailrecord;
```

Now the path name would need to recognize this additional level of structure. For example, if mailinfo were declared to be a variable of type dailymail:

| | |
|---|---|
| *mailinfo[3].address.zipcode* | —the zip code of the third parcel |
| *mailinfo[100].name.last[1]* | —the first character in the last name on the one hundredth parcel |

This latter example illustrates an extremely important point that will be developed and stressed throughout the book. All Pascal data structures are recursive and can utilize each other in their definitions. Very complex and interesting data types can be built up by combining in a hierarchical fashion the basic data structures presented in these sections (arrays of arrays, arrays of records, records containing records, etc.). So far we have introduced only the basic syntax of these data structures. Complex structured data types will be covered in detail in Chapters 7 to 9.

**1.4.3.5    The Set Data Type.**    A *set* is a collection of identically typed objects. However, unlike the array, individual elements of a set are not accessed and manipulated. Instead, the set is treated as an indivisible unit.

The members of a specific set are of the same data type, called the *base type*. To create a set, we use the following **type** declaration.

**Set** type definition:

For example:

**type**
    *letterset*    = **set of** *'a' . . 'z'*;
    *colorset*    = **set of** *colors; {see Sec. 1.4.3.1}*

Constants of set type are indicated by enclosing the set elements in brackets: [ ]. For example, if *l1* and *l2* are variables of type letterset, and *c1* and *c2* are variables of type colorset, the following are valid statements.

    *l1* := ['a', 'e', 'i', 'o', 'u'];
    *l2* := ['a', 'b', 'c'];
    *c1* := [red, yellow, blue];
    *c2* := [  ] {the empty set}

There are seven operators that can be applied to sets. (For the following, assume that a and b are sets of the same base type and x is a variable or constant of the same type.)

    +    Set union. An element is contained in the *union* of two sets, denoted a+b, if and only if it is a member of set a or set b or both.
    −    Set difference. An element is contained in the *difference* of two sets, denoted a−b, if and only if it is a member of set a but not of set b.

*    Set intersection. An element is contained in the *intersection* of two sets, denoted a*b, if and only if it is a member of both sets a and b.

=    Set equality. The relation a=b is true if and only if every member of a is a member of b and every member of b is a member of a.

<>   Set inequality. The relation a<>b is true if and only if the relation a=b is false.

<=   Set inclusion. The relation a<=b is true if and only if every member of a is also a member of b.

**in**   Set membership. The operation x **in** a is true if and only if the element x is a member of the set a.

For example, using the previous assignments:

| | |
|---|---|
| *l1 + l2* | *is ['a', 'b', 'c', 'e', 'i', 'o', 'u']* |
| *l1 − l2* | *is ['e', 'i', 'o', 'u']* |
| *l1 * l2* | *is ['a']* |
| *c1 = c2* | *is false* |
| *l2 = (l1*l2) + ['b', 'c']* | *is true* |
| *red* **in** *c1* | *is true* |
| *'z'* **in** *l2* | *is false* |

**1.4.3.6    The Pointer and File Data Types.**    Pointer and file are complex structured data types that are discussed fully later in the book. For now we will limit ourselves to the proper syntax for the corresponding **type** statements.

Pointer type definition:

The pointer data type and the linked structures that can be created with it are discussed in Chapter 8.

**file** type definition

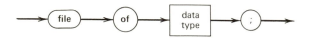

The file data type will be discussed in Chapter 9.

The hierarchy of data types in Pascal is summarized in Figure 1.6.

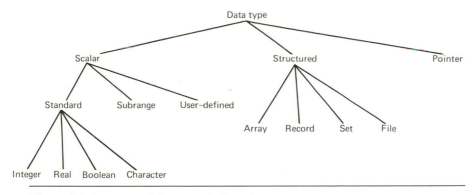

*Figure 1.6.* Summary of data types in Pascal.

### 1.4.4  The Var Declaration

Pascal is a *strongly typed language*. This means that the data type plays a very explicit role in the design and use of the language. Before any operation is performed, the data type of each operand is checked for consistency and correctness. *Every variable in a Pascal program* must be explicitly declared as to its data type. This is done through the **var** declaration.

Var declaration:

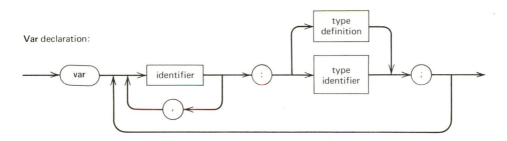

The data type identifier can be standard (e.g., integer) or a user-supplied name created through the **type** declaration discussed in the previous sections.

```
var
    customerlist    :   twod; { See Section 1.4.3.3 }
    i               :   integer;
    j               :   integer;
    loss            :   real;
    profit          :   real;
    years           :   integer;
```

All variables, without exception, must be so declared; failure to do so is always a fatal error. Be certain that you clearly understand the difference between a data type and a variable of that data type. In the previous example, twod is a data type name (like real or integer). Customerlist is a variable declared to be of that type.

In the previous sections we assumed the existence of certain variables in order to present some examples. We can now show exactly how these variables would be declared.

```
var
    z           : example1;           {Section 1.4.3.3}
    mike        : studentrecord;      {Section 1.4.3.4}
    parcel      : mailrecord;         {Section 1.4.3.4}
    mailinfo    : dailymail;          {Section 1.4.3.4}
    l1, l2      : letterset;          {Section 1.4.3.5}
    c1, c2      : colorset;           {Section 1.4.3.5}
```

Pascal allows a shorthand notation for combining the role of the **type** and **var** declarations. The ''data type identifier'' in a **var** declaration can be replaced by the entire data type definition.

```
(a) type
        twod    = array [1. .50, − 5. .5] of integer;
    var
        x       : twod;
(b) var
        x       : array [1. .50, − 5. .5] of integer;
```

are identical except that in (b) the new data type does not have an explicit name. (However, be aware that even though the preceding types match by inspection, some compilers may not always treat them identically.)

## 1.4.5   The Function and Procedure Declaration

Pascal provides two classes of subprograms—the *procedure* and the *function*. These subprograms are created through the **procedure** and **function** declarations.

A procedure or function subprogram is identical in structure to the Pascal program shell in Figure 1.5 with two exceptions.

The first line of a subprogram is the **procedure** or **function** declaration, as shown in the examples on the top of page 26.

**Procedure** heading:

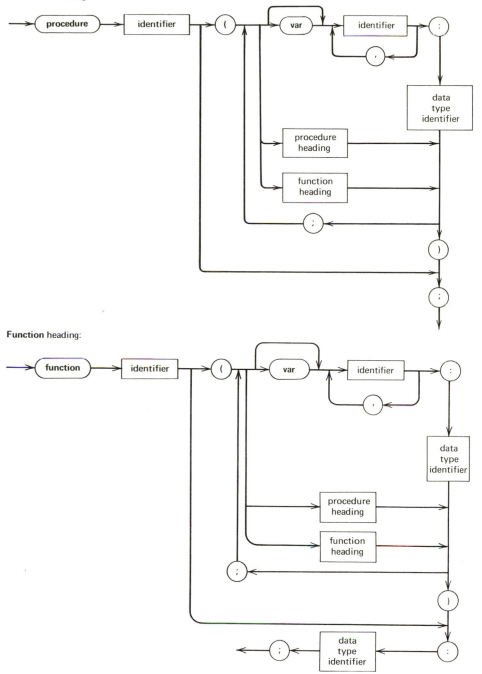

**Function** heading:

Examples:

```
procedure sample(x : real; y : char);
procedure loop(var a,b,c : integer);
procedure search(var list : twod; var n : integer);
procedure check(procedure validate; var ok : boolean);
function average (var scores : arraytype; size : integer) : real;
function legal(val : integer) : boolean;
```

The identifier immediately following the reserved identifier **procedure** or **function** is the name of this subprogram. The identifiers inside the parentheses specify the names and data types of all formal parameters. (The reserved word **var,** which may optionally precede any of the parameters, determines the class of formal parameter it is. These classes of parameters will be discussed fully in Chapter 3.) The last type identifier in a **function** declaration determines the data type of the result returned by the function. For example, function average above would return a real result, while function legal would produce a boolean one. A function or procedure name can itself be a parameter within a function or procedure call.

The second major syntactic difference between a program and a subprogram is that a function or procedure ends with a semicolon, not a period.

To invoke a procedure, we write out its name followed by the list of actual parameters. For example, assume the following declarations.

```
var
    ch        : char;
    delta     : real;
    int1      : integer;
    int2      : integer;
    int3      : integer;
    int4      : integer;
    z         : twod; {twod is the type declaration of Section 1.4.3.3}
```

Then the following are all proper procedure invocations of the preceding procedures.

```
sample(delta,ch);
loop(int1,int2,int3);
search(z,int4)
```

A function is invoked by using it as an element of an arithmetic or logical expression anywhere in the program, exactly as if it were a variable, constant, or standard function of the corresponding type. Examples of function invocations will be shown in Section 1.5 after we have introduced expressions. In addition, Chapter 3 will expand on the proper design and use of both types of subprograms.

These, then, are the six classes of declaration available on Pascal. Figure 1.7 presents a program heading and declaration section that illustrate many of the features discussed in Sections 1.1 to 1.4. (The complete program will be presented at the end of this chapter.)

```
program example(input,output);
{ sample declaration section }
const
      low        = 0;          { lowest possible test score }
      high       = 100;        { highest possible test score }
      testcount  = 10;         { maximum number of quizzes per student }
type
      grades     = (a,b,c);
      cutofflist = array [grades] of real;
      id         = 0. .999999;
      range      = low. .high;
      testlist   = array [1. .testcount] of range;
var
      average    : real;       { average of all test scores }
      bad        : integer;    { total number of illegal scores }
      cutoff     : cutofflist; { cutoffs for grade of a, b, or c }
      goodflag   : boolean;    { flag set for valid scores }
      grade      : char;       { letter grade for a student }
      i          : integer;    { loop counter }
      idnum      : id;         { student id number }
      n          : integer;    { number of test scores }
      score      : integer;    { holds a single score during validation }
      scores     : testlist;   { scores on a single student }
      sum        : integer;    { sum of all test scores }
{ this procedure assigns the letter grades a, b, c, or f to the average score 'aver'
based on standards set by the user. The grade is returned as 'finalgrade.' }
procedure assigngrade(aver : real; var finalgrade : char; cutoff : cutofflist);

begin
      if aver >= cutoff[a] then finalgrade := 'a'
      else if aver >= cutoff[b] then finalgrade := 'b'
      else if aver >= cutoff[c] then finalgrade := 'c'
      else                      finalgrade := 'f'
end; { of procedure assigngrade }
      .

      .
      . { other procedure or function declarations will follow }
```

*Figure 1.7.* Sample declaration section.

## 1.5  EXECUTABLE STATEMENTS IN PASCAL

### 1.5.1  The Assignment Statement

The syntax of the assignment statement is:

Assignment statement:

The expression on the right side of the assignment operator (: =) can be of arbitrary complexity and is composed of constants, variables, operators, functions (standard or user written), and grouping marks ( ). The operands of the expression must all be of the *same* data type, except for the acceptable mixing of integers and reals mentioned in Figure 1.3.

To determine the order of evaluation of operators, we use the precedence rules summarized in Figure 1.8. Within a precedence group, we evaluate operators from left to right.

The following are examples of assignment statements in Pascal. (We will assume that all variables have been properly declared.)

| *Mathematical Notation* | *Pascal Translation* |
|---|---|
| $x = \sqrt{(a + b)/c}$ | x := sqrt((a+b)/c) |
| $Root = ax^2 + bx + c$ | root := a * sqr(x)+b*x+c |
| Increment i by 1 | i := i + 1 |
| $z = \mu + \dfrac{1}{2\sigma}$ | z := mu + (1.0/(2.0 * sigma)) |
| $\dfrac{Pi\ (1 + i)^2}{(1 + i)^2 - 1}$ | value := p*i*sqr(1.0+i)/ (sqr(1.0+i)−1.0) |

### 1.5.2  Input/Output Statements

In Pascal most input is handled by the two statements read and readln.[4]

Input statements:

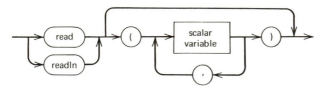

[4]Read, readln, write, and writeln are not actually statements but standard procedures provided by the language. It is common, however, to view these 4 operations as basic statements of the language.

| Highest precedence | parenthesized expressions |
| | **not**, (unary) —, (unary) + |
| ↓ | * / **div mod and** |
| | + − **or** |
| Lowest precedence | < <= > >= <> = **in** |

*Figure 1.8*. Precedence rules in Pascal.

For example:

```
read (x, lower, root)
readln (delta, bound, i, j)
```

The list of variables inside the parenthesis is referred to as the *read list*.

The operation of the two statements is almost identical. The following description applies to both forms. (We will only be describing input from the standard file input. Chapter 9 will expand that to include input from arbitrary files.)

Both input statements begin scanning the input stream (punched cards or terminal lines) from the point where the previous input operation ended in order to find a value of the appropriate type for each variable in the read list. That is, if the next variable in the read list is of type integer, we will scan the input stream until we come to a data item, which should be an integer constant. If a value of the wrong data type is encountered a fatal error will occur, with the exception of an integer being accepted and converted to type real. Blanks are skipped (except for type char) and, if necessary, the scan will continue to succeeding cards or lines until all variables have a value. The input operation is complete when all variables in the read list have been defined. The difference between the two input commands is that the read leaves the input pointer exactly where it was when the input operation ended. The next input command will begin from that point. The readln, however, discards the remainder of the current record (card or line) and resets the input pointer to column 1 of the next card or line.

For example, given the following data cards:

```
              Input pointer
                  ↓
Card 1:         100.7    +4.5
Card 2:          25.6   − 9.800*
```

the following operations will have the results shown. (Assume *a* and *b* are integers, *x* and *y* are real, and *z* is char.)

(a)  `readln(x, y)`          $x = 100.7$     $y = 4.5$
(b)  `readln(x); readln(y)`    $x = 100.7$     $y = 25.6$

(c)  *read(x); read(y)*                                      *x = 100.7     y =    4.5*
(d)  *readln; read(x,y);*                                    *x =   25.6     y = −9.8*
     *readln(z)*                                                            *z = '*'*
(e)  *readln(a,b)*                                           *fatal error*

Data lines in Pascal are free format with the following exceptions.

1.   There must be at least one blank between numbers.
2.   Numbers cannot be broken between two cards or lines. A value started on one line must be completed on that line.
3.   The syntax of numeric values entered as data is identical to the syntax of those values as they would appear in a program.

Two predicates (boolean functions) that are very useful during input processing are:

*eof*—(*End-of-F*ile). eof is true if we are currently at the end of the input file and false otherwise.

*eoln*—(*End-of-Lin*e). eoln is true if we are at the end of the current input line and false otherwise.

The values of eof and eoln are set by the system. We will have much more to say about these two important predicates later.
     The two basic output commands in Pascal are the write and writeln statements (see footnote[4]).

Output statements:

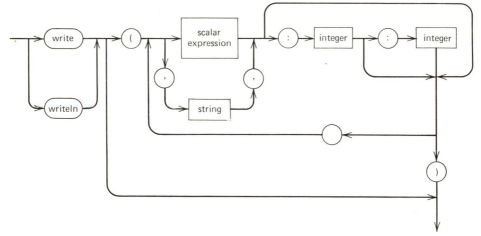

The writeln command, the more frequently used of the two, will output either of two types of values: (1) the result of evaluating a scalar expression, or (2) a string (a character array). If the object to print is a scalar expression it is evaluated and the value is printed. The number of columns that will be used to print the value will be an automatic default value in effect at your installation. If the object is a string, the characters between the apostrophes are printed exactly as is. Any object to be printed can be followed by one (or, in the case of type real, two) *field width designators*. The first designator is the total field width to be used in printing the value and overrides the default value. If the field is larger than needed, blanks are added at the left to fill out the field. The second designator, if present, forces reals to be printed in decimal notation and specifies how many decimal places to print. Both of these field designators are extremely important in producing legible, elegant output—a topic discussed at length in Chapter 3. (For example, assume $i = 4$, $j = -30$, $x = 51.87$, $z = 0.038$, and $c = $ '$*$'; also assume the default field width for integers is 10 columns, for reals 17 columns, and for characters 1 column.)

| *Statement* | *Output ( __ is a blank character)* |
|---|---|
| writeln(i,x); | __ __ __ __ __ __ __ __ __ 4 __ __ __ __ 5.18700000e + 1 |
| writeln(' j = ', abs(j):6); | __ j __ = __ __ __ __ __ 30 |
| writeln(x:10:1); | __ __ __ __ __ __ 51.9 |
| writeln(z:10:4); | __ __ __ __ __ 0.0380 |
| writeln(' x sin(x)'); | __ x __ sin(x) |
| writeln; | (blank line) |
| writeln(i+j:5); | __ __ − 26 |
| writeln(c:4) | __ __ __ * |

The only difference between the two forms of output command is that after the writeln command has been completed, it will force a carriage return and line feed and go to the next line. The write command will leave the printer exactly where it was when the output operation terminated. The next output command will begin from that point. Thus, using the write will allow you to continue printing on the same line.

| | *Result* |
|---|---|
| for i := 1 to 5 do<br>    write(name[i]) | smith |
| for i := 1 to 5 do<br>    writeln(name[i]) | s<br>m<br>i<br>t<br>h |

### 1.5.3  Compound Statements

In Pascal a compound statement is any group of statements bracketed by the reserved identifiers **begin** . . . **end**.

Compound statement:

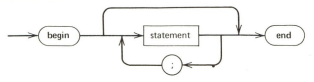

Any place in Pascal that you can put a single statement you can also put a compound statement. Syntactically, the compound statement, regardless of its length and the number of individual statements it includes, is always considered a single statement. (Looking at Figure 1.5, we see that all Pascal programs are composed of a single, large, compound statement.)

```
begin {this is an example of a compound statement}
    readln(x,y);
    writeln(' x = ', x:5, ' y = ',y:5)
end
```

### 1.5.4  Conditional Statements

There are two forms of conditional statements in Pascal: **if/then/else** (and the **if/then** variant), and the **case/end**.

**1.5.4.1  The If/Then/Else Statement.**  The format of the **if/then/else** is shown here, and the semantics of the statement are summarized in Figure 1.9a.

If statement:

First, the boolean expression is evaluated. If it is true, the **then** clause is executed and the **else** clause is skipped. If it is false, the **then** clause is skipped and the **else** clause is executed. If there is no **else** clause (in the **if/then** variant), we proceed to the statement following the **if**. In either clause the statement may be anything at all, including a compound statement that performs a great many operations.

(a)  **if** *hours* <= *40.0* **then**
        *pay* : = *hours* * *payrate*
    **else**
    **begin**
        *overtime* : = *hours* − *40.0;*
        *pay* : = *(40.0* * *payrate)* + *(overtime* * *(payrate* * *1.5))*
    **end** *{of the else clause}*

(b)  **if** *(theta* = *pi / 2.0)* **or** *(theta* = *(3.0* * *pi) / 2.0)* **then**
    **begin**
        *writeln(theta);*
        *writeln('*** tangent is infinite ***')*
    **end**
    **else**
    **begin**
        *tangent* : = *sin(theta)/cos(theta);*
        *writeln(theta, tangent, ' (in radians)')*
    **end**

(c)  **if** *a* = *0* **then**
    **if** *b* = *0* **then** *writeln('no answer possible')*
    **else** *root* : = *c* **div** *b {example of the dangling else}*

Example (c) illustrates an important point. Without an additional rule, that statement is ambiguous. Does the **else** clause belong to the outer (a = 0) or inner (b = 0) **if** statement? The rule for this "dangling **else**" problem is that an **else** clause always belongs to the closest **if**. Thus the **else** clause in Example (c) will be executed only if both a = 0 and b <> 0.

**1.5.4.2  The Case Statement.**  A common construct with the **if/then/else** is the staircase structure (s$_i$ is any valid statement).

**if** *a* = *1* **then** s$_1$
**else if** *a* = *2* **then** s$_2$
    **else if** *a* = *3* **then** s$_3$
        **else** .
            .

This structure occurs when we are checking a variable for one specific value out of many. For each possible value there is a specific action to be taken — exemplified by the preceding s$_i$. The structure shown here is clumsy and cumbersome and, to replace it, Pascal provides the **case-end** control construct.

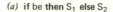

(a) if be then S$_1$ else S$_2$

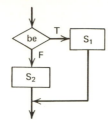

(b) case expression of
    C$_{11}$, C$_{12}$, . . . : S$_1$ ;
    C$_{21}$, C$_{22}$, . . . : S$_2$ ;
             $\vdots$
    C$_{n1}$, C$_{n2}$, . . . : S$_n$
    end

(c) while be do S

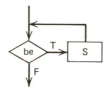

(d) repeat S$_1$, S$_2$, . . . until be

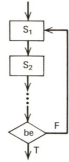

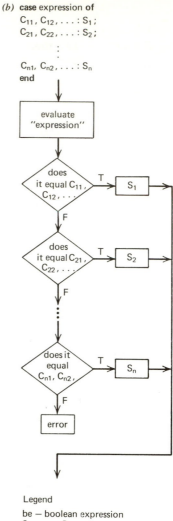

(e) for v : = e$_1$ to e$_2$ do S

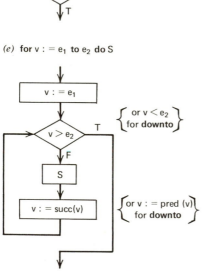

Legend

be — boolean expression
S  — any Pascal
        statement
e  — any ordinal
        expression
v  — any ordinal
        variable
C  — any ordinal
        constant

*Figure 1.9.* Semantics of the Pascal control statements.

Case statement:

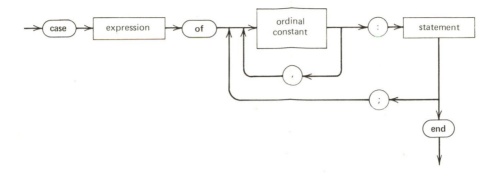

The expression is evaluated to produce an ordinal result. Then the set of ordinal constants contained in the **case** statement, which must be of the same type as the expression, is searched to find a match with the value of the expression. When the match is found, the statement immediately following the colon is executed. The **case** is a way to select exactly one statement out of many for execution. The semantics of the **case** are summarized in Figure 1.9*b*. For example, if the integer variable yearinschool had the values 1-4, 5 (for graduate students) and 9 (for special students), the following would correctly produce a count by student status.

```
case yearinschool of
    1, 2, 3, 4 : undergrad := undergrad + 1;
            5 : grad := grad + 1;
            9 : begin
                    special := special + 1;
                    writeln('special student')
                end {of special student processing}
end {of case statement}
```

A serious problem arises when the value of the expression is not contained in the case list (e.g., yearinschool = 6). The actions taken are implementation dependent, and we cannot say for sure what will happen. Some versions of Pascal consider the situation a fatal error and terminate the program, while others continue on without doing anything, as though the statement were a "no operation." The best thing to do is to prevent that from occurring by methods we will discuss later.

### 1.5.5   The Goto Statement

The **goto** statement in Pascal is the standard unconditional branch.

goto statement:

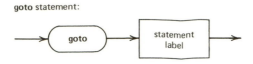

The next statement to be executed is the one with the designated label.

Misuse of the **goto** is one of the most common programming errors. With the control constructs we have introduced so far (or will introduce in Section 1.5.6), there is rarely a need for the low-level and elementary **goto** statement in any Pascal program. When the situation does arise where a **goto** can aid the clarity of a program, we will not avoid using one, but we must stress that these situations are rare. Most programs presented here will have very few or no unconditional branches.

### 1.5.6   Looping Constructs

There are three statements in Pascal for constructing *loops* (groups of statements to be repeated until a certain condition is met). They are the **while, repeat,** and **for**.

#### 1.5.6.1   The While Statement

While statement:

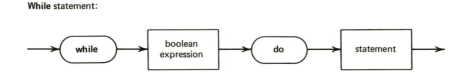

The **while** executes a statement (which may be anything, including a compound statement or another **while**) as long as the boolean expression in the **while** header remains true. The boolean expression is evaluated at the beginning of the loop. If the expression is false initially, the statement is not executed. The semantics of the **while** are shown in Figure 1.9*c*.

```
(a)  sum := 0;
     counter := 0;
     while not eof do
     begin
         readln(score);
         sum := sum + score;
         counter := counter + 1
     end {of the while loop }
```

(b) *read(ch);*
   *i := 0;*
   **while** *(ch <> '.')* **and** *(i <= linesize)* **do**
   **begin**
      *i := i + 1;*
      *line[i] := ch;*
      *read(ch)*
   **end** *{of the while loop}*

The user must include statements within the body of the **while** that will eventually cause the boolean condition to become false and allow the loop to terminate. Failure to guarantee this results in an *infinite loop*, a common programming error. For example:

   *readln(count); {tells us how many times to execute loop}*
   **while** *count <> 0* **do**
   **begin**
      *processdata; {procedure to solve problem}*
      *count := count - 1*
   **end** *{of while loop}*

The preceding loop will work correctly only as long as the initial value of count is greater than or equal to 0. If count < 0, we will never reach the condition (count = 0), and we have an infinite loop. The correct way to implement this loop is to write:

   **while** *count > 0* **do**

The loop will now terminate properly.

### 1.5.6.2 The Repeat Statement

Repeat statement:

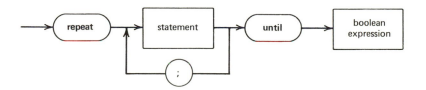

The **repeat** statement serves the same purpose as the **while**—to repeat a group of Pascal statements. Aside from the syntactic differences between the two statements, there are two other basic differences between the **repeat** and the **while**.

1.  The condition for termination of the **repeat** is the reverse of the **while**: the **repeat** executes a block **until** the condition becomes true. The **while** repeats a block **while** the condition stays true (i.e., until it becomes false). The following examples illustrate this difference.

    (a) **repeat**                                   **while not** *eof* **do**

        .                                             .
        .                                             .
        .                                             .

        **until** *eof*
    (b) **repeat**                                   **while** *(count* $<=$ *80)* **do**

        .                                             .
        .                                             .
        .                                             .

        **until** *(count* $>$ *80)*
    (c) **repeat**                                   **while not** *(a* **or** *b)* **do**

        .                                             .
        .                                             .
        .                                             .

        **until** *a* **or** *b*

2.  The termination condition on the **repeat** is tested at the end of the loop, not the beginning. (Compare Figures 1.9*c* and 1.9*d*.) This guarantees that the body of the **repeat** loop will always be executed at least once, even if the condition is true initially.

    (a) **repeat**
            *x* := *x* + *delta*; *{take a step along the x-axis}*
            *f* := *func(x)*; *{compute the function value at the point}*
            *writeln('function value at ', x, ' is ', f)*
        **until** *x* $>=$ *final*

    (b) **while not** *eof* **do** *{loop to process all the students}*
        **begin**
            *i* := *0*;
            **repeat**     *{the loop to read in quiz scores of a student}*
                *i* := *i* + *1*;
                *read(quiz[i])*
            **until** *i* $>=$ *quizcount*; *{criteria to terminate reading quiz scores}*
            *processdata(quiz)*; *{procedure to process array we just read}*
            *readln* *{discard remainder of the current line}*
        **end** *{of while loop}*

As with the **while,** the user is responsible for guaranteeing that the conditions for termination are eventually met.

**1.5.6.3 The For Statement.** The **for** statement, unlike the **repeat** or **while** statements, bases its looping on a counter value as opposed to a boolean condition.

For statement:

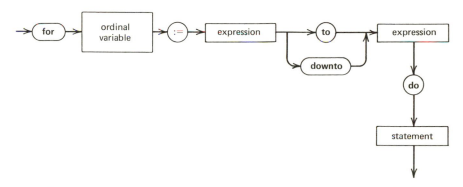

The ordinal variable (which may be of any scalar type except real) is initialized to the value of the first ordinal expression. It is then tested to see if the current value of the variable (usually called the *control variable*) is greater than (if the **to** option is selected) or less than (if the **downto** option is selected) the second expression. If not, the statement is executed, the control variable, v, is set to either succ(v) for **to** or pred(v) for **downto**, and the loop is repeated. Thus the number of times the loop is repeated is determined by how many times it takes to get from the initial value of the control variable to the final value by repeated applications of succ or pred. The semantics of the **for** loop are summarized in Figure 1.9*e*.

(a) **for** *i* := *100* **downto** *0 do*
    *list[i]* := *0*

(b) **for** *ch* := *'a'* **to** *'z'* **do**
    **begin**
        *write('the number of occurrences of ', ch:2);*
        *writeln(letters[ch]:5)*
    **end**

(c) **for** *i* := *1* **to** *row* **do**
    **begin**
        *sum* := *0;*
        **for** *j* := *1* **to** *column* **do**
            *sum* := *sum* + *a[i,j];*
        *writeln('sum of row', i, ' is ', sum:10)*
    **end**

These examples all show that the **for** loop control variable (i, j, ch) can be referenced or used within the body of the loop. It is always illegal, however, to attempt to redefine or *change* the value of the control variable inside the loop in any way.

## 1.6 SAMPLE PASCAL PROGRAM

We will complete this section by working through a complete program that uses many of the language features introduced in this chapter.

Assume we have a number of data cards on students. Each card contains a student identification number (integer) and a series of quiz scores in the range 0 to 100 (also integer). The student cards are preceded by two special data cards. The first card tells us what the cutoffs are for the grades of A, B, and C. The second card specifies exactly how many test scores there will be for each student. What we would like is a report listing for each student, their identification, test scores, overall average, and a letter grade—A, B, C, or F, based on the criteria of 85 for A, 70 for B, 55 for C, and F for any score below 55. Any erroneous input should result in an appropriate error message, and the total number of input errors should be tallied. The input and output requirements are represented in Figure 1.10. The declaration section for this problem was presented in Figure 1.7. The program uses two subprograms.

1. *legal*. A function that validates the quiz scores of each student.
2. *assigngrade*. A procedure that assigns a letter grade of A, B, C, or F based on the overall average of all test scores.

The entire program to solve this problem is shown in Figure 1.11.

We have summarized most of the Pascal language and presented a complete Pascal program. Readers should study this program carefully to be sure that they understand all the statements and declarations used as well as all those presented in

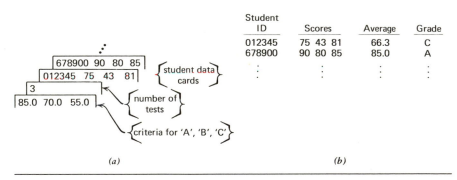

(a)                                                         (b)

**Figure 1.10.** Input/output specification of the sample program. (a) Input. (b) Output.

```
program example(input,output);
{ a sample program to illustrate the statements of pascal }

const
      low           = 0;                  { lowest possible score }
      high          = 100;                { highest possible score }
      testcount     = 10;                 { maximum quizzes per student }

type
      grades        = (a,b,c);
      cutofflist    = array[grades] of real;
      id            = 0..999999;
      range         = low..high;
      testlist      = array[1..testcount] of range;

var
      average       : real;               { mean quiz score for a student }
      bad           : integer;            { illegal quiz score count }
      cutoff        : cutofflist;         { a,b,c cutoff scores }
      goodflag      : boolean;            { true if all scores were valid }
      grade         : char;               { letter grade }
      i             : integer;            { for loop index }
      idnum         : id;                 { student id number }
      n             : integer;            { actual number of quizzes }
      score         : integer;            { holds a score before validation }
      scores        : testlist;           { scores of a single student }
      sum           : integer;            { sum of quiz scores by a student }

{ procedure assigns a letter grade of a,b,c,f to the overall average quiz score. the
  cutoff points are stored in the cutoff array that was read from the first input line.}
procedure assigngrade(aver : real; cutoff : cutofflist; var finalgrade : char);
begin
      if aver >= cutoff[a] then finalgrade := 'a'
      else if aver >= cutoff[b] then finalgrade := 'b'
            else if aver >= cutoff[c] then finalgrade := 'c'
                  else finalgrade := 'f'
end; { of procedure assigngrade }

{ a function to check to see if a quiz score is within the proper range. the
  function returns true if the score is valid, and false otherwise. }
function legal(valu : integer) : boolean;
begin
      legal := true;
      if (valu < low) or (valu > high) then
      begin
            legal := false;
            bad := bad + 1
      end
end; { of function legal }
```

```
begin { main program }
    { read grade cutoffs first }
    read(cutoff[a], cutoff[b], cutoff[c]);
    writeln(' grade assignments: ',high:4,' >= a >= ',cutoff[a]:4:1,
            ' > b >= ',cutoff[b]:4:1,' > c >= ',cutoff[c]:4:1,' > f >= ',low:3);
    if not ( (high > cutoff[a]) and (cutoff[a] > cutoff[b]) and
             (cutoff[b] > cutoff[c]) and (cutoff[c] > low) ) then
        writeln(' *** error --cutoff list must be decreasing and ',
                'be between ',low,' and ',high,' processing halts')
    else
    begin
        readln(n);
        writeln(' each student has ',n,' quiz scores');
        if (n <= 0) or (n > testcount) then
            writeln(' *** error -- number of quiz scores per student ',
                    'is negative or exceeds the limit of ',testcount, 'processing
                    halts')
        else
        begin
            bad := 0;
            while not eof do { repeat this for all students }
            begin
                sum := 0;
                goodflag := true;
                read(idnum);
                for i := 1 to n do { repeat this for each quiz score }
                begin
                    read(score);
                    if legal(score) then
                    begin
                        scores[i] := score;
                        sum := sum + scores[i]
                    end { of good score }
                    else
                    begin
                        goodflag := false;
                        scores[i] := low;
                        writeln(' *** error in score of quiz no. ',i, '
                                of student ',idnum,'. bad one was ',score)
                    end; { of processing a bad score }
                end; { of for loop }

                { echo the input data }
                write(idnum:10);
```

```
for i := 1 to n do write(scores[i]:5);

{ now if there was no error we will compute average and the letter
    grade }
if goodflag then
begin
    average := sum / n; { average will be real }
    assigngrade(average,cutoff,grade);
    write(average:10:1,grade:5)
end
else
{ we get here if there was an error in a quiz score. we will put in
    the appropriate places }
    write('*****':10, '***':5);
writeln; { finally print the line constructed }
readln { and get a new input line }
end; { of outer while loop }

{ finish up by printing out how many errors were encountered during
    processing }
writeln;
writeln(' total number of quiz score errors encountered ',
        'was ',bad);
end { number of quiz scores was valid }
end { cutoff list was valid }
end. { program example }
```

*Figure 1.11.* Sample Pascal program.

this chapter. In the next few chapters we will be concerned with the proper and elegant use of Pascal. Specifically, we will develop evaluation criteria so that we may look at programs, such as the one in Figure 1.11, and decide not only if the program is correct but if it is *good*. We want to develop rules and guidelines for good *programming style*.

## EXERCISES FOR CHAPTER 1

1.   Which of the following are valid scalar constants in Pascal? If invalid, state why.

(a)  $-345$

(b)  $+345$

(c)  truth

(d)  $-$maxint

(e)  123,456

(f)  $-6.000000000$

(g)  $-5.$

(h)  $3 + 2i$

(i)  31e17
(j)  − 31e − 17
(k)  1.0e5.5
(l)  1.e10
(m)  − 0.0

(n)  ' '
(o)  '1'
(p)  'character'
(q)  $10^{35}$

2.  Assume the following declarations.

**var**

    a,b      : integer;
    x,y      : real;
    c1       : char;
    l        : boolean;

For each of the following, state whether they are valid Pascal expressions. If valid, state what the data type of the result will be.

(a)  sqr(a) ∗ sqr(b/2) (1.0 − a)

(b)  sqrt(sin((abs(x + y)) / 4.0) + 1)

(c)  chr(ord(succ(c1) + 1))

(d)  a < 5 **and** b = 1 **or** 5

(e)  sqr(9 + chr(17))

3.  Write a single **const** declaration to create the following four symbolic constants.

(a)  e, the base of the natural logarithms (2.71828...).

(b)  π (3.1415927).

(c)  End-of-the-sentence is a '.'.

(d)  Current year is 1981.

4.  Write a single **type** declaration to create the following four user-defined data types.

(a)  The states of Minnesota, Wisconsin, Iowa, and New Jersey.

(b)  The grains hay, wheat, rye, barley, soybeans, and corn.

(c)  The people Sandy, Buttons, Ruth, Benjamin, and Rebecca.

(d)  The primary colors.

Write a **var** declaration to create a variable of each type.

5.  Write a single **type** declaration to create the following subrange data types.

(a)  The integers − 500 to 500.

(b)  The characters 0 to 9.

(c)  The states of Iowa and New Jersey (Exercise 4a).

Write a **var** declaration to create variables of each type.

6.  Write a single **type** declaration to create arrays with the following structure.

(a)  A 200-element real array indexed by the integer values 0, 1, ... .

(b)  An array, indexed by the commodities in Exercise 4b, that contains the price of each commodity in dollars and cents.

(c)  An array, indexed by the names in Exercise 4c, that contains the age, in years, of each person.

(d)  A 50 × 50 array of real values. The indices are all integers that begin at 1.

(e)  A 10 × 10 × 10 array of true/false values. Each index should be an integer and begin at 0.

Write a **var** declaration to create variables of each type.

7.  Write a single **type** declaration to create the following record structures.

(a)  *payroll record*
     name (last, first, middle initial)
     department (building number, room number)
     social security number
     hourly payrate

(b)  *inventory record*
     part name (1 to 10 characters)
     part number
     supplier (name, address, city, state, zip code)

Write a **var** declaration to create a record variable of each type.

8.  Write a Pascal **procedure** called

**procedure** *readrecord(***var** *inv : inventoryrecord)*

that reads input cards containing the following information and stores these values into the correct fields of inv—the record variable passed in as an input parameter. Inv has the structure shown in Question 7b.

(a)  Columns   1 to 10—part name.

(b)  Columns 11 to 20—part number.

(c)  Columns 21 to 30—supplier's name.

(d)  Columns 31 to 40—supplier's address.

(e)  Columns 41 to 50—supplier's city.

(f)  Columns 51 to 60—supplier's state.

(g)  Columns 61 to 65—zip code.

9.  Translate the following into valid Pascal assignment statements.

(a)  c = W ln (1 + s/n)

(b)  payment = $(1 + i)^2 - P$

(c)  $ax^3 + bx^2 + cx + d$

(d)  y = $\sin^2\Theta - \cos^2\Theta$

(e)  Increment z by 100

10.  Given the values a = 1.0, b = 7.0, c = 5.5, d = 35, e = 12, f = 14, g = true, and h = false, what are the values of the following expressions?

(a)  (a + sqrt(−b + 2.0 ∗ c)) / b − 2.0

(b)  e **mod** d **div** 5 ∗ f

(c)  −a ∗ b/c ∗ d

(d)  (d = 3 ∗ e) **and** g **or** h

11.  Write the necessary read or readln commands to read data properly in the following format.

(a)  An 80-element character string entered on a single line.

(b)  A variable length character string entered on a single line, ending with an eoln.

(c)  Three integer values on three separate lines.

(d)  Four real values entered as two lines of two values each.

12.  How should one enter the data pairs x, y if the input commands are written as follows?

(a)  read(x); read(y)

(b)  readln(x); readln(y)

(c)  readln(x); readln; readln(y)

13.  Assume the following declaration.

**var** x,y : real;

Show the output commands needed to produce the following results.

(a)  value of x = nnn.nn      value of y = nnn.nn

(b) x is n.nnnenn
    y is n.nnnenn

(c)

| x | y |
| --- | --- |
| nnn.nn | nnn.nn |

14. Write both a **repeat** and **while** loop to do the following operations. Do not use the **goto** to achieve the desired results.

(a) Sum up the values in a 100-element list until you reach the end of the list or the sum exceeds 10,000.

(b) Read input lines with pairs of integers until you come to the end of the data or both values are 0.

(c) Look through a table of length n, counting the number of elements equal to a − 1.

(d) Look through a table of length n for the *last* occurrence of a special value called key and save its position. If key never occurs, set position to −1.

15. Write a Pascal **function** called

**function** *prime(n : integer;* **var** *factor : integer) : boolean;*

The function accepts an integer n > 0 and determines whether or not n is a prime number. The function value is true if n is prime and is set to false otherwise. Additionally, if n is not prime, the parameter factor is set to a factor of n.

16. Write a Pascal procedure as follows:

**type** *table :* **array** *[1..10;1..10]* **of** *integer;*

**procedure** *magic(square: table; n : integer;* **var** *ismagic: boolean;*
    **var** *value : integer);*

The procedure takes an n x n table(n <= 10) and determines whether or not it is a magic square. (A magic square is one in which all rows, columns, and diagonals add up to the same value.) The variable "ismagic" is set to true if the square is magic and is false otherwise. If the square is magic the parameter "value" holds the numeric sum of the rows and columns.

17. Write a complete Pascal program that uses the function prime from Exercise 15 to solve the following problem.

Read in a value k, k >= 2 and determine for every value l, l = 1,2,..,k whether or not l is prime. For every value l the output of the program should be either nnn is prime or nnn is not prime, mmm is a factor.

18. Write a complete Pascal program that reads in a series of data lines and computes the frequency of the 26 letters of the alphabet. The program should read in all data until the end of file and produce as output the following report.

| Letter | Count | Percentage |
|--------|-------|------------|
| a | XXX | XX.X |
| b | XXX | XX.X |
| . | | |
| . | | |
| . | | |
| z | XXX | XX.X |
| Totals | XXX | XX.X |

19. Write a complete Pascal program to handle a bank's savings transactions. For each savings account, the first card is the initial balance card containing the account number and the starting balance for this period. All succeeding cards are transactions and have the following information:

(a)  transaction code  0 : deposit
                       1 : withdrawal
                       2 : end of transactions for this account

(b)  transaction amount (real)

Process all transactions for a single account and produce as output the following report.

acct no: xxxxxx      starting balance $xxx.xx
nn withdrawals with a total of $xxx.xx
nn deposits with a total of      $xxx.xx
        current balance          $xxx.xx

Do not let a balance become negative. If a withdrawal would overdraw an account, do not process it or count it as a legal transaction. Instead, write out:

** illegal withdrawal   **
    withdrawal amount $xxx.xx
    current balance      $xxx.xx

and go to the next transaction.

# BIBLIOGRAPHY FOR PART I

Bowles, K. *Microcomputer Problem Solving with Pascal*. New York: Springer-Verlag, 1977.

Conway, R., D. Gries, and E. Zimmerman. *A Primer on Pascal*, Englewood Cliffs, N.J.: Winthrop, 1976.

Findlay, W., and D. A. Watt. *Pascal—An Introduction to Methodical Programming*. Potomac, Md.: Computer Science Press, 1978.

Grogono, P. *Programming in Pascal*. Reading, Mass.: Addison-Wesley, 1978.

Hoare, C. A. R., and N. Wirth. "An Axiomatic Definition of the Programming Language Pascal." *Acta Informatica*, Vol. 2, 1973.

Holt, R. C., and J. Hume. *Programming in Standard Pascal*. Englewood Cliffs, N.J.: Reston Publishing House, 1980.

Jensen, K., and N. Wirth. *Pascal—User Manual and Report*. New York: Springer-Verlag, 1974.

Kieburtz, R. *Structured Programming and Problem Solving with Pascal*. Englewood Cliffs, N.J.: Prentice-Hall, 1978.

Ravenel, B. "Toward a Pascal Standard." *Computer*, Vol. 12, No. 4, April 1979.

Schneider, G. M., S. Weingart, and D. Perlman. *Programming and Problem Solving with Pascal*. New York: Wiley, 1978.

Webster, C. *Introduction to Pascal*. Rochelle Park, N.J.: Hayden, 1976.

Wirth, N. "The Programming Language Pascal." *Acta Informatica*, Vol. 1, 1971.

*Part II*

# PROGRAMMING STYLE AND EXPRESSION

# Chapter 2

# PROGRAMMING STYLE

## 2.1 INTRODUCTION

Chapters 2 and 3 will introduce and develop an important concept called *programming style*. This refers to the entire set of conventions, guidelines, aids, and rules that make computer programs easier *for people* to read, work with, and understand. The phrase *for people* is central to the definition of programming style. Computer programs are used by both machines *and* people—a fact that is too often overlooked. The long litany of syntactic rules and restrictions presented in Chapter 1 insures that a *computer* can properly accept and execute a computer program. The style rules and restrictions that we will now present insure that *people* can properly and easily read, understand, and work with that same program. We should stress that, from the machine's point of view, these style rules are completely unnecessary. As long as we adhere to all the language rules of Pascal, it is immaterial to the computer (actually, to the *compiler*) how that program was prepared or how it originally looked. As an extreme example, Figure 2.1 shows two program fragments that utilize very different styles. From the computer's viewpoint, these two fragments behave identically and, given the same input data, would produce the same results. However, for a person trying to understand the purpose of these program units, they are hardly ''identical''; the second example would almost certainly be considered more difficult to read and understand. Although this is an extreme case, it illustrates the importance, *to people*, of developing and using pleasing, helpful, and effective stylistic guidelines.

It is often difficult for students in an academic environment to appreciate the importance of good programming habits. The academic environment is very different from the real-world, production programming environment, where programming style and expression are critical. In an academic environment, working programs are typ-

```
found := false;
i := 0;
while (i < tablesize) and (not found) do
begin
    i := i + 1;
    if table[i] = key then found := true
end;
```

(a)

```
a := 0; i := 1; 97: if t[i] = k then
goto 98; if i >= tsz then goto 99;
i := i + 1; goto 97; 98: a := 1;
99:
```

(b)

*Figure 2.1.* Example of two identical program pieces using differing styles.

ically run only once and are almost never changed or modified after that final run. These programs are rarely examined in detail by anyone other than those already familiar with the problem—the teacher who designed it or the student who coded it. Usually neither of these people leaves the project (i.e., the class) prior to completion of the program. Finally, no one is paying any salaries or "real money" to develop this piece of software, and there are no budget restrictions to worry about. Taken together, these characteristics make for a most unrealistic situation.

In a production environment, computer programs are used for fairly long periods of time—possibly longer than the employment period of the original author of the program. In addition, programs are most definitely *not* static; they change frequently in response to changes in things such as tax laws, federal or state regulations, accounting policies, management priorities, new product lines, and consumer demands as well as the inevitable selection of a new computer. These two characteristics—program longevity and frequent change—mean that *program maintenance*, the process of keeping programs current and correct, is commonly done by someone who is unfamiliar with the original program. In this environment, the importance of clear, readable, and understandable programs becomes obvious.

Likewise, most production programs are large software systems, not the small 100- or 200- line "toy" programs used in programming classes. These large, real-world projects are usually not the effort of one person but of a *programming team* of 3 to 10 people. The team members cooperate in program development, design, and coding and can pick up the work of another teammate who has left the project. Again, this necessitates software that is clear and readable and can be understood quickly and easily by various people. (We will say more about team programming in Chapter 5.)

Another reason that good program design and style habits are important is the enormous *decrease* in hardware costs and a concomitant *increase* in "people costs":

salaries and benefits. Ten or twenty years ago, when every computer was a multi-million dollar investment (not even counting the expensive computer rooms, air conditioning, etc.), the guiding philosophy in programming was to avoid wasting precious computer time. Today, however, complete microcomputer systems sell for as little as $5000, and even a sophisticated, medium-sized computing facility capable of handling 16 simultaneous users may cost only $50,000 to $100,000. If we assume a life expectancy of 5 years, the average annual cost for the hardware will be $10,000 to $20,000—less than the salary of a single programmer! Studies on the overall cost of computing systems show that the people costs of developing and maintaining software are now 50 to 70% of the overall costs, and they are *still* increasing!

So the new philosophy in computer programming is to avoid wasting precious programmer's time by writing unintelligible programs. The way to achieve that is to develop program design methods and style guidelines for creating software that is *initially* free of bugs, easy to read and understand, and subsequently easy to maintain. We always want to avoid software like that shown in Figure 2.1*b*. Such programs, even though they may be technically correct, are extremely conducive to errors that will cost time and money to locate and correct; the inevitable updating of such programs will be an expensive and time-consuming task as we try to interpret intricate, obtuse, and confusing logic.

There is one final point that we must mention. A discussion of programming style and expression can present only guidelines, not rules. Unlike the syntax rules of Chapter 1, there are no absolute rights or wrongs of programming style— only personal judgments and subjective evaluations. While we can say with absolute certainty that i = i + 1 is an *incorrect* assignment statement and that i := i + 1 is *correct*, there are no clear-cut answers to stylistic questions. For example, which of the two following indentation schemes for the **if/then/else** is superior:

```
if boolean-expression then
begin
   . . .
end
else
begin
   . . .
end
```

or

```
if boolean-expression
      then begin
         . . .
            end
      else begin
         . . .
            end
```

Of course, there is no answer to that question. Like many issues we will raise, this is a debatable point based on personal taste.

We will be presenting numerous examples of what we feel are useful stylistic guides. However, as readers become more experienced, they will develop personal programming styles that may differ somewhat from ours but with which they feel more comfortable. Remember that the "bottom line" on programming style is *not* to make your programs look exactly like ours but to make them easy for people to read, understand, and work with.

Programming style and expression will be considered in the next three chapters. The remainder of this chapter will discuss how to develop programs that *look* nice (i.e., that are easy to read and understand). Chapter 3 will talk about programs whose execution time behavior is beneficial to either the end-user or other programmers who must work with that program. Finally, Chapter 4 will discuss how programs can be *structured* nicely so that they are composed of a sequence of properly nested, single-entry, single-exit control blocks.

The following chapters will discuss how to write *individual program units* (a procedure, a function, a main program) that are stylistically and behaviorally good. However, any realistic program will be composed of dozens of subprogram units that, when properly but together, will solve the problem. The task of organizing, identifying, and developing all the subprograms needed to solve a problem is called *program design*, and it will be covered in Part III. Initially, we will concentrate on learning how to write the individual units properly and elegantly. Then we will look at the bigger picture of putting these separate units together to solve the problem.

## 2.2 STYLISTIC GUIDELINES

### 2.2.1 Clarity and Simplicity of Expression

Often programmers pride themselves on having designed a particularly intricate piece of code. The feeling is that a solution that looks difficult is better than one that is simple and clear. Nothing could be more untrue. For example, look at the program in Figure 2.2 on the next page.

Twenty-six lines and four loops—surely something important must be going on! But eight lines and one loop are all that is really needed to sum the positive and negative values in the list as shown below.

```
positivesum := 0;
negativesum := 0;
for i := 1 to n do
    if list[i] >= 0
    then positivesum := positivesum + list[i]
    else negativesum := negativesum + list[i];
writeln(' sum of all positive values = ',positivesum);
writeln(' sum of all negative values = ',negativesum)
```

```
sw := false;
while not sw do
begin
     { sort list using bubble sort }
     sw := true;
     for i := 1 to n − 1 do
          if list[i] > list[i + 1] then
          begin
               temp := list[i];
               list[i] := list[i + 1];
               list[i + 1] := temp;
               sw := false
          end
end;
sum := 0;
i := 1;
repeat
     sum := sum + list[i];
     i := i + 1
until list[i] > 0;
writeln (' sum of all negative values =',sum);
sum := 0;
repeat
     sum := sum + list[i];
     i := i + 1
until i > n;
writeln(' sum of all positive values =',sum)
```

*Figure 2.2.* Example of unnecessary complexity in a program.

The above example was actually submitted as an example of "programmer expertise" and cleverness. The student who wrote it was proud of his exotic approach; it served his personal purposes and did wonders for his ego. To him it did not matter that the code was confusing and *unnecessarily* complicated. (And it also did not matter that the code was *wrong*! The program in Figure 2.2 will not work properly unless the list contains at least one positive and one negative value. This points out another key fact: it is very easy for errors to slip unnoticed into confusing and complex programs.)

This type of programming behavior should no longer be admired or even tolerated. Programmers must always remember that the code they develop will be read and studied by others and must therefore clearly and directly reflect the operation that it is carrying out. When choosing among alternate methods of coding a programming task, *never* sacrifice clarity of expression for cleverness of implementation. Do it the simple and straightforward way. If you are tempted to break this rule, be sure you are doing it for a good reason (e.g., unusual conditions, special circumstances).

Another example of misplaced priorities is the effort to reduce execution time at all costs. If we were to learn that, on our computer, a floating-point addition could be carried out in 1.6 microseconds while a floating-point multiplication took 18.0 microseconds, some programmers might actually be tempted to code the operation a = b * n as:

```
a := 0.0;
if n <= 11
then for i := 1 to n do a := a + b
else a := b * n
```

The gains to be made with this optimization (if any) will usually be measured in thousandths or millionths of a second. But the loss in programmer time caused by repeated use of confusing code fragments is measured in hours, days, and weeks! *Never sacrifice clarity for minor reductions in machine execution time.* (However, this does not imply that we would tolerate *gross* inefficiencies only to achieve elegance of code. The trade-offs between machine efficiency and programming style will be discussed more fully in Chapter 11.)

Finally, look at the following statement:

```
y := x − (x div n) * n
```

You will probably determine, after some effort, that it says the same thing as:

```
y := x mod n
```

but with considerably more bombast. Where possible, avoid the use of programming tricks whose intent will not be immediately obvious to readers.

In summary, programmers will need to learn how to suppress their egos. The biggest or the most complex is not necessarily the best. The goal is not to impress fellow programmers but to enlighten the end-users and facilitate program maintenance. The overriding and motivating factor in all programs that we will discuss is *simplicity* and *clarity*: say exactly what you mean.

*Style Review 2.1* —————————————————————————————

### Clarity of Expression

Never sacrifice clarity of expression for cleverness of expression.
Never sacrifice clarity of expression for minor reductions in machine execution time.
Avoid confusing programming tricks.
Always strive for simplicity and clarity.

### 2.2.2 Names

Nothing contributes as much to a program's clarity and readability as the simple expedient of choosing good *mnemonic names*. For example:

$m := s / (n - b)$

gives no clue to the purpose of the operations that we are performing. However:

*average := sum / (count − invalid)*

is quite clear, even when presented out of context. The syntax of identifiers in Pascal (Section 1.2.3) does not limit the number of characters in a name, nor does it require the selection of a specific initial character to guarantee a specific data type. The only requirement is to select a name that is immediately recognizable and helpful to someone looking at the code. In most cases this is quite easy.

*root, profit, sumsquared, mean, stepsize, score, date*

are all easily recognized and quite appropriate. A programming technique that can be helpful is to use a common prefix (or suffix) to identify a group of variables that are logically related within the program. For example, if we were computing individual weekly payrolls and then carrying the totals along in a year-to-date (ytd) file, we might choose to preface our variables with weekly and ytd to identify this relationship.

| | |
|---|---|
| *weeklygross* | *ytdgross* |
| *weeklyfica* | *ytdfica* |
| *weeklynet* | *ytdnet* |

Likewise, the suffix "file" can quickly identify all files used in the program.

*masterfile   transactionfile   updatefile*

Pascal (and other languages) does not allow blanks, underscores, or hyphens to be included as part of a name, as in:

*cost-per-fluid-ounce*

Because of this some names, when written without spacing, can be very difficult to read.

*costperfluidounce*

A nice stylistic aid is to utilize the uppercase and lowercase abilities of the character set and to use capitalization as a visual delimiter to help in reading long names.

*AmountInStock*
*CostPerFluidOunce*
*AverageScore*

There are some very important *do nots* related to name selection.

1. Do not use nonstandard, personally invented abbreviations that are not immediately clear:

   *dlds* (days lost due to sickness)
   *inclmtx* (increment limit of x)
   *mxdspn* (maximum dispersion)

   It is well worth it to use the few extra characters needed to create meaningful names. Abbreviated or shortened identifiers are acceptable *only* as long as they are unambiguously recognizable and still carry sufficient mnemonic value. Therefore:

   *freqtable*    instead of *frequencytable*
   *maxvalue*    instead of *maximumvalue*

   seem to be reasonable names, while ft, mv, ftab, or maxv would most likely be unacceptable. Remember, when selecting a name, your goal is not to minimize keystrokes but to maximize clarity.

2. Be very careful about choosing two names that might be confused. Be especially cautious about confusing 0 and o, 1 and i, 2 and z.

   *positionmax*    *xxyz*    *kkk*     *interestrate*
   *positionmax2*  *xxy2*    *kkkk*    *interstate*

   The eye can easily gloss over the minimal differences in these pairs.

3. In some languages you are forced into misspellings because of restrictions on data types or identifier length. Pascal does not impose these restrictions so there is no need to create ''cute'' misspellings or unnecessary abbreviations that can lead to trouble when the person attempts to use the correct spelling.

   *kount*      (instead of count)
   *averag*     (instead of average)
   *matrx*      (instead of matrix)
   *xnumber*   (instead of number—a hangover from other languages)

One bad habit closely related to this is the selection of cute names that form a joke or a play on words.

**if** *(puss* **in** *boots)* **then** . . .
**while** *boolexp* **do** *wackadoo* *: =* . . .

It may be funny, but it is not good programming.

Finally, there are two places where the selection of a name is particularly critical as a legibility aid. The first is the assignment of a symbolic name to a scalar constant. By themselves, constants have no mnemonic value (except for a few special ones such as 3.14159 and 2.71828). Therefore the appearance of a constant in an expression does not in any way assist us in understanding the purpose of the expression. For example:

*credit : = (0.0525 * balance) − 5.00*

may leave us bewildered as to the purpose of the constants 0.0525 and 5.00. In Pascal, the **const** declaration allows us to overcome this limitation by assigning a helpful name to a scalar constant:

**const**
    *interestrate*      *= 0.0525;*    { *the annual interest paid on the account balance* }
    *annualcharge*    *= 5.00;*    { *the fixed $5.00 annual charge against all acccounts* }

This assignment statement becomes:

*credit : = (interestrate * balance) − annualcharge*

From the program's point of view we have added nothing, since interestrate and annualcharge are still constants. But, from the user's point of view, we have made the purpose of the assignment much more obvious.

The second place where name selection is especially critical is with procedures and functions. If we are careful in our choice of procedure names, we should be able to determine what a program does merely by looking at the main program, which calls the various procedures. The example at the top of the following page illustrates this point quite clearly.

```
readdata(list);
validate(list,ok);
if ok then
begin
    computemean(list,mean);
    computevariance(list,variance);
    printresults(mean,variance)
end
else errorhandler
```

This fragment of a main program is extremely lucid, partly because of our choice of descriptive names for the various procedures: readdata, validate, errorhandler, and the like. It should be sufficient to look at the main program to determine *what* the program is doing. The only reason we have to scan the procedures is to determine *how* they actually perform that task. We will stress this point more in Chapters 5 and 6 when we discuss program design.

*Style Review 2.2* _____

*Names*

*Do:*

Pick good mnemonic names for all variables.
Pick good mnemonic names for all procedures and functions.
Use standardized prefixes and suffixes for related variables.
Assign names to scalar constants where it will help clarity and readability.

*Do Not:*

Make up cryptic or unclear abbreviations for variables.
Limit the length of names solely to save keystrokes.
Use "cute" names that do not have mnemonic value.

## 2.2.3 Comments

Most students treat comments as something to be added after a program has been completed, in order to guarantee that they receive full credit. Again, this is because of the environment in which the students operate; the programs, usually quite short, are seen only by the authors, and only when they are fresh in their minds. In addition, the program never requires maintenance. Taken together, these characteristics result in an unnatural situation. Contrast this with the production environment discussed at the beginning of this chapter, where *proper* commenting is absolutely essential for correct program utilization. Notice the emphasis on the word *proper*. Not all comments are helpful, and certain commenting styles can even be detrimental to program understanding.

The single most important comment in a program is the *preface* or *prologue* comment, which should appear near the beginning of each program unit. Just as the preface of a book introduces its contents and purpose, the prologue comment should introduce the program unit to follow. The exact nature of the information in a prologue comment depends on the specific program or on specific company policy, but it should contain the following information.

1. A brief summary of what the program does and the method it uses.
2. The programmer's name(s).
3. The date written.
4. A reference to the written technical and user documentation manuals that give additional information about this program.
5. A brief history of all modifications to the program and why they were necessary.

For a procedure or function subprogram, a useful prologue comment should probably include the following two points.

1. *The Entry Conditions.* The initial state of the subprogram and what initial values are passed into it.
2. *The Exit Conditions.* What values are returned on completion and whether the state of the program has changed in any other way.

Figure 2.3 shows a typical prologue comment for the sample program of Figure 1.11.

```
program example(input,output);

{

    this program reads in student quiz scores, computes the mean score, and
    outputs the test scores, the mean, and the letter grade a, b, c, or f based on
    the mean. the program does extensive error checking and will detect improper
    student scores.

    author: g. michael schneider

    date : january 1, 1980

    refer to the book advanced programming and problem solving in pascal,
    john wiley and sons, pp. 40 to 43 for further specifications and details. }
```

**Figure 2.3.** Sample prologue comment.

In a way, the declaration section itself can be thought of as a prologue or introduction to the program. Therefore closely related to the prologue comment are comments explaining each individual item in either a **const** or **var** declaration, making these declarations act almost as a table of contents for the rest of the program.

```
const
    high          = 999;              { highest valid flight number }
    low           = 1;                { lowest valid flight number }
    max           = 1000;             { the maximum length of the flight
                                        tables }
type
    flightrange   = low .. high;
    list          = array [1 .. max] of flightrange;
var
    available     : integer;          { the number of seats currently
                                        available on this flight }
    flights       : list;             { list of all flight numbers going in or
                                        out today }
    found         : boolean;          { switch indicating whether flight was
                                        found }
    i             : integer;          { loop index }
    j             : integer;          { loop index }
    key           : flightrange;      { flight number we are searching }
    tally         : integer;          { if found was true, tally will be the
                                        number of passengers holding
                                        confirmed reservations }
```

If the declarations section has been commented in this fashion, we can determine the purpose of any variable by referring to the **var** declaration (always in a fixed and known location) and reading the comment about that variable. The declaration section of the sample program in Figure 1.11 also uses this technique.

To make the declaration section even easier to use, it is good stylistic practice to alphabetize the names in the **const, var**, and **type** declarations. As we have stressed repeatedly, real-world programs can be large, containing thousands of statements and hundreds of variables. Alphabetizing this list can significantly reduce the time needed to locate a specific declaration and its comment. We will follow this practice in our sample programs.

Another very useful set of comments are those that *paragraph* the program (i.e., identify and introduce blocks of code that perform a single task). The comment can explain the purpose of the upcoming segment and mark the beginning and end of the individual segments.

```
{ data input section }
n := 0;
while not eoln do
begin
    n := n + 1;
    read(string[n])
end;
readln;
read(ch);

{ guarantee that the character just read is alphabetic }
if ch in ['a' . . 'z'] then
begin
    { count frequency of character in the input text }
    count := 0;
    for i := 1 to n do
        if ch = string[i] then count := count + 1;

    { output the character and its frequency count }
    writeln(' input character = ',ch);
    writeln(' frequency count = ',count)
end
else writeln(' *** error - - character must be alphabetic, please resubmit *** .')
```

When dividing a program into paragraphs, it is helpful to have an explanatory comment to introduce each new paragraph. However, even a series of comments consisting of nothing more than *blank lines* can be helpful. Even though they do not explain anything, they still visually isolate groups of related statements.

```
read(x,y,z);
read(theta);

first := sqr(x) + theta;
second := sqr(y) + theta;
third := sqr(z) + theta;

writeln;
writeln(first,second,third)
```

Another very helpful class of comments are those that identify matching *pairs* of reserved words. This is especially important in matching the beginning and end of nested loops or large, compound statements. It is very common to have a number of compound statements (within loops or conditionals) end near one another.

Although the indentation will give us a clue, a comment attached to each **end** can also help us identify the matching **begin-end** or **case-end** pairs, as in the following.

> **end** { *of inner while loop* }
>     **end** { *of else clause for negative parameters* }
>     . . .
> **end**; { *of procedure readfile* }

Surprisingly, the least useful types of comments are those that annotate a single line of code. If we adhere to the stylistic guidelines that are being presented here, most individual lines of code will be quite clear and "self-documenting." Comments that simply restate the obvious intent of a Pascal statement are usually worthless. For example:

> { *input the date* }
> *read(day,month,year);*
>
> { *validate day and month* }
> **if** *(day > 31)* **or** *(month > 12)* **then**

However, this does not mean that all such comments are unnecessary. The purpose of some individual statements may not be immediately obvious and, for those, a comment can be quite useful, even essential. For example, if we were searching for a match between a string of length m within some text of length n (m $<=$ n), the last possible starting position for such a match (assuming the array was indexed from 1) would be the subscript n $-$ m $+$ 1. However, to a first-time reader, the reason for the upper limit in this loop construct — **for** i $:=$ 1 **to** (n $-$ m $+$ 1) **do** — would probably not be immediately clear. It would have been much better to say:

> { *position n $-$ m $+$ 1 is the last possible place a successful match can start* }
> **for** *i $:=$ 1* **to** *n $-$ m $+$ 1* **do**

Closely related to the unnecessary comment is the *cryptic* comment. The code should tell us *how* an operation is to be done. The comment should enlighten us about *why* we are doing it. Avoid the unclear, technical, and jargon-filled comment.

> { *check the 2 link fields in the doubly linked list to see if they are pointing to the same node element* }
> **if** *head↑ = tail↑* **then** . . .

How much nicer it is to say:

> {*determine if we have come to the end of the list*}
> **if** *head↑ = tail↑* **then** . . .

So far we have concentrated on *which* comments should (and should not) be placed in a program. Another important stylistic point is that comments should be written along with the code, as it is being developed, *not after*. Programmers should treat the comment as a Pascal statement (much as the assignment or **if/then/else**) and "code" with it right from the start. If you fail to do this, you will have to comment an entire (possibly large) program all at once. What will probably result is a grossly undercommented program, with brief comments that are only marginally helpful. Also, if there is a significant time lapse between the writing of a code section and the commenting of that section, the comment may be either inappropriate or less detailed than it would have been if we had done it immediately. Time can quickly dim even the best of memories.

Finally, after all comments are written, our job is still not finished. As we perform the job of program maintenance, it is critical that we modify the comments to reflect the new intent of any modified code. Programmers rely heavily on comments to explain what is going on. Incorrect or outdated comments can thus be ruinous to a proper understanding of the program. If we say:

{ *here we compute the median score of all who completed the test* }

but then later rewrite the code to calculate the arithmetic mean (or average), we could completely mislead someone who is trying to understand what our program does. When you change a section of a program, simultaneously change all necessary comments to reflect the change. Incorrect or misleading comments are much worse than no comments at all.

*Style Review 2.3* _____

### Commenting

*Do:*
> Use prologue comments.
> Comment each line in a **var** and **const** declaration.
> Paragraph a program with either comments or blank lines.
> Use comments to help match **begin-end** pairs.
> Comment any statement whose intent is not immediately obvious.
> Include comments from the very beginning of the coding phase.

*Do Not*
> Simply reiterate the obvious intent of a statement.
> Let comments get out of date with the code.
> Write cryptic comments.

### 2.2.4  Indentation and Formatting

As we indicated in Figure 2.1*b*, Pascal is a free-format language (with the minor exception that it does not allow splitting constants and identifiers between two lines). The selection of guidelines for both horizontal and vertical spacing of programs is totally a matter of style.

The question of vertical spacing—blank lines—was treated in the previous section under the topic of blank comments. The question of horizontal spacing, usually termed *indentation*, will be treated here.

As we have stressed repeatedly, Pascal programs will always be written as a series of properly nested control constructs. The depth of this nesting is usually called the *level* of a statement. All programs initially begin at level 0.

Each time we enter either a loop or an alternative of a conditional statement, we enter the next level. The major use of indentation guidelines is as a *level indicator*— as a visual aid in determining the nesting structure of the program we are reading or writing.

Although there are a number of different indentation styles (and their proponents still have heated arguments about which is best), the best style is still determined by which indentation scheme most clearly highlights the hierarchical structure of a program. Figure 2.4 shows the indentation guidelines that we feel are most helpful and most succinct; they will be used in the remainder of this text. In all cases the process of proceeding from level i to level i + 1 is clearly delineated by an appropriately indented space; all statements at the *same* level will always be aligned. The result is good visual interpretation. The actual number of spaces to indent between levels is a matter of personal taste, and this is usually based on things such as the printing characteristics of a particular output device or the width of the paper being used. Three to five spaces usually give enough good visual clues and avoid one major problem—indenting so far to the right margin that there is barely enough room left for the statement itself. (However, if you have nested yourself this many levels deep in a single program unit, you probably have other programming problems, which we will discuss in Chapter 5.)

The procedure sort in Figure 2.5 clearly illustrates the advantages of a good indentation scheme. There is no difficulty at all in determining the scope of any of the loops or conditions within the procedure, even though the procedure reaches a nesting depth of five levels.

It is important to remember, however, that indentation only *highlights* structure, it does not *cause* it. Regardless of whether or not we indent statements, execution will be unaffected, and the program will have the same number of levels. This point is demonstrated by the program fragment at the top of page 69.

```
while bool-exp do
begin
      s1;
      s2;
      . . .
      sn
end { of while }

repeat
      s1;
      s2;
      . . .
      sn
until bool-exp

for j := initial-exp to final-exp do
begin
      s1;
      s2;
      . . .
      sn
end { of for }

if bool-exp then
begin
      s1;
      s2;
      . . .
      sn
end { of then clause }
else
begin
      s1;
      s2;
      . . .
      sn
end { of if }

case exp of
      caselabels : s1;
      caselabels : s2;
      . . .
      caselabels : sn
end { of case }
```

**Figure 2.4.** Indentation guidelines on Pascal control statements.

|  | *Level* |
|---|---|
| **while not** *eof* **do** | 1 |
| **begin** | 2 |
| *readln(n);* | 2 |
| **for** *i := 1* **to** *n* **do** | 2 |
| **begin** | 3 |
| *read(ch);* | 3 |
| *write(ch)* | 3 |
| **end** | 3 |
| **end***;* | 2 |
| *write(n)* | 1 |

This fragment represents a nesting depth of two levels, regardless of the fact that we have (improperly) chosen not to indicate that.

```
{ sort

    this procedure sorts a list of n integers into descending order using the process
    called a selection sort. if the sort was successful then success is set true,
    otherwise it is false.

    note: we assume the following have been previously declared
const
    lowerlimit          = ?;   (lower bound of array)
    upperlimit          = ?;   (upper bound of array)
type
    arrayindex          = lowerlimit .. upperlimit;
    arraytype           = array [arrayindex] of integer;

    entry conditions:
    list                the array of integers to be sorted
    firstelement        the index of the first element to sort
    lastelement         the index of the last element to sort
    lowerbound          the lower array bound for list
    upperbound          the upper array bound for list

    exit conditions:
    success             true, if sort was successful; false otherwise
    list                if success is true, then the list of values is returned in
                        sorted order

    note: a selection sort is appropriate for use on relatively small lists only. for a
    discussion on sorting efficiency, see chapter 11 }
```

```
procedure sort (var list : arraytype; firstelement : arrayindex;
        lastelement : arrayindex; lowerbound : arrayindex;
        upperbound : arrayindex; var success : boolean);
var
    big         : integer;      { largest item of sublist }
    i           : arrayindex;   { loop index }
    j           : arrayindex;   { loop index }
    location    : arrayindex;   { position in sublist of big }
    temp        : integer;      { used to exchange two array elements }

begin
    if (firstelement > = lowerbound) and (lastelement < = upperbound)
        and (firstelement < = lastelement) then
    begin
        success : = true;
        for i : = firstelement to lastelement − 1 do
        begin
            big : = list[i];
            location : = i;
            { find the largest item in sublist beginning at position i + 1 }
            for j : = i + 1 to lastelement do
                if list[j] > big then
                begin
                    big : = list[j];
                    location : = j
                end; { of if and for j }

            { now interchange the largest item with the one at position i }
            temp : = list[i];
            list[i] : = list[location];
            list[location] : = temp
        end { of for i }
    end { of then clause }
    else success : = false
end; { of procedure sort }
```

*Figure 2.5.* Selection sort procedure—an example of indentation.

Aside from the use of indentation to indicate levels, the other indentation guidelines are of significantly less importance. Two that should be mentioned concern multiple statements per line and multiple lines per statement.

Pascal does allow you to place more than one statement on a line; however, there are two good reason to avoid this. First, we may miss a statement altogether as our eye naturally follows down the margin of aligned statements.

```
initialvalue := 0; i := 1;
read(final);
median := (initialvalue + final) / 2.0
```

We may, on a quick and casual reading of the preceding statements, miss the initialization of i. Second, when a statement appears on a single line, its nesting level is not indicated by indentation level. This can be important if you attempt to place two control statements on a single line—usually a bad practice.

**if** *a* = *0* **then if** *b* <> *0* **then** *c* := *b* **else** *c* := *a*

However, we will not adhere to this restriction too rigidly and will occasionally place more than one *noncontrol* statement per line, especially if they are logically related or if we wish to describe both with a single comment.

```
z := 0; y := 0;    { initialize the cartesian coordinates }
read(x);
writeln(' x = ',x)
```

If a statement is too long to fit on a single line, we should position the next (or continuation) line so that structurally related parts of the statement are aligned.

**if** ( (*a* < *0*) **or** (*a* > *99*) ) **and**
    ( (*b* < *0*) **or** (*b* > *99*) ) **and**
    ( (*c* < *0*) **or** (*c* > *99*) )
**then** . . .

*Style Review 2.4* _____

*Indentation*

*Do:*
    Indent to highlight the nesting depth of a group of control statements.
    Indent multiline statements so that structurally related clauses are
        aligned.

*Do Not:*
    Place more than one control statement per line.

### 2.2.5   Unburdening the User

In Figure 1.8 we listed a set of precedence rules that allow us to determine unambiguously the order of execution of arithmetic or logical operations within an expression. Using these rules, we could determine, for example, that in the assignment statement:

$b := x - a$ **div** $2$

a **div** 2 is to be evaluated before the subtraction. However, we want to avoid, whenever possible, writing a program that requires recall of specific syntactic or semantic rules—in this case, the precedence rules of the language. Instead, we would add our own stylistic rule that says to *parenthesize for both order of evaluation and clarity of expression*. Parenthesize expressions so the order of evaluation is obvious on a first reading.

$b := x - (a$ **div** $2)$

Nothing has been changed (from the machine's point of view), but now readers of the program cannot possibly misinterpret the meaning of the statement. Of course, there is a point of diminishing returns, where overparenthesizing can actually make an expression *more* difficult to read. However, this usually does not occur until the nesting depth reaches three or four levels. In most cases, the intelligent use of a limited number of extra grouping symbols will enhance a program's clarity.

A second example of the same problem is the "dangling **else**" problem referred to in Section 1.5.4.1. The statement:

**if** $x > 0$ **then**
    **if** $y > 0$ **then** *root* $:= y$
    **else** *root* $:= x$

although indented reasonably, could still confuse a reader as to which **then** the **else** belongs to. An "extraneous" **begin-end** could completely alleviate this problem and avoid having a program that relies on the user's (possibly faulty) memory.

**if** $x > 0$ **then**
**begin**
    **if** $y > 0$ **then** *root* $:= y$
    **else** *root* $:= x$
**end** { *of if* $x > 0$ }

There are numerous other situations where this can occur; the rule is to *help users by unburdening them from having to remember a specific syntactic rule to interpret*

*properly a statement or piece of data.* If it means having to add unneeded symbols—
( ), **begin-end**, or blank spaces — that is insignificant. What is important is what we
have been stressing and striving for in this chapter: clarity of expression.

## 2.3  EXAMPLE PROGRAM

We will now write a program to illustrate the stylistic characteristics that we have
been stressing throughout this chapter. The program will read in examination scores,
punched one per card (or entered one per line), and produce seven values.

1.  A list of scores with illegal scores flagged as errors.
2.  The number of valid scores.
3.  The arithmetic *mean*, or average.
4.  The *mode* and its frequency. The mode is the individual score that occurred
    with the greatest frequency. (If two or more scores have the same frequency,
    print the highest mode.)
5.  The lowest and highest scores.
6.  A *histogram*, giving for each valid score a bar whose length is proportional
    to the number achieving that score.
7.  The total number of illegal scores.

For the following input:

```
 30
100
 20
100
 50
```

the desired output (for an assumed exam score range of 0 to 100) is:

the list of all scores:
```
 30
100
 20
100
 50
```
number of good scores = 5
mean = 60.00
mode = 100 frequency = 2
lowest score = 20 highest score = 100

```
score frequency
    100 **
     99
     98
     97
     96

      .
      .
      .
     50 *

      .
      .
     30 *

      .
      .
     20 *
number of illegal scores = 0
```

The complete program to produce the preceding output is shown in Figure 2.6 on pages 75-77. It should be used as a guide for the syntactic style points that we have discussed in this chapter.

```
program statistic(input,output);
{ this is a program to input test scores and produce as output the median, the
  mode, the lowest and highest scores, and a histogram of all scores.

authors   : g. michael schneider
            steven c. bruell
            university of minnesota
            minneapolis, minn

date      : january 1, 1980 }

const
     headersize    = 10;                  { length of header for histogram }
     high          = 100;                 { highest test score }
     linesize      = 132;                 { maximum output line width in characters
                                            for our line printer }
     low           = 0;                   { lowest test score }
     printchar     = '*';                 { histogram printing character }

type
     scorerange    = low .. high;

var
     count         : integer;             { count of all valid scores
     freq          : array [scorerange] of integer;   frequency count of test
                                                        scores }
     i             : scorerange;          { for loop index }
     illegal       : integer;             { count of all invalid scores }
     j             : integer;             { for loop index }
     maxcount      : integer;             { used to find the mode
     maxscore      : scorerange;          { highest legal score read }
     mean          : real;                { average of all valid scores
     minscore      : scorerange;          { lowest legal score read }
     mode          : scorerange;          { score with the highest frequency }
     printsize     : integer;             { number of spaces left for printing the
                                            histogram after we have printed the
                                            header }
     score         : integer;             { an individual quiz score
     sum           : real;                { sum of all valid scores }

begin
     { initialization section }
     for i := low to high do freq[i] := 0;
     count := 0;
     illegal := 0;
     sum := 0.0;
     printsize := linesize – headersize; { the header for each line of the histogram is
                                           headersize characters }
     writeln(' the list of all scores:');
```

```
{ main loop to process all test scores }
while not eof do
begin
      readln(score);
      writeln(score:headersize);
      if (score >= low) and (score <= high) then
      begin
            sum := sum + score;
            count := count + 1;
            freq[score] := freq[score] + 1
      end { of then }
      else
      begin
            illegal := illegal + 1;
            writeln('the above score is illegal and is not included ',
                      'in the following statistics')
      end { of else }
end; { of while }
```

```
{ now we compute the mean and mode of the scores, if there were any valid
  scores encountered. }
if count > 0 then
begin
      writeln(' number of good scores = ',count:headersize div 2);
      mean := sum / count;
      writeln('mean = ',mean:headersize:2);

      { now compute mode }
      maxcount := 0;
      for i := low to high do
            if freq[i] >= maxcount then
            begin
                  maxcount := freq[i];
                  mode := i
            end; { if and for }
      writeln(' mode = ',mode:headersize div 2, ' frequency = ',
                freq[mode]:headersize div 2);

      { find min and max scores }
      minscore := low;
      while freq[minscore] = 0 do minscore := minscore + 1;
      maxscore := high;
      while freq[maxscore] = 0 do maxscore := maxscore - 1;
      writeln(' lowest score = ',minscore:headersize div 2, ' highest score = ',
                maxscore:headersize div 2);
```

```
{ now set up to print the histogram }
writeln(' score ':headersize, 'frequency');
for i := maxscore downto minscore do
begin
      { print the header for this line }
      write(i:headersize div 2, ' ':headersize div 2);

      { we must first check to see if the length of the bar to be printed on the
         histogram exceeds the length of a line. if so, we will have to print it on
         multiple lines. }
      while freq[i] > printsize do
      begin
            { print a full line }
            for j := 1 to printsize do write(printchar);
            writeln;
            write(' ':headersize); { skip header on continuation line }
            freq [i] := freq[i] – printsize
      end; { of while }

      { print the partially filled last line }
      for j := 1 to freq[i] do write(printchar);
      writeln
end; { of for i }

{ finish up the report with the illegal score count }
writeln(' number of illegal scores  =  ',illegal:headersize div 2)
end { of then clause for valid scores }
else writeln(' sorry - - no valid scores were processed')
end. { of program statistic }
```

**Figure 2.6.** Sample program for elementary statistics.

## EXERCISES FOR CHAPTER 2

1.  The following program reads a value for x and computes the value $e^x$ using the following formula.

$$e^x \approx 1 + x + \frac{x^2}{2!} + \frac{x^3}{3!} + \ldots \cdot \frac{x^n}{n!}$$

```
program ex(input,output);
var i : integer;
    x,s,n,d : real;
begin
readln(x);
s := 0.0;
n := 1.0;
d := 1.0;
i := 0;
repeat
s := s + n / d;
n := n * x;
i := i + 1;
d := d * i
until (n/d > (− 0.01) ) and (n/d < 0.01) or (i > 10);
writeln(s)
end.
```

Discuss what you feel are poor stylistic habits displayed in this program. Rewrite the program so that it achieves the same result but in a way that is easier to read and understand.

2.   The following procedure computes the first and third quartile of a list of examination scores. Discuss which comments you feel are helpful to understanding the program and which you feel are not very helpful. Discuss which comments actually interfere with your understanding of the program. Rewrite the procedure to meet your own commenting standards and other stylistic concerns.

```
procedure quartile(list : arraytype; n : integer;
        var fq, tq : integer; var fail : boolean);
```

{ *This procedure takes a list of examination scores, 'list,' of length 'n,' and returns the first and third quartiles in 'fq' and 'tq,' respectively. The first quartile is the score at which 25% did worse and 75% did better. The third quartile is the score at which 75% did worse, 25% did better. If we were unable to compute the quartiles, the parameter 'fail' is set to true; otherwise, it is set to false.* }

```
var errflag :   boolean;   { a flag returned by the sorting routine to determine
                              whether or not it worked properly }
    fqindex :  integer;    { the subscript position in list of where the first quartile
                              score is located }
    tqindex :  integer;    { the subscript position in list of where the third quartile
                              score is located }
```

```
begin
    { first we must sort the list into ascending order }
    sort(list, n, errflag);
    { check the flag to see if there was any error encountered during the sorting
      operation }
    if errflag then
        { set the switch to true }
        fail := true
    else
    begin
        { set switch to false }
        fail := false;
        { now we will compute the quartiles. first accurately determine their position
          in the list }
        fqindex := round (n/4.0);
        { determine the third quartile index }
        tqindex := round (3.0 * n/4.0);
        { now set the parameters to the appropriate value. first fq }
        fq := list [fqindex];
        { and now tq }
        tq := list [tqindex]
    end { of the else no error clause }
end; { of procedure quartile }
```

3. Modify the statistics program in Figure 2.6 so that, in addition to what it now does, it also computes and prints the *standard deviation*. The standard deviation, sd, of a list of n values, $x_i, \ldots, x_n$, is defined as:

$$sd = \sqrt{\frac{\sum_{i=1}^{n} (x_i - M)}{n}}$$

where M is the mean of the scores $x_i$. Discuss how you feel the style of the program helped or hindered you in understanding and modifying the program.

4. Modify the procedure sort in Figure 2.5 so that it:

(a) Sorts into ascending order.

(b) Only sorts the first m items into place, $m <= n$. The remaining $n - m$ items can be left unsorted at the end of the list. That is, if $m = 7$, $n = 10$, procedure sort will correctly sort the first seven items in the list. The values in list[8] . . . list[10] are not affected by the procedure.

Again, discuss how you feel the presentation format of the program helped you with the maintenance task.

5.  Write a complete Pascal program to solve the following problem. Given a series of data cards on the physical characteristics of men and women in the following format:

(a)  Sex code: "m" or "f".

(b)  Age (in years).

(c)  Height (in inches).

(d)  Weight (in pounds).

produce a report in the following format.

| Age | Average Height for Men (in.) | Average Height for Women (in.) | Average Weight for Men (lb) | Average Weight for Women (lb) |
|---|---|---|---|---|
| 0–17 | XX.X | XX.X | XXX.X | XXX.X |
| 18–29 | . | . | . | . |
| 30–39 | . | . | . | . |
| 40–49 | . | . | . | . |
| Over 50 | . | . | . | . |

Discuss the stylistic aspects of your program with regard to mnemonic names, indenting, commenting, and parenthesizing. If you have developed personal stylistic habits that differ from those presented in this chapter, discuss why you feel your presentation format is an improvement.

6.  Write a Pascal procedure *roman* that takes a character array containing Roman numerals and returns the corresponding decimal value. The end of the Roman numeral is indicated by a blank. The valid characters that may appear in the Roman numeral are I (1), V (5), X (10), L (50), C (100), D (500) and M (1000). Thus, if the array contained

```
roman[1] = 'C'
roman[2] = 'C'
roman[3] = 'L'
roman[4] = 'I'
roman[5] = 'X'
roman[6] = ' ' (blank)
```

the procedure should return the decimal value 259.

7.  Take a program you wrote in an earlier programming class before you were introduced to the material in this chapter. Review your own early programming style and discuss its weak and strong points. In what ways did you follow the guidelines presented here? In what ways was your style different? How do you think it can be improved?

# Chapter 3

# PROGRAMMING STYLE: THE RUN-TIME BEHAVIOR OF PROGRAMS

## 3.1 INTRODUCTION

In Chapter 2 we presented guidelines for developing programs that are *clear* and *readable*. But clarity alone is not enough. To maximize utility and flexibility, the run-time characteristics of programs must also conform to certain stylistic constraints and guidelines.

Too often, however, programmers are concerned only with *correctness*. Of course, a program must produce proper results, but correctness is not the only important run-time characteristic. This chapter will discuss many other desirable qualities.

## 3.2 ROBUSTNESS

A *robust* program will produce meaningful results from any data set, regardless of how illegal, implausible, improper, or "pathological" it may be. Notice that we said meaningful results, not correct answers. For most problems there will be data sets for which we cannot produce an answer or apply an algorithm. But we can always perform certain operations that are meaningful and helpful to the user. This could include printing an appropriate error message, explaining why the operations cannot be executed and, if possible, supplying details about how to correct and resubmit the data. Under certain conditions, the program could also reset data items within allowable limits (after informing the user).

```
if n > maximum then
begin
    n : = maximum;
    writeln ('the list was too large. it will be reset to',maximum)
end { of if }
```

Robustness guidelines require that under *no* circumstances will a program terminate abnormally during execution. Terminations normally should occur only at the end of the main program, and the output produced should be a meaningful result directly related to the input data provided.

### 3.2.1  Input Validation

The most important operation a programmer can perform to insure robustness is a complete and thorough validation of all input. Errors in input data are common and can be caused by any of the following.

1. Data entry errors when typing in the data (hitting a 7 instead of a 1).
2. Improper ordering of the data items on a line or of the lines themselves.
3. A misunderstanding of what type of value to enter [e.g., entering a date as June 10, 6/10, or 161 (the number of days since January 1)].

There is a popular phrase in programming: GIGO (garbage in, garbage out). If improper data values are accepted by a program, unpredictable and meaningless operations will result, regardless of the care we have taken to write a good program.

Input values must first be validated for *legality*. They must be within the bounds set by either the problem specification or physical reality. For example, the specifications of a payroll program may state that the following bounds apply.

$$0 < \text{ID number} <= 99999$$
$$0 <= \text{Dept number} <= 99$$
$$\$1.90 <= \text{Payrate}$$
$$0 <= \text{Exemptions}$$

To this we could add the following constraints based on natural physical limits.

$$0 <= \text{Daily Hours Worked} <= 24$$
$$0 <= \text{Days Worked This Week} <= 7$$

and check that all input falls within this range.

```
type
    errortype = (iderror, depterror, payerror, exemerror, hourserror, dayerror);

var
    error      : set of errortype;
        .
        .
        .
begin
    error := [ ];
    if (id <= 0) or (id > 99999) then
        error := error + [iderror];
```

```
if (dept < 0) or (dept > 99) then
      error : = error + [depterror];
if payrate < 1.90 then
      error : = error + [payerror];
if exemptions < 0 then
      error : = error + [exemerror];
if (hours < 0) or (hours > 24) then
      error : = error + [hourserror];
if (days < 0) or (days > 7) then
      error : = error + [dayerror];

      .
      .
      .
if error = [ ] then "process the data"
else "invoke error procedure"
```

This example illustrates another important point about input validation; in case of an error, we should gather enough information to be able to tell the user exactly what the error was and where it occurred. In the preceding example we made a separate entry in the variable error for each of the six possible types of error. Now the error procedure we write can produce detailed messages of the form:

```
** error in the hours-worked field of this record
** value must be an integer in the range 0-24 **
```

simply by checking the variable error:

```
if iderror in error then
      writeln (' ** error in id field. value must ',
            'be in the range',low,'to',high);
if hourserror in error then
begin
      writeln(' ** error in the hours-worked field of ',
            'this record.');
      writeln(' ** value must be an integer in ',
            'the range 0-24 ** ')
end
```

If we had made error a boolean value with, say, false signifying no errors and true signifying one or more of the preceding six error types, the most we could say is:

```
** Error on this data record **
```

We would have no additional information about the error because of the data type selected. Always try to relate an error message to the mistake that caused it. All-inclusive error messages should be avoided.

Validating input for legality is not sufficient. We must also validate for *plausibility*. When a data value is highly unlikely (although not impossible), we should identify it for special (manual) processing. Many values provided to a program have no preestablished upper bound that we can check. For instance, in the previous example, what is the maximum payrate allowed? (Some consultants may earn $200 per hour.) What is the maximum number of exemptions allowed? If there were no way to validate this type of value, we would develop programs that accept hourly payrates of $123,456 or a claim of 35,000 dependents—certainly suspicious! To prevent this, during the problem specification phase we must develop *plausibility bounds* in conjunction with the end-user. Then we can verify within the program that all values are legal and also plausible.

```
{ do not accept pay rate over $30 }
if payrate > 30.00 then error : = error + [overpayerror];

{ do not accept 20 or more dependents }
if exemptions > = 20 then error : = error + [overexempterror]
```

These cases cannot be automatically considered input errors because the values might be correct, albeit unusual. However, users will at least be made aware of the suspicious value and can rely on their own judgment instead of on the program to determine if the data item is correct.

Finally, even if improper input slips through our checks for legality and plausibility, we have one final checkpoint—the user. By always *echo printing* the input data, we allow the user to locate incorrect data before utilizing the output. Always include, somewhere in the output, the input data that produced those results.

```
read(x);
writeln(' x-coordinate = ',x)
```

Exactly where this is to be printed will naturally depend on the problem specifications. However, regardless of exactly where it is to be placed, the output should include, as one of its reports, an exact copy of the input file.

Between echo printing and validating for legality and plausibility, the input validation phase can be quite long. It is not unusual, especially for commercial and business-oriented problems, for the input and input checks to account for 20, 30, or even 40% of the statements in the program. Your program must be protected against *all* improper data, no matter how pathological. The number of lines it takes to achieve that level of security is not important. A well-known maxim of programming states that the error you do not check for is the one that will occur!

*Style Review 3.1* _____

> *Input Validation*
>
> Always validate input for legality.
> Always validate input for plausibility.
> Always echo print the input.
> If an error is detected, try to capture enough information to identify the exact cause of the error.
> Your goal should be to write a program that is protected against all improper data.

### 3.2.2 Protecting Against Run-time Errors

A *run-time error* causes a program to terminate abnormally during execution. It may be caused by improper input data, or even by data that are perfectly legal and plausible. For example, even if we validate that $0° <= \theta <= 360°$, $\tan(\theta)$ is still undefined at $\theta = 90°$ and $270°$, and any attempt to use those values will cause the program to abort. Obviously, protection against run-time errors is necessary in addition to input validation to guarantee program robustness.

In Pascal, the most common run-time errors are caused by:

1. Array subscript out of bounds.
2. Subrange out of bounds.
3. Real to integer conversion with the absolute value of the real > maxint.
4. Case statement expression not matching one of the case labels.
5. Improper argument for a function.

   | | |
   |---|---|
   | *sqrt(x)* | *x < 0* |
   | *ln(x)* | *x <= 0* |
   | *chr(x)* | *x < 0 or x > maximum number of characters* |
   | *succ(x)* | *where x is the last entry in the scalar data type* |
   | *pred(x)* | *where x is the first entry in the data type* |

6. Input errors: illegal input formats or attempts to read past the end-of-file.
7. Referencing a data object before it is defined.
8. Division by zero.

To prevent an operation from aborting (even in spite of our input checks), we must include sufficient tests in the program before performing that operation. Inclusion of these precautions is usually termed *defensive programming*—anticipating potential trouble spots and guarding against them. [There is one notable exception. Most implementations of the read and readln procedures will automatically abort the program if they encounter an error during input (e.g., a nonnumeric character punched in an integer field). Users cannot prevent this loss of control except by choosing not to use the read or readln procedures.]

For example, the following assignment statement computes one root of a quadratic equation using the well-known quadratic formula: (Assume that we have already checked that a ≠ 0.)

```
root1 := (−b + sqrt(sqr(b) − 4.0*a*c)) / (2.0*a)
```

However, if sqr(b) − 4.0*a*c < 0.0, there is a complex root, and the preceding statement will attempt to find the square root of a negative value, causing the program to terminate. It is better to say:

```
discriminant := sqr(b) − (4.0 * a * c);
if discriminant >= 0.0 then
begin
      root1 := (−b + sqrt(discriminant)) / (2.0 * a);
      root2 := (−b − sqrt(discriminant)) / (2.0 * a)
end
else complexroots { procedure to compute complex roots }
```

Another example is the following method for counting the frequency of each letter in a text.

```
var
      freq      : array ['a'..'z'] of integer;
          .
          .
          .
begin
      read(ch);
      freq[ch] := freq[ch] + 1
```

This code works properly only as long as we can guarantee that all characters in the text are in the range 'a' to 'z'. The occurrence of a single out-of-range character will abort the entire program with an array subscript out of bounds. A better way to code that operation would be:

```
read(ch);
if (ch < 'a') or (ch > 'z') then illegal := illegal + 1
else freq[ch] := freq[ch] + 1
```

Another common cause of run-time errors is the failure to handle the null case properly. The *null case* is the empty data set (e.g., no input, no legal values, or a list of length 0). If we are not careful, it is easy to write code that works properly only when there is one or more valid pieces of data. This harmless-looking loop:

```
i := 0;
repeat
    i := i + 1;
    readln(table[i])
until (i = listsize) or eof
```

will "blow up" if the input file is *initially* empty. We will attempt to perform a readln operation while eof is true and will terminate with a fatal error. Likewise:

```
sum  := 0;
i     := 0;
repeat
    i      := i + 1;
    sum  := sum + table[i]
until i = listsize
```

is correct only if listsize > 0. If we are ever presented with a table of length 0, this fragment will abort with an array index out of bounds (i.e., table[1]).

Referring to the code fragment in Figure 2.2, we can see one more example of the failure to handle the null case. That program attempted to sum up separately the positive and negative values contained in a list. However, it fails to handle the special case of a list without either one positive or one negative value. (It either produces a wrong answer or terminates abnormally with an array reference out of bounds.)

When writing a loop, immediately ask yourself if any null cases are possible and, if so, check to see if the loop as written will properly handle those cases.

Finally, be particularly careful with the **case** statement in Pascal. Failure to match a case label with the value of the selector expression can be a fatal error.

```
case yearinschool of
    1,2 : processlowerclassman;
    3,4 : processupperclassman;
    5   : processgraduate
end { of case }
```

If we cannot guarantee that "yearinschool" will always be an integer in the range 1 to 5 (possibly from an earlier input check), the preceding statement is very risky. It is much safer to write:

```
if (yearinschool >= 1) or (yearinschool <= 5) then
begin
    case yearinschool of
          1,2 : processlowerclassman;
          3,4 : processupperclassman;
          5   : processgraduate
    end { of case }
end { of then }
else errorhandler
    .
    .
```

As the previous examples clearly indicate, programmers must think defensively by anticipating and preventing problems. A run-time error is a sure sign of a poorly designed and poorly written program.

### 3.2.3   Protecting Against Representational Errors

On any machine, we can represent exactly all scalar data types except real. With any positional numbering system—binary or decimal—we may incur a *round-off error* in attempting to represent a real value with a finite number of decimal places. We are all familiar with this limitation in our own decimal notation.

$$1/3 = 0.33333. . .$$
$$1/7 = 0.1428571428. . .$$

The same thing happens in the *binary,* or *base 2,* numbering system used by almost all computers.

$$\frac{1}{5}_{10} = 0.001100110011. . ._2$$

Because of this round-off error we should never expect a real number to be exactly equal to a particular value. For example, when $\frac{1}{5}$ is represented on a computer with eight binary place accuracy, we get the following representation.

$$\frac{1}{5} \doteq 0.00110011_2 \doteq 0.1992$$

Therefore the apparently true relationship:

$$\frac{1}{5} + \frac{1}{5} + \frac{1}{5} + \frac{1}{5} + \frac{1}{5} = 1$$

would actually be false. (It would be approximately 0.996 on our eight-place machine.)

The implications of this in our programming is that we should never test for exact equality between elements of the real data type. We should either use an ordinal data type that does not suffer from round-off error (integers, character, user-defined) or write our programs in such a way that they are not sensitive to round-off errors in the real values. For example, the following code was intended to process values of x in the range 0.0 to 2.0 in steps of 0.2, a total of 11 iterations.

```
x := 0.0;
while x <> 2.2 do
begin
    .
    .
    .
    { process this value of x }
    .
    .
    .
    x := x + 0.2
end { while loop }
```

However, this code is highly dangerous and could very well result in an infinite loop. Because of round-off errors the value of x after the eleventh iteration may not be exactly 2.2 but $2.2 - \epsilon$, where $\epsilon$ is some small, positive value. This slight difference will cause the inequality test to remain true and the loop to execute repeatedly. A better way to structure the above code would be:

```
x := 0.0;
while x <= 2.0 do
begin
    .
    .
    .
    { process this value of x }
    .
    .
    .
    x := x + 0.2
end { while loop }
```

Now the loop will correctly execute 11 times, even in the presence of a small round-off error in the value of x.

Another source of errors in real numbers is termed *truncation error*. These are errors caused by approximating an infinite mathematical process with a finite number of programming operations. The minute we stop our program we have induced an error into the final result. For example, the roots of the equation:

$$x^2 - 3x - 10 = 0 \qquad \text{(roots are } -2, +5\text{)}$$

can be found by a repeated execution of the following iterative formula:

$$x_{i+1} = x_i - \left( \frac{x_i^2 - 3x_i - 10}{2x_i - 3} \right) \qquad i = 0,1,2,\ldots$$

The formula (based on a technique called *Newton's method*) will ultimately converge to one of the roots of the preceding formula (which root it finds will depend on the starting point $x_0$). If $x_0 = 10$ the formula will generate the following sequence of values (to three-place accuracy):

$$x_0 = 10.000$$
$$x_1 = 6.471$$
$$x_2 = 5.218$$
$$x_3 = 5.006$$

and we are obviously converging on the root $+5$. However, if we stop the iteration after three applications of the formula, we will have induced a truncation error of about 0.006 (5.006 − 5.000). If we were to run the process a great many more times we would get much closer to the correct answer but, at least theoretically, would never get the exact result. Therefore, whenever programming any type of iterative process that converges on the desired result, we cannot use the criteria of exactness as the termination of the process. Instead, we must check to see if the last two values in our sequence are "close enough": that is, if the *absolute truncation error* is acceptably small. For example:

```
{ see if we have converged to the root }
if abs (x[i+ 1] − x[i]) <= epsilon then . . .
```

would work given an appropriate value for epsilon. An alternate way to perform the check is to test the size of the *relative truncation error:*

```
if abs ((x[i+ 1] − x[i]) / x[i]) <= epsilon then . . .
```

This latter way is preferred. This is because it may be extremely difficult in certain cases to achieve an absolute level of accuracy. If epsilon were 0.0001 and our root were near 1 then we would need about 4 significant digits to satisfy the termination test. However, if the root were near 1 million we would need about 10 significant digits. This may be beyond the accuracy capabilities of our machine. A relative accuracy test with $\epsilon = 0.0001$ simply says that the last two values must be within 0.01% of each other, regardless of their absolute magnitudes.

In summary, one must be very careful when using the real data type. Unlike the ordinal types, all real values cannot be represented exactly. Checking for equality of real quantities or expecting exactness of real results can produce a program that may work properly for certain cases but exhibits incorrect or abnormal run-time behavior on others, and this is precisely the flaw we want to avoid.

This has been only a brief introduction to an important but complex subject. For a more technical discussion of finite representations, errors, and error propagation, refer to References 3 and 6 at the end of this chapter.

### 3.2.4  Graceful Degradation

So far this section has been concerned with detecting errors and preventing them from ruining our program. If an error does occur, we should do as much as possible to recover.

The worst thing about run-time errors is that they cause the program to lose control. A cryptic error message is produced (see Section 10.1.2) and the program is terminated whether or not there is still useful work that can be done. By checking for run-time error conditions, we can maintain control and take the actions we feel are appropriate, not the actions provided automatically by the system.

Aside from producing a meaningful error message, a program should always apply the following rules in the following order.

1.  Make a reasonable assumption about the erroneous item that allows continued processing of that item. Inform the user about this assumption.
2.  Discard the erroneous item but continue if possible to process the remainder of the current record.
3.  Discard the current record containing the erroneous item but, if possible, continue to process the current file, until eof.
4.  Discard the file containing the erroneous item but, if possible, process any subsequent files until the end of all information.

Only if it is impossible to apply one of these rules do we come, as a last resort, to rule 5.

5.  Die gracefully. Produce useful, meaningful messages that will allow the user to properly rerun the program and then exit by the end of the main program.

### 3.2.5  Example

Figure 3.1 shows a program that converts dates in the form mm/dd/yy into dates of the form nnn—01/01/80 = 001, 01/02/80 = 002, . . . ,12/31/80 = 366. (The program should probably be developed as a procedure, but we are writing it this way to make a point.)

Notice that we validate $00 <=$ year $<= 99$; $1 <=$ month $<= 12$; and $1 <=$ day $<= 28$, 29, 30, or 31, depending on the month and year. If an error makes it impossible to compute the result, we terminate that data case and go on to the next one. We also check the format of the data to insure that a '/' appears between the month, day, and year. However, since this error does not preclude a correct date (it may have been entered as mm-dd-yy), we will continue to process that data set after warning the user.

The program in Figure 3.1 clearly illustrates a point we made earlier. Almost 50% of the code in that program is there to prevent incorrect results. The program is robust. Regardless of how many statements it takes, make your code impervious to bad data.

*Style Review 3.2*

### Defensive Programming

1.  Prevent run-time errors by checking for them before any risky operation.
2.  Never assume exact equality of real values.
3.  Produce a meaningful error message directly related to the specific error. Whenever possible, keep going in order to get as much additional information and/or results as you can.
4.  Never stop if something useful can be done.
5.  Check to see that you have handled the null cases properly.

## 3.3    PROPER USE OF PROCEDURES AND FUNCTIONS

The proper use of subprograms (procedures and functions) is critical to the development of programs of any reasonable size and complexity. In later chapters we will discuss the development of large programs in a modular, hierarchical fashion using procedures. (Except when otherwise indicated, we will use "procedure" in a generic sense, meaning either a procedure or function subprogram.) The ability to write each type of subprogram is central to good programming. Section 1.4.5 discussed correct Pascal syntax for procedures and functions; this section will discuss the stylistic guidelines for usable, effective subprograms.

**program** *dateconversion(input,output);*
*{ this program converts dates of the form mm/dd/yy into an integer date of the*
*form nnn, where nnn is between 001 and 366 }*

**const**
    *separator*      = *'/';*          *{ separator between portions of date }*

**type**
    *months*      = *1..12;*

**var**
    *ch1*          *: char;*          *{ separator between month and day }*
    *ch2*          *: char;*          *{ separator between day and year }*
    *day*          *: integer;*          *{ input data }*
    *daycount*      *: 0..366;*          *{ the date as an integer }*
    *days*          *:* **array** *[months]* **of** *1..31; { number of days in each month }*
    *error*          *: boolean;*          *{ error flag for input data }*
    *leapyear*      *: boolean;*          *{ true if year is a leap year }*
    *m*          *: months;*          *{ loop index }*
    *month*          *: integer;*          *{ input data }*
    *year*          *: integer;*          *{ input data }*

**begin**
    *{ initialization section }*
    *days[1] := 31; days[2] := 28; days[3] := 31; days[4] := 30;*
    *days[5] := 31; days[6] := 30; days[7] := 31; days[8] := 31;*
    *days[9] := 30; days[10] := 31; days[11] := 30; days[12] := 31;*

    **while not** *eof* **do**
    **begin**
        *error := false;*
        *readln(month,ch1,day,ch2,year);*
        *writeln(' input date: ',month,ch1,day,ch2,year);*

        *{ now validate the input }*
        **if** *(ch1 <> separator)* **or** *(ch2 <> separator)* **then**
            *writeln (' illegal date format. it should be mm',*
                *separator,'dd',separator,'yy');*

        **if** *(month < 1)* **or** *(month > 12)* **then**
        **begin**
            *error := true;*
            *writeln(' error in month field -- must be 1..12. it will be set to 1');*
            *month := 1*
        **end;** *{ of if }*

```
if (year < 0) or (year > 99) then
begin
    error := true;
    writeln(' error in year field -- must be 0. .99. it will be set to 0');
    year := 0
end; { of if }

leapyear := ((year mod 4) = 0) and (year <> 0);
if leapyear then days[2] := 29
else days[2] = 28;

if (day < 1) or (day > days[month]) then
begin
    error := true;
    writeln(' error in day field -- must be 1. .',days[month],
            ' for month ',month,'. it will be set to 1');
    day := 1
end; {of if }

{ start date computation }
if not error then
begin
    daycount := 0;
    for m := 1 to (month − 1) do
        daycount := daycount + days[m];
    { finally add on days in current month }
    daycount := daycount + day;
    writeln(' the corresponding integer date = ',daycount)
end { of if not error }
end { of while not eof }
end. { of program dateconversion }
```

*Figure 3.1.* Program to do date conversion.

### 3.3.1 Parameter Passing Mechanisms

In some programming languages (most notably FORTRAN and BASIC), there is only one way to pass parameters between a calling program and a procedure. In Pascal, however, there are two ways to manage this exchange, and it is extremely important to select the proper technique for each parameter.

The first type of parameter passing mechanism is *call-by-reference*, indicated by the reserved word **var** before the formal (or dummy) parameter. This method functions as if every occurrence of the formal parameter in the procedure were replaced by the actual parameter in the procedure invocation.

```
procedure sample(var x,y : integer);
begin
     x := x + 1;
     y := y + 1
end;

        .
        .
        .

begin
     a := 5;
     b := 10;
     sample(a,b);
     writeln(a,b)
        .
        .
        .
end.
```

In this example, every occurrence of the name "x" is replaced by the name "a," and every occurrence of "y" is replaced by "b." (Actually, *addresses* are being replaced, but we can think in terms of names.) The assignment statements in the procedure actually modify the variables a and b, and the output produced by the writeln command would be:

6   11

The second method of parameter passing is *call-by-value* and is indicated by omitting the reserved word **var** before the parameter name. With call-by-value, we make *local copies* of all dummy parameters and initialize those local copies to the current value of the actual parameters.

```
procedure sample(x,y : integer);
begin
     x := x + 1;
     y := y + 1
end;

        .
        .

begin
     a := 5;
     b := 10;
     sample(a,b);
     writeln(a,b)

        .
        .

end.
```

Now the dummy parameters x, y are call-by-value. The call to sample will create a local variable x that is automatically initialized to the current value of a and a local variable y that is automatically initialized to the current value of b. Except for this initialization, the variables a and b are inaccessible to the procedure. Now the assignment statements in the procedure modify not a and b, but the local copies of x and y. The variables a and b will be *unaffected* by the procedure, and the output produced by the writeln will be:

*5  10*

By way of analogy, we can think of two different ways to pass a file of information to an office worker. First, we could simply pass the key to the filing cabinet from one person to another. Now that person has complete access to the original information and can add, change, or delete any details as desired. Any changes made to the file are permanent. This method is analogous to *call-by-reference*.

Second, we could open the file cabinet, remove the needed information, duplicate it, put the original back in the file cabinet (and lock it), and give the *duplicate* to the other person. That person now has access to the needed information, but any changes affect only that duplicate copy. The original is inaccessible and therefore immune to change. This is analogous to *call-by-value*.

Call-by-reference creates a *two-way path* between a calling program and a procedure. Information can flow both into and out of a procedure. Thus *result* or *output parameters,* which will be returned to the calling program, must be made call-by-reference. *Input parameters* (the values that only need to be passed into a procedure) should be made *call-by-value*. This will make the value available to the procedure but will prevent any accidental (or intentional) change to that value.

The proper use of the call-by-value feature will guarantee the security of a variable and enhance the robustness of a program. Avoid the tendency to make all parameters call-by-reference simply because that is the method you learned first. (FORTRAN supports only call-by-reference.)

There is one important exception to these guidelines. As we indicated, with call-by-value the system will make a local copy of the dummy parameter. This is fine for scalar variables or small structured types. But for large, complex data types such as:

```
var
    x : array [0. .5000] of integer;
```

this copy operation can be very expensive in terms of memory. Passing the preceding array by value will require the system to create a local 5001-element integer array, possibly exceeding the total memory capacity of the computer. So, even though security will always be our primary concern, we may also have to consider the availability of memory when selecting parameter passing mechanisms. If memory is a critical resource, we may have to pass large data structures such as the previous one by reference, not by value.

If we wanted to write a procedure to scan one line of alphabetic text (beginning at some designated starting position) to find and return the next word and its length, we might set up the parameter passing mechanism in the following way.

| Parameter | Purpose | Passing Mechanism |
|---|---|---|
| Line | The array of characters comprising one line | Value (or reference if the array is very large) |
| Start | The index into line of where to start the scan | Value |
| Word | The array of characters containing the next word we found on the line | Reference |
| Length | The length of the word | Reference |
| Ending | The index into line of where the last scan ended | Reference |

The procedure heading would look like this.

```
procedure scan(line : linetype; start : integer; var word : linetype;
    var length : integer; var ending : integer);
```

### 3.3.2 Global Variables versus Parameters

The *scope* of a variable is the set of all blocks in which that variable is known and can be accessed. In Pascal the scope of a variable is the block in which it is declared and all blocks contained therein. Referring to Figure 3.2, the *scope* of the variable x is blocks a, b, c, and d; the scope of y is b and c; and the scope of z is d.

A variable (or any declaration) is said to be *global to* a procedure if it was not declared inside that procedure but inside a procedure (or program) declared at a higher level. In Figure 3.2, x is global to b, c and d, and y is global to c. Looking back at Figure 3.2, procedures b and d could exchange information through the global variable x to which both have access. In fact, carried to the extreme, parameters could be omitted entirely by declaring all variables in the outermost block of the program and letting each procedure access these global variables directly.

However, whenever possible, *avoid* using global variables for exchanging information between program units. Global variables impart some extremely negative behavior characteristics to programs that use them.

1. The procedure is *bound* to use of a specific variable name that cannot be replaced by other actual parameters at invocation time. That is, if we write

```
program    a;
var
     x        : integer;

procedure  b;
var
     y        : integer;

     procedure c;
     begin { of procedure c }
          .

          .

          .
     end; { of procedure c }
begin { of procedure b }
     .

     .

     .
end; { of procedure b }


procedure  d;
var
     z        : integer;
begin { of procedure d }
     .

     .

     .
end; { of procedure d }


begin { of main program a }
     .

     .

     .
end. { of main program a }
```

*Figure 3.2.* Scope of a variable.

a procedure "sort" to sort a list called table, where table is a global variable, we are bound to sorting only that object. To use sort on other lists with other names would first involve copying into the variable called table.

2.   The use of global variables requires close coordination between the person writing the outer block and the person writing the procedure. Specifically, they both must be aware of and agree to the name to be used for a variable.

3.   Since we access by common name, a change to an outer block declaration

could require changes to all procedures that access that object (and vice versa).

4.   Global variables cannot be used for external procedures.

In general, the use of global variables risks program anarchy. We effectively open our entire program name space to every procedure, allowing any and all changes with no limits imposed. We lose all security. The seemingly innocuous call:

*proc;*

can erroneously change any variable. The use of global variables is considered stylistically so bad that the pejorative term "side effect" is used for the modification of any global variable by a procedure during its execution.

In general, it is best to write procedures that operate without any side effects. Pass all information into or out of a procedure using either value or reference parameters.

### 3.3.3  Local Variables

A variable is said to be *local to* a block if it is declared within that block. Thus, referring again to Figure 3.2, we would say that y is local to procedure b, and z is local to procedure d.

In almost any procedure, we will need to use variables that are not passed in or out as parameters but are used only within the procedure itself. For example:

| | |
|---|---|
| Temporary variables for operations (temp) | *temp := a;*<br>*a := b;*<br>*b := temp* |
| Loop indices (i) : | **for** *i := 1* **to** *n* **do** . . . |

Theoretically, these variables could be declared in any block containing the preceding statements, including the outermost block—the main program. The variables temp and i would then be global to the procedure containing these statements. However, that is a very poor practice to follow. All variables should be declared *local* to the block where they are used. Specifically, temporary variables should always be declared local to the procedure in which they are needed to prevent the accidental modification of any variable outside the procedure.

As an example, look at the following fragment, which sums each row of an n-x-n matrix called table:

```
for i := 1 to n do
    rowsum(table, n, i, sum)
```

If we write rowsum the following way, with i as a global variable:

```
procedure rowsum(matrix : tabletype; size,row : integer;
     var sum : integer);
begin
    sum := 0;
    for i := 1 to size do
        sum := sum + matrix[row,i]
end; { of procedure rowsum }
```

then the loop index i referenced within the procedure rowsum is the same variable as the loop index used in the calling program. Each call to rowsum will improperly modify the loop count being kept by the outer **for** loop—a disastrous side effect. If we instead had written the procedure using a local variable as the index:

```
procedure rowsum(matrix : tabletype; size,row : integer;
     var sum : integer);
var
    i      : integer; { local variable used for loop index }
begin
    sum := 0;
    for   i := 1 to size do
        sum := sum + matrix[row,i]
end; { of procedure rowsum }
```

there would be no problem because of Pascal's *name precedence rule*. When two variables of the same name are declared in different blocks, any reference to that name will always refer to the *innermost,* or *most recent,* declaration of that variable. In procedure rowsum, any reference to i will be to the local i declared within the procedure. Outside of procedure rowsum, references to i will be to the other (global) i. By always declaring variables local to the current block, programmers are free to code modules independently, without worrying about name conflicts or accidental modifications of variables outside the scope of the procedure they are writing.

In order to eliminate side effects and minimize errors caused by the interaction of two or more program units, all data either defined or referenced by a procedure should be one of the following three types of objects.

1. Call-by-value parameter.
2. Call-by-reference parameter.
3. Local variable.

Local variables in Pascal do *not* maintain their values between calls of a procedure. Each time a procedure is activated, the space for local variables in that procedure is created anew and their values are undefined. Some languages (not Pascal) support a special class of local variables that maintain their value from one procedure call to the next.

### 3.3.4  Signal Flags

In response to our demand for robustness in programs (Section 3.2) we would probably not write rowsum as we did in the previous section. If rowsum were given a row index outside the range 1. .size, the program would "blow up" because of an array index out of bounds. Our first reaction might be to add the following statement as the first line of procedure rowsum.

```
if (row >= 1) and (row <= size) then
begin
        .
        .
        .
end { of if }
else writeln('error in row index to rowsum')
```

While this would prevent a run-time error in rowsum, it represents an extremely poor programming technique and a misunderstanding of the purpose of procedures. While rowsum knows that the procedure did not complete properly, the calling program does *not*. If the calling program contained the following code segment:

```
for i := first to last do
begin
        rowsum(table,n,i,sum);
        writeln(' sum of row ',i,' = ',sum)
end { of for }
```

we would again have created the possibility of a run-time error caused by attempting to write out a value (sum) that was not properly defined.

One of the most important output values produced by a procedure is a *signal flag*. This is an output parameter that indicates the *status* of the computation carried out by the procedure (i.e., whether or not it was successful) and which of several special cases occurred. The signal flag parameter can be tested by the calling program to determine the outcome of a procedure *before* attempting to use the results of that procedure.

```
procedure rowsum(matrix : tabletype; size,row : integer;
    var sum : integer; var success : boolean);
var
    i        : integer;
begin
    if (row >= 1) and (row <= size) then
    begin
        success := true;
        sum := 0;
        for i := 1 to size do
            sum := sum + matrix[row,i]
    end { of then }
    else success := false
end; { of procedure rowsum }
```

Any procedure that might fail to complete the operations for which it was invoked must be able to return that information to the calling program through a signal flag passed as a call-by-reference parameter. Likewise, to insure robustness, any program that invokes a procedure that returns a signal flag should check that flag *before* referencing any of the other output parameters.

```
rowsum(table,n,i,sum,correct);
if correct then { carry on with the computation }
else { invoke error recovery methods }
```

Since we are now able to return the status of a computation, we can eliminate two other approaches to handling errors within procedures that actually represent bad programming habits:

```
if "error condition" then halt
```

or

```
if "error condition" then writeln "error message"
```

The termination of execution within a procedure violates the graceful degradation characteristic discussed earlier. If execution is halted within the procedure, the calling program cannot analyze what caused the error and, if possible, recover. In addition, by deciding at this point to produce a specific error message, we lose the flexibility of having the calling program analyze the output and determine if an error message should be produced and what type of message it should be. In fact, a procedure should generally not do output of any kind unless it is specifically an output procedure. When we include writeln commands directly within the procedure, we again lose the flexibility of allowing the calling program to decide if we will print a value and, if so, in what format.

A well-written procedure should do no more than perform its computation and *return* the results and status of that computation to the calling program. All further decisions about how to use those results and how to continue should be done outside the procedure. This style of writing procedures will be critical in the implementation of large and complex computer programs (to be discussed in Chapter 5).

*Style Review 3.3* —————————————————————————————

### Program Libraries

Before you begin to design and code a procedure, first see if it already exists in a program library on your computer. There is no sense in "reinventing the wheel." Build on the work of others.

For most common tasks—sorting, merging, basic mathematical and statistical operations—many good commercial packages already exist. For example, these routines are in the Minnesota Subprogram Library: IMSL (The International Mathematical and Statistical Library); SPSS (Statistical Package for the Social Sciences); PLOTS (General Purpose Plotting Package); QSORT (Highly efficient "Quicksort" procedure); and TEKLIB Procedures for performing graphical displays on a terminal.

Check on the availability of these packages before you start to code a procedure.

*Style Review 3.4* —————————————————————————————

### Use of Procedures

1.  Protect input parameters using call-by-value.
2.  Avoid global variables and procedures with side effects.
3.  Make all temporary variables local to the procedure where they are used.
4.  Never halt in a procedure.
5.  Avoid producing output within a procedure (unless the sole purpose of the procedure is output).
6.  Where appropriate use signal flags to return the status of a computation to the calling program.

### 3.3.5  Example Program

Figure 3.4 shows a procedure to *integrate* a function numerically using the *trapezoidal rule*. The integral of a function f on the interval [a,b] is the area under the graph of that function between the two points, a and b, and is written as:

$$\int_a^b f(x)dx$$

This is illustrated in Figure 3.3*a*. The trapezoidal rule approximates that area by placing a set of trapezoids under the curve f and summing up their area as shown in Figure 3.3*b*. The more trapezoids used, the better the approximation.

The formula for the area, i, under the curve, f, using the trapezoidal rule is given by the following formula.

$$i \ = \ \Delta *[(1/2)f(a)+f(a+\Delta)+f(a+2\Delta)+ . . +f(a+(n-1)\Delta)+(1/2)f(b)]$$

where a and b are the end-points of the integration, delta ($\Delta$) is the width of each individual trapezoid, n is the number of trapezoids to use in the approximation, and f(x) is the value of f at the point x. Given a, b, and n, delta can be determined by the following formula.

$$\Delta = (b - a) / n \qquad for \ b > a \ and \ n > 0$$

Our procedure, called trapezoidrule, uses four value parameters: the function f, and a, b, and n. The procedure returns two reference parameters: the result and a boolean flag. The flag is set to false upon detection of the one fatal error condition:

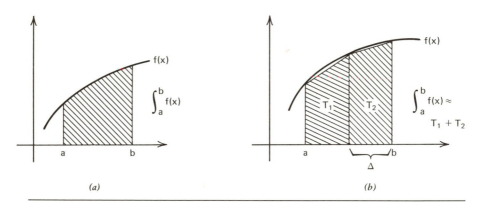

*(a)*             *(b)*

**Figure 3.3.** The meaning of integration. (*a*) The value of $\int_a^b$ f(x). (*b*) Its approximation using two trapezoids.

n < 1. If the procedure detects the condition a > b, it will simply interchange the two values, since the area from a to b is the same as the area from b to a. (Remember, wherever possible, keep going and get meaningful results.) Finally, all other values that are not parameters but are needed by the procedure are declared *local* to the procedure. The complete procedure is shown in Figure 3.4.

```
{ trapezoidrule
  procedure to integrate functions using the trapezoidal rule.

  entry conditions:
      f                 : the formal function to be integrated
      a                 : lower bound of integration
      b                 : upper bound of integration
      n                 : the number of trapezoids to use (n > 0)

  exit conditions:
      result            : the result of integration
      flag              : returned true if operation was successful, false otherwise
}

procedure trapezoidrule(function f(x : real) : real; a : real;
        b : real; n : integer; var result : real; var flag : boolean);
var
      delta           : real;          { the step size along the axis }
      i               : integer;       { loop index }
      sum             : real;          { accumulates intermediate results }
      temp            : real;          { temporary variable }
      x               : real;          { will range from a to b in steps of delta }

begin
     if n >= 1 then
     begin
         flag := true;
         if a <> b then
         begin
             if b > a then
             begin
                 { the integral from a to b = the integral from b to a.
                   let's switch the limits. }
                 temp := a;
                 a := b;
                 b := temp
             end; { of if b > a }

             delta := (b − a) / n;
             sum := 0.0;
             x := a;
```

```
            { first sum the intermediate terms f(a + i*delta),
                i=1. .n−1 }
            for i := 1 to n−1 do
            begin
                x := x + delta;
                sum := sum + f(x)
            end; { of for }

            { add in the end terms 0.5*f(a) + 0.5*f(b) and multiply by delta to
                complete the computation }
            result := (0.5 * (f(a) + f(b)) + sum) * delta
        end { of if a <>b }
        else result := 0.0 { integral from a to a of f(x) is 0 }
    end { of if n >= 1 }
    else flag := false
end; { of procedure trapezoidrule }
```

*Figure 3.4.* Procedure to integrate using the trapezoidal rule.

## 3.4  GENERALITY

*Generality* frees a program from dependence on any specific data set. That is, the program can operate properly on many different data sets without requiring change to the code itself. Generality is a very desirable property because it reduces the need for program modification as the needs of the end-user change. Modifications take time and can induce errors. As an example of inflexible, nongeneral code, look at the following.

```
count := 0;
sum := 0;
for i := 1 to 25 do
    if (score[i] >= 0) and (score[i] <= 100) then
    begin
        count := count + 1;
        sum := sum + score[i]
    end; { of if }
mean := sum / count
```

(The code is also not very robust, since the last statement could possibly cause a division by 0.)

In order to work properly, this code requires a data set containing exactly 25 items, and it will produce the correct values only when the legal range of the items is exactly 0 to 100. Any change to those two conditions will require a search through the code to find every occurrence of the constants 25, 0, and 100 and change them to the appropriate values. If even one occurrence of any of those constants is over-looked, it will almost certainly introduce a bug into the code. (Actually, the problem is even worse, since we only want to change the occurrence of the constants 0, 25, and 100 used for this purpose. A value of 25 that refers to some other aspect of the code must be left unchanged. The likelihood of inducing errors into the program during this type of modification is obviously very high.) A first pass at making the code more general might be:

```
const
    datasets    = 25;           { number of data items }
    high        = 100;          { max range of valid scores }
    low         = 0;            { min range of valid scores }

        .

        .

        .

count := 0;
sum   := 0;
error := false;
for i  := 1 to datasets do
    if (score[i] >= low) and (score[i] <= high) then
    begin
        count := count + 1;
        sum   := sum + score[i]
    end;
if count > 0 then mean := sum / count
else error := true
```

This approach is an improvement because it localizes the change to a single statement. It is no longer necessary to read through the entire program searching for constants. Now we need only change a single declaration. This significantly reduces the possibility of introducing an error during the modification process.

However, the preceding change may not be sufficient. If items such as datasets, high, and low change often enough, declaring them as constants requires the program to change on virtually every run. Values such as these that may change from one data set to another should always be made *variables,* not constants. This will make the code independent of any specific values.

```
         .
         .
         .
readln(datasets,high,low)
         .
         .
         .
count := 0;
sum := 0;
error := false;
for i := 1 to datasets do
    if (score[i] >= low) and (score[i] <= high) then
    begin
        count := count + 1;
        sum := sum + score[i]
    end; { of if and for }
if count > 0 then mean := sum / count
else error := true
```

This fragment will now find the sum of any number of scores over any range of values. The generality has been increased significantly.

There is a similar stylistic guideline for writing generalized procedures. Any value that may change from one procedure invocation to another should be made a *parameter* to the procedure instead of a local variable or constant.

The following procedure for reading strings of characters into a packed array exhibits the same lack of generality as our earlier example.

```
type
    string    : packed array [1. .80] of char;
    .
    .
    .
procedure stringread(var list : string);
var
    ch      : char;          { input character }
    i       : integer;       { loop index }
begin
    i := 0;
    read(ch);
    while (not eoln) and (i < 80) and (ch <> '.') do
    begin
        i := i + 1;
        list[i] := ch;
        read(ch)
    end { of while }
end; { of procedure stringread }
```

The procedure will read a set of characters until either end-of-line, the specific terminating character '.', or an upper bound of 80 characters is encountered. There is no way, short of changing the procedure, for the user to specify either a maximum character string length other than 80 characters or a terminator character other than '.'. By adding the appropriate parameters, however, we can generalize the procedure and eliminate these shortcomings.

```
procedure stringread(var list : string; var size : integer;
    max : integer; terminator : char);
const
    maxstringlength    = 80;
begin
    if max > maxstringlength then max := maxstringlength;
    size := 0;
    read(ch);
    while (not eoln) and (size < max) and (ch <> terminator) do
    begin
        size := size + 1;
        list[size] := ch;
        read(ch)
    end
end; { of procedure stringread }
```

The procedure will now allow the user to read any string of characters up to a length of 80 or until the first occurrence of any user-designated character. This procedure could now be used, *without change,* to process any of the following fields.

A last name of up to 20 characters ending with a ','.

A first name of up to 20 characters ending with a ' '.

A comment of up to 80 characters ending with a '}'.

Generality is not merely *coded* into a program; it is *designed* in. During the development of the problem specification, we should always attempt to anticipate variations that will occur with future modifications or different end-users. We should incorporate sufficient generality into our design to accommodate these variations.

As a final example of program generality, look at the following skeleton of a program.

```
begin
    initialize all values;
    read data;
    if data is valid then process data
    else error;
    write out results
end
```

This is a reasonable outline for a program that operates on a single data set to produce a single set of results. But if we wish to run the program on more than that single data set, we must modify it as follows.

```
begin
    while not eof do
    begin
        initialize for this data set;
        read this data set;
        if data is valid then process data set
        else error;
        write out results
    end { of while }
end
```

It now works on any arbitrary number of data sets—0, 1, or more. This marked increase in generality was achieved with virtually no increase in program complexity and little change to the code itself.

Figure 3.5 shows a generalized pattern matching procedure that will attempt to locate a pattern (up to length n) within a string of text of length $<=$ n. The attempted match can begin at any location within the text and will return either the location of the match or 0 if no match is found. In addition, the special character '$\wedge$' will match any single character within the text. Given the following input stream:

    Text: "d x f h e f g d z f g c e k"
    Pattern: "d $\wedge$ f g"

the procedure will return an 8, the index of the first column of the match between "pattern" and "text."

This example illustrates a truly generalized program, but it could be further generalized. For example, we could include yet another parameter, direction, of type (left,right) that would indicate to the procedure the direction in which we would attempt to perform the match, left to right for text in English, French, Russian, and so forth, or right to left for languages such as Hebrew and Arabic. Naturally, we would need to know more about the proposed use of this procedure to determine whether this feature (or others) would be useful enough to warrant the additional effort. From the very first stages of program design, always look for useful ways to increase generality in the programs and procedures you develop.

## 3.5  PORTABILITY

While generality makes a program independent of any particular data set, *portability* implies program independence from the hardware, or a specific operating system or

{ *patternmatch*

*procedure to do generalized pattern matching.*

|  |  |
|---|---|
| *text* | : *the string of characters in which we will look for the match. (arraytype)* |
| *size* | : *the actual length of text.* $1 <= size <= max.$ |
| *start* | : *the starting location in text to begin the match.* $1 <= start <= 1 + size - length.$ |
| *pattern* | : *the object string of the search. (arraytype)* |
| *length* | : *the length of the character stream in pattern.* $1 <= length <= max.$ |
| *found* | : *if no match is found,* − *found* − *will return 0. if the pattern is matched, the first position of the match* $(1..size)$ *will be returned.* |
| *error* | : *returns* − *errnone* − *if all input parameters are correct, otherwise returns* − *errbound* −, − *errsize* −, *or* − *errstart* − *to indicate the presence of one of these classes of errors.* |

*the following global declarations are assumed:*

*const    max = nnn;    / for some integer nnn /*
*type    arraytype    = packed array [1..max] of char;*
*        error        = (errnone, errbound, errsize, errstart);*

}

**procedure** *patternmatch(text : arraytype; size : integer; start : integer;*
    *pattern : arraytype; length : integer;* **var** *found : integer;*
    **var** *error : errortype );*
**const**
    *anymatch        = '/\';*                { *the character that matches anything* }
**var**
    *lastmatchplace  : integer;*              { *last possible position in text where a successful match can occur* }
    *matched        : boolean;*
    *p              : integer;*               { *index into pattern* }
    *t              : integer;*               { *index into text      * }
**begin**
    *error := errnone;*
    *lastmatchplace := size − length + 1;*

    { *error checking* }
    **if** *(size < 1)* **or** *(size > max)* **or** *(length < 1)* **or** *(length > max)* **then**
        *error := errbound;*
    **if** *length > size* **then** *error := errsize*
    **else**
    **if** *(start < 1)* **or** *(start > lastmatchplace)* **then** *error := errstart;*

```
if error = errnone then
begin
    repeat { loop to look through all possible match locations }
            p := 1;              { to index through pattern }
            t := start;          { to index through text    }
            matched := true;

            repeat { loop to check a match at a specific location }
                    if (pattern[p] <> anymatch) and (pattern[p] <> text[t]) then
                        matched := false
                    else
                    begin
                        p := p + 1;
                        t := t + 1
                    end
            until (p > length) or not matched;

            if not matched then start := start + 1
    until (start > lastmatchplace) or matched;

    if matched then found := start
        else found := 0
    end { of else error = errnone }
end; { of procedure patternmatch }
```

**Figure 3.5.** A generalized pattern matching procedure.

compiler. A portable program is independent of any particular *machine environment*. Because computer changes are so rapid and frequent and because programmers exchange and share programs, portability is a critical characteristic. We would like to avoid significant reprogramming efforts for new machines in order to keep costs down and minimize new errors in the programs during the change. Programs that require little or no change as environments change will certainly cause fewer problems during their lifetime.

The most important guideline in writing portable programs is strict adherence to the *standard version* of a programming language. Many people are unaware that some programming languages, like electrical voltages, radio frequencies, and screw sizes, are standardized by international conventions. The group that directs most of these standardization efforts is the International Standards Organization (ISO), an agency of the United Nations. ISO is itself made up of the standards agencies of each UN member nation that votes on and approves international programming language standards. Figure 3.6 lists some of these member agencies.

ISO publishes a programming language standard that specifies the *syntax* of statements in the language and the action (called the *semantics*) taken by each of those statements. A *standard compiler* for a language accepts and correctly compiles the standard language. It may also accept additional nonstandard features (called *exten-*

ISO—International Standards Organization (UN Agency)
ANSI—American National Standards Institute
BSI—British Standards Institute
CSA—Canadian Standards Agency
JISC—Japanese Industrial Standards Committee
DNA—Deutscher Normenausschuss (German Standards Organization)
AFNOR—Association Francaise de Normalisation (French Standards Association)
NIU—Ente Nazionale Italiano de Unificazione (Italian National Standards Agency)
SAA—Standards Association of Australia

**Figure 3.6.** Standards agencies around the world.

*sions*). In Figure 3.7 compilers A and B would be standard compilers, since they include the entire language standard as a subset.

By using only the standard version of a language, we insure that our program will be immediately movable to any other standard compiler for that language. But if we use a language feature that is a local extension to our compiler (e.g., point x in Figure 3.7), there is no guarantee that our programs will run when moved.

Most Pascal compilers contain at least a few local extensions, and it is often very tempting to use them; after all, they are usually added to provide a shorter way of performing some common operation. All of the following (and others) have been proposed and implemented as extensions to standard Pascal.

1. A *value* declaration (like a "data" statement in FORTRAN).

2. An *otherwise* clause in the **case** statement as the default match.

3. An exponentiation operator.

4. Dynamic array bounds.

5. A "string" data type with the basic string operations—concatenate, pattern match, delete substring.

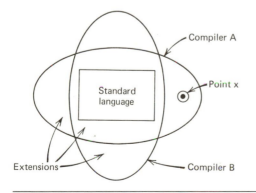

**Figure 3.7.** Language standards.

6. Extensions of the procedure read to read arrays, records, sets, and the like.
7. Random-access files.

When using any such nonstandard language feature, be aware that it may not be possible to transfer your program to other systems without a massive reprogramming effort. This fact must be carefully weighed against the programming advantages to be gained from using these features.

All examples presented in this book will be written in the standard Pascal proposed by the ISO and described in Reference 9 at the end of Chapter 1. (However, as of this writing, this "standard" version of Pascal has not been officially approved by the ISO. It is currently only an "informal standard.")

We must also try to avoid machine-dependent features that could prevent our program from running on another computer system. Where possible, try to avoid:

1. Machine-dependent constants. Avoid constants that are specific to one machine. Use maxint instead of $2^{48} - 1$ and ord('a') instead of 141 (base 8). These values may not be the same on a different system.

2. Machine-dependent collating sequences. Do not make any assumptions about the collating sequence of a particular code set except the following:

   'A' < 'B' < . . . < 'Z'
   'a' < 'b' < . . . < 'z'
   '0' < '1' < . . . < '9'

   You must check all other assumptions. For example, do not assume that 'a' . . 'z' are contiguous without gaps (BCD and ASCII-true, EBCDIC-false). Instead, check whether:

   ord('z') = ord('a') + 25

   and take the appropriate action for each case.

At times it will be impossible to avoid using machine-dependent features. When that occurs, *localize* the machine-dependent information to a single place within the program and identify it clearly as information that must be changed when the program is moved, as in the following.

```
{ the following parameters apply to the CDC CYBER/74. they should be reset
  when moved to a new machine. }
const
    wordsize  = 60;
    bytesize  = 6;
    codeset   = 'display';
```

*Style Review 3.5*

---

**Program Independence**

Try to develop programs that are independent of:

*The Machine Environment* (Portability)
1. By using the standard version of a language.
2. By avoiding machine dependent constants.
3. By avoiding specific collating sequences.
4. By localizing and identifying unavoidable machine-dependent information.

*The Data Set* (Generality)
1. By making key data values *variables,* not constants.
2. By making key data values *parameters* to procedures, not local variables.
3. By using flexible data formats.
4. By considering how to achieve generality from the earliest stages of program design and specification.

## 3.6  INPUT/OUTPUT BEHAVIOR

Most of the stylistic guidelines we have described in Chapters 1 to 3 have been directed at helping the people who must read, understand, and modify a computer program—that is, the *maintenance programmers*. Much of what we have discussed (indentation rules, naming conventions, parameter passing mechanisms) is irrelevant to the end-user.

To end-users, a program is not unlike a black box. They know what goes in and what comes out but nothing about what goes on inside.

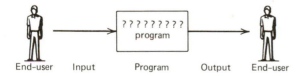

To end-users, the stylistic guidelines we follow, the elegance of our design, and the beauty of our code may be unimportant. What does matter is the interface between the users and the program, the program's *input* and *output behavior*. A beautifully structured program that produces confusing and poorly formatted output will probably end up in the wastebasket. We should have as much concern for elegance of input/output as for elegance of code.

When working interactively, we should identify the input we are requesting, either with a *prompt:*

*Please Enter ID:*

or a *screen template*—preset information on a CRT screen identifying what must be provided.

```
name: _____
address:
city:
state:
 zip:
```

Never write a program that simply waits for some unidentified input:

*?*

First-time users of this program will be thoroughly confused.

However, once familiar with a program, users may wish to avoid the wordiness of either of the preceding user aids. A good program allows users to *suppress* aids, such as prompts, if desired.

```
{ assume that we have previously read a command string from the user telling
  us whether or not to supply prompts for input data. the variable promptmode is
  set true if prompts are desired }
      .
      .
      .

  if promptmode then write('please enter user id');
  readln(userid)
      .
      .
```

This facility will allow us to develop a program useful to both the novice and the expert.

Be careful not to write a program with an infinite "input-error" loop.

```
? 123
error -- please reenter
? 123.
error -- please reenter
? 123,
error -- please reenter
? help
error -- please reenter
```

These loops are usually caused by code organized as follows.

```
read data
if data is invalid then
    repeat
        write a vague or unhelpful error message
        read data
    until data is valid
```

Once in the loop, we cannot extricate ourselves until we correct the mistake and, in some cases, we may not know how. We should provide users with either some sort of useful error message or a "help mode" that they may request or that is activated automatically when they are at a loss about what to do.

```
switch := false;
repeat
    read data
    if data is invalid then
        if switch then helpmode
        else
        begin
            write out an error message
            switch := true
        end
until the data is valid
```

The specifics of the procedure "helpmode" will, of course, depend on the nature of the input and the program. It should at least provide users with information and advice about what to do or where to go when they are lost.

There are a number of I/O programming styles that force end-users to perform operations that the computer can and should do for them. *Never* terminate data files (except the very smallest ones) by an explicit user-supplied count field. Always use either *eof* or a special *signal card*. Why ask a person to count a set of data cards (and risk the chance of an erroneous count) when computers can count much more quickly and reliably?

```
count := 0;
while (not eof) and (count < arraylimit) do
begin
    count := count + 1;
    read(item[count])
end
```

Similarly, do not require users to employ the internal representation scheme of the computer. The following illustrates this kind of thinking.

*enter 0=yes 1=no*

Wouldn't it make more sense simply to say:

*enter yes or no*

and let the program perform the translation?
The line:

*smith,john      85      93      −1      75      60*

from a finished report of student test results might leave us wondering how a student could possibly receive a −1 on a test. Only after looking at the code would we discover that a −1 was used by the program as a flag to indicate "missing data." (The −1 was probably chosen so that we could distinguish between a missing score, −1, and a valid test score (0. .n). However, just because the program uses a value internally does not mean that we must output the value in precisely that format, especially if it will be unclear or misleading to end-users. The following statement could greatly improve the clarity of the output.

**if** *item[i]* = *−1* **then** *write(' * missing * ')*

.

.

.

The output would then look like this.

*smith,john      85      93      * missing *      75      60*

Request and provide information in a format that is best for end-users.

In Section 3.2.1 we talked about validating input for both legibility and plausibility. However, there are incorrect values that are both legal and plausible. Our last line of defense against erroneous input are the users themselves. *Echo printing* the input allows users to check input and locate improper data values. Avoid programs that produce output such as:

the result is 11

but does not identify the input values that led to this result. Users have no way of knowing if the input is correct or if the output can be trusted. (Remember: garbage in, garbage out.)

Be sure to select input formats that are easy and natural to prepare, proofread, and verify. Input fields that are run together, broken between lines, or encoded with some cryptic representation are highly prone to errors and extremely difficult to proof-read, verify, and correct. Input data prepared in the following format:

    smith10189341301101011020203201
    jones13896231501102012010103201

will almost certainly be the cause of significant grief during the life of the program. Although it may be a convenient format for the program, it is highly unnatural to typical end-users and will be difficult to verify visually. The programmer has im-properly traded a reduction in programming effort for an increase in user effort. This is another example of catering to the machine instead of to the people using it. How much nicer to prepare the preceding data in the following way.

| | | | |
|---|---|---|---|
| smith | 101893 | sr | 13 |
| math | 101 | | |
| math | 102 | | |
| cs | 201 | | |
| / | | | |
| jones | 138962 | jr | 15 |
|     . | | | |
|     . | | | |
|     . | | | |

Input errors will be much easier to locate and correct.

Finally, there is the problem of requiring users to provide too much data. In Section 3.4 we encouraged the design of generality into all programs. One way that generality can be achieved is by allowing greater user control over input values. However, carried to the extreme, generality would require users to provide literally dozens of input values in order to customize a very generalized program to their specific needs. This would result in much duplication of effort. *Default values* can reduce the user's burden. A default value is a value set by the program to be used *unless* the user specifies otherwise. For example, the following assignments:

```
speed := 300;            { terminal speed in bits per second }
linesize := 80;          { characters per line }
pagesize := 57;          { lines per page }
code := ascii;           { code set }
eol := '.';              { end of line character }
missingdata := −1;       {the missing data indicator flag }
rubouts := 0;            { used for delays during carriage
                           returns }
    .                        .
    .                        .
```

set default values for terminal description parameters needed within a typical word processing program. Instead of providing dozens of parameters, users need only supply the values for which the defaults do not apply. This could be done, for example, with *self-identifying input*.

linesize = 132

The preceding data card indicates that the linesize parameter is to be reset to 132, while all other values remain set to their defaults.

Reading a data card in this format cannot be done with a single statement; it involves a fairly complex procedure to locate the name and its value and update them to the appropriate values. (We will be discussing this in the following chapters.) This time we have traded an increase in program size, complexity, and running time for increased end-user services and ease of use. In most cases, it is worthwhile.

The placement of this section on I/O behavior at the very end of the discussion on programming style is very appropriate. During the program design, coding, and debugging phases, we really should not be concerned with the elegance of our input or output as long as we can understand what the program is doing. End-users do not see the program during this phase, so our *first* concern should be with correctness, readability, code structure, robustness, generality, proper procedure usage, and portability. Most books on programming and style include a rule that says:

Make it correct before you make it pretty.

In following the guidelines we have presented, we would add the following related rules, among others.

Make it readable before you make it pretty.

Make it robust before you make it pretty.

Make it general before you make it pretty.

But we would add one final important corollary.

Eventually, *do* make it pretty!

*Style Review 3.6* _____

*I/O Behavior Guidelines*

Always identify the input you are requesting—either with a prompt or screen template.

Avoid programs that trap users in an infinite loop demanding correctly structured input, without offering assistance in preparing it.

Use a "help" mode to aid users who are confused about how to prepare input.

Always terminate data through eof or a signal card, never user-supplied count fields.

Have users supply input in a form most natural for them, not the program.

Select input formats on the basis of ease of preparation and legibility, not ease of programming.

Use defaults to reduce the amount of input data required for generalized programs.

Use self-identifying input to modify defaults.

Above all, spend sufficient time in producing "finished" output that is elegant, legible, and directly usable by intended end-users.

## 3.7  SUMMARY

Chapters 2 and 3 have discussed guidelines for producing programs that are easy to use for the *programmers,* who will work with and maintain them, and for the *end-users,* who will use the results from them. Programs and procedures that follow these guidelines (Figures 1.11, 2.4, 2.6, 3.1, 3.4, and 3.5) should help reduce the sky-rocketing costs associated with software development and maintenance. They will also nurture an idea that should be a commandment to every programmer. The two most important elements of any computer system are the *people* who use the computer (not the computer itself), and the useful *information* we provide to those people (not the raw data on which we perform the computations).

Part III will deal with the overall management of the programming process.

There is one final comment about the guidelines presented in Chapters 2 and 3. In an *academic* environment, many of the techniques we have discussed may seem unnecessary, excessive, or out of place. This is not because the techniques are wrong, but because the environment in which students work is different from the *production* environment in which they will someday be working. This point was raised in the beginning of Chapter 2 and is presented in greater detail in Table 3.8. A quick glance through the table indicates one major difference: in a production environment, a wide range of people (with a wide range of ability and familiarity with the problem) will come in contact with that program during its lifetime. These people will be looking at, reading, correcting, updating, and modifying the program and will need to understand its purpose quickly and correctly. The programming guidelines we have been discussing will insure that these people can perform their work properly and keep costs down. Even if you do not immediately see the need to develop such "beautiful" homework assignments, remember that the programming habits you are developing now will be of critical importance throughout your career.

*Table 3.8*
*Differences Between Academic and Production Programming Environments*

| Academic Assignments | Production Programs |
|---|---|
| 1. Once working are typically run once or twice and then discarded. | 1. Once working are run on a regular basis for months or years. |
| 2. Are short, rarely exceeding 500 to 1000 lines. | 2. May be quite long, 1000 to 5000 lines is not unusual and even 25,000 lines is not unheard of. |
| 3. Are done alone. | 3. Are done in groups or teams of 3 to 10 members. |
| 4. Are started and completed by the same individual. | 4. May have many different programmers in charge over the life of the program. |
| 5. Are rarely changed after working correctly. | 5. May be updated frequently over the life of the program. |
| 6. If maintenance is ever required it is typically done by the original programmer. | 6. Maintenance may be handled by someone other than the original programmer. |
| 7. The program is never looked at except by those intimately familiar with the problem—the instructor and the student. | 7. The code may be looked at by levels of management not immediately familiar with the structure of the program. |
| 8. The end-users of the program are only those who are intimately familiar with the problem. | 8. The end-users may be totally naive in the area of computers or the specific program. |
| 9. No salaries are paid to the student programmers when they do their homework. Lower productivity does not result in decreased profits. | 9. Real money is being paid to the programmers for all work performed, whether useful or not. |
| 10. If the program does not perform up to specifications, the student will receive a lower grade. | 10. If the program does not perform up to specifications, a significant financial loss may be suffered, possibly including the programmer's job. |

## EXERCISES FOR CHAPTER 3

1. Under what conditions could the following fragments terminate abnormally? Rewrite the fragments to avoid this problem.

(a)  { *compute the average of 'scorecount' number of exams in the range*
        *low. .high* }
      *bad*  := 0;
      *i*     := 0;
      *sum* := 0.0;
      **while** *i* < *scorecount* **do**
      **begin**
            *i* := *i* + 1;
            **if** *(exam[i] < low)* **or** *(exam[i] > high)* **then**
                  *bad* := *bad* + 1
            **else**
                  *sum* := *sum* + *exam[i]*
      **end;** { *of while loop* }
      *average* := *sum/(scorecount − bad)*

(b)  **var**
            *salaryclass* : *(salaried, hourly, piecework, temporary)*

                  .

                  .

                  .

      { *get payroll information on all regular employees* }
      *readln(salaryclass,dept,name,payrolldata);*
      **case** *salaryclass* **of**
            *salaried*    : *salariedproc(dept,name,payrolldata);*
            *hourly*      : *hourlyproc(dept,name,payrolldata);*
            *piecework* : *pieceproc(dept,name,payrolldata)*
      **end** { *of case* }

            .

            .

            .

(c)  { *The end of data is marked by the special signal card 999. The signal value*
        *will be stored in the last position of the item list* }
      *i* := 0;
      **repeat**
            *i* := *i* + 1;
            *readln(item[i]);*
            *processdata(item[i])*
      **until** *(item[i]* = 999)

(d) { *read all the transactions that occurred this week and merge them into the master file* }
>
> **repeat**
>> *readln(number,name,amount);*
>> *merge(number,name,amount);*
>> *writeln(number,name,amount)*
>
> **until** *eof*

(e)  *p := succ(p)   { p is some user-defined data type }*

(f)  *upperbound := 0.5 * sqr(ln(0.25 * (a + b)))*

2. In Exercise 6 of Chapter 2 you wrote a program to translate Roman numerals to decimal quantities. Discuss the robustness of that program with regard to the following problems.

(a) The total maximal length (in characters) of the Roman numeral.

(b) An invalid character (i.e., not I, V, X, L, C, D, or M) encountered during translation.

(c) An illegal sequence of characters encountered (e.g., LLL instead of CL).

(d) The handling of the null case—a Roman numeral with 0 characters.

Rewrite the procedure Roman so that it correctly handles the special circumstances.

3. Write a procedure *validate* that, given a real matrix M, of size n × n, n >= 1, validates that:

(a) All entries of M are nonnegative.

(b) At least one diagonal element of M is nonzero.

(c) The matrix is not upper triangular (i.e., all elements to the right of the diagonal are not zero).

If any of these conditions occur, the procedure should set the parameter called errorflag to one of the following:

(negative, diagonal, triangular)

Otherwise errorflag should be set to the value ok.

4. Write a program to generate *mailing labels*. The input to the program will be data cards in the following format.

Last, first                                     Where first and last names are 1 to
                                                20 characters in length.

City, state zip

Where city and state are 1 to 10 characters in length, and zip is an integer 0 to 99999.

Month/year

Month and year of expiration.
$1 <=$ month $<= 12$
$80 <=$ year $<= 99$

Your program should output a four-line mailing label of the form:

expiration date
firstname lastname
city, state
zip

if and only if the month and year of expiration are *after* the current month and year. After writing the program, determine what percentage of this typical "business-oriented" program is devoted to:

(a) Input and input validation.
(b) Processing.
(c) Output.

5. Rewrite the program in Figure 3.1 so that the input is in the form:

mmm. dd, 19yy

where mmm is the three-letter abbreviation for the current month and is selected from the following data type:

**type** *months = (jan, feb, mar, apr, may, jun, jul, aug, sep, oct, nov, dec);*

Perform the same types of validation on this new input that you did on the input in Figure 3.1.

6. Given the following procedure:

**procedure** *example(a : integer;* **var** *b : integer);*
**var** *c : integer;*
**begin**
    *a := 20; b := 30; c := 40*
**end;**

what is the result of executing these statements in another program unit?

**var**
    *a,b,c : integer;*
    .
    .
    .

*a := 5; b := 10; c := 15;*
*example(a,b,c); writeln(a,b,c);*
*example(c,b,a); writeln(c,b,a)*

7. Rewrite the date conversion program in Figure 3.1 so that it is a Pascal *function*. Be sure to consider and handle the problems of:

   (a) Signal flags.
   (b) Output statements.
   (c) Local variables/global variables.
   (d) Parameters and proper parameter-passing mechanisms.

   Now show a mainline program, which uses the date conversion function, to read in dates entered from a terminal in mm/dd/yy format and print them out in Julian format. The main program should continue until end-of-file.

8. Write a complete Pascal procedure to compute the determinant of a 3-x-3 matrix M. The determinant of M, called D(M) is defined as:

$$M = \begin{bmatrix} a_{11} & a_{12} & a_{13} \\ a_{21} & a_{22} & a_{23} \\ a_{31} & a_{32} & a_{33} \end{bmatrix}$$

$$
\begin{aligned}
D(M) = &\ (a_{11}\, a_{22}\, a_{33} + a_{12}\, a_{23}\, a_{31} + \\
&\ a_{13}\, a_{21}\, a_{32}) - \\
&\ (a_{13}\, a_{22}\, a_{31} + a_{12}\, a_{21}\, a_{33} + \\
&\ a_{11}\, a_{32}\, a_{23})
\end{aligned}
$$

   If the determinant is 0, M is *singular*. Return a special flag called singular that is true if this condition exists and false otherwise.

   Use the preceding procedure to find the solution to the 3-x-3 system of equations:

$$a_{11}x_1 + a_{12}x_2 + a_{13}x_3 = b_1$$
$$a_{21}x_1 + a_{22}x_2 + a_{23}x_3 = b_2$$
$$a_{31}x_1 + a_{32}x_2 + a_{33}x_3 = b_3$$

   using *Cramer's rule*.

   Cramer's rule states that the solution to the preceding problem can be found by building an *augmented matrix* $M_i$ by replacing the ith column of M with the constants $b_1$, $b_2$, and $b_3$. The value of $x_i$ at the solution point is simply:

$$x_i = \frac{D(M_i)}{D(M)}$$

For example, to find the value of $x_1$:

$$x_1 = \frac{D\begin{bmatrix} b_1 & a_{12} & a_{13} \\ b_2 & a_{22} & a_{23} \\ b_3 & a_{32} & a_{33} \end{bmatrix}}{D\begin{bmatrix} a_{11} & a_{12} & a_{13} \\ a_{21} & a_{22} & a_{23} \\ a_{31} & a_{32} & a_{33} \end{bmatrix}}$$

[if D(M) is not singular]

and similarly for $x_2$ and $x_3$. Write a program that reads in the values for $a_{ij}$ and $b_i$ and solves the resulting 3-x-3 system of equations. Be concerned with the style and organization of all procedures and functions used.

9. Write a function called power(a,b) that will compute, for any real values a,b: $a^b$ by using the relation $a^b = \exp(a*\ln(b))$. However, this will only work for b>0. Write a function that also works for b<=0 by noting that $a^{-b} = 1/a^b$.

10. Look at Exercise 5 in Chapter 2. Is there any reason for which you would criticize the program for a lack of *generality?* If so, rewrite it to make it more generally applicable to a wider range of problems.

11. What would happen to the procedure trapezoidal if it attempted to evaluate the function f(x) at the point where it is undefined? For example:

$$f(x) = \frac{1}{x - 2}$$

at the point x = 2. Rewrite the example from Figure 3.4 assuming that the function f(x) returns the value ($\pm$ maxint) if it is undefined at the point x. Your procedure should now take the proper recovery action and inform the user about what has happened.

12. Rewrite the generalized pattern matching procedure from Figure 3.5 so that:

(a) It can match from right to left or left to right based on the value of a parameter called direction.

(b) It will delete the string if it is found and if a parameter called ''delete'' is set to true.

13. The program fragment:

```
var
        x : alfa;
        y : integer; value 1;
     w,z : integer ;
        .
        .
        .
     readln(x,z);
     if x = 'all done ' then
     begin
          writeln('program completed');
          halt
     end;
     w := z + y;
     case w of
          1,2 : proc1;
          3,4 : proc2;
          else : proc3
     end { of case }
end. { of program }
```

uses the following nonstandard Pascal features.

(a)  A type alfa that is predefined to be a **packed array** [1. .10] **of** char.

(b)  A **value** clause that initializes a variable to a value at compile time.

(c)  A procedure halt that terminates program execution.

(d)  An **else** clause in the case to handle the situation in which the selector expression did not match any label.

(e)  A direct comparison between a variable of type alfa and a 10-character string constant.

Rewrite this fragment so that it does not use any nonstandard features and would be acceptable to any standard Pascal compiler.

14. Write a complete Pascal program to generate a concordance. A concordance is an alphabetized list of every word that appears in some text and every line or page number on which it appears. For example:

| | |
|---|---|
| see spot run. | (line 1) |
| see spot bite. | (line 2) |
| see jane run. | (line 3) |
| run jane run. | (line 4) |

would produce the following concordance:

| bite | 2 |
| jane | 3,4 |
| run | 1,3,4 |
| see | 1,2,3 |
| spot | 1,2 |

Write a program to produce such a concordance given the following assumptions.

(a)  Words will always be separated by one or more blanks.

(b)  Words will never be broken between two or more lines of text.

(c)  The only punctuation marks used are the period and comma.

(d)  The end of the text is marked by an end-of-file condition.

Follow all the guidelines presented in this chapter in terms of the proper run-time behavior of programs.

# BIBLIOGRAPHY FOR PART II

Kernighan, B. W., and P. J. Plauger. *The Elements of Programming Style*. Second Edition. New York: McGraw-Hill, 1978.

Kernighan, B. W., and P. J. Plauger. *Software Tools*. Reading, Mass.: Addison-Wesley, 1976.

Knuth, D., *The Art of Computer Programming*, Vol. 2, "Seminumerical Algorithms." Reading, Mass.: Addison-Wesley, 1969.

Ledgard, H., J. Hueras, and P. Nagen. *Pascal With Style*. Rochelle Park, N.J.: Hayden, 1979.

Ledgard, H. *Programming Proverbs*. Rochelle Park, N.J.: Hayden, 1975.

Ralston, A., and P. Rabinowitz. *A First Course in Numerical Analysis,* New York: McGraw-Hill, 1978.

Scowens, R. S., and B. A. Wichman. "The Definition of Comments in Programming Languages." *Software Practice and Experience,* April 1974.

Schneiderman, B. *Software Psychology*. Englewood Cliffs, N.J.: Winthrop, 1980.

Van Tassel, D. *Programming Style, Design, Efficiency, Debugging, and Testing*. Second Edition. Englewood Cliffs, N.J.: Prentice-Hall, 1978.

Wetherell, C. *Etudes for Programmers*. Englewood Cliffs, N.J.: Prentice-Hall, 1978.

# PROGRAM DESIGN

*Chapter 4*

# STRUCTURED CODING

## 4.1 INTRODUCTION

This chapter introduces a concept called *structured coding*. A primary purpose of structured coding is to create highly readable and understandable programs. To appreciate the superior legibility of structured code, compare the unstructured program fragment in Figure 4.1*a* to the structured program fragment in Figure 4.1*b*, which achieves exactly the same result (to find the largest of three values).

Since readability and clarity are a major reason for using structured code, this upcoming discussion could have properly been placed in Chapter 2 under the topic "Programming Style." However, because structured coding is important for reasons beyond just readability, we have placed this topic in a separate chapter. Remember, however, that there are few things more helpful in creating a readable program than structuring the control statements according to the guidelines that will be presented.

## 4.2 STRUCTURED CODING

*Structured coding* involves constructing individual programs or subprograms with only three types of statements: sequential, conditional, and iterative. A *sequential statement* performs an operation and then continues on to the next statement in the program. The flow of control of any sequential statement, s, can be represented visually by the diagram shown in Figure 4.2*a*. Examples of sequential statement types in Pascal are the assignment, input, output, compound, and procedure calls.

A *conditional statement* performs a test to decide which statement to execute next. In general, the model of a conditional statement is as diagrammed in Figure 4.2*b*. The **if/then/else** (Figure 1.9*a*) and **case/end** (Figure 1.9*b*) are examples of two-way and multiway branch conditionals, respectively.

132

```
8:      read(x,y,z);
        if x >= y then goto 1;
        if y >= z then goto 5;
        goto 4;
1:      if x >= z then goto 2;
        goto 4;
5:      big := y;
        goto 3;
2:      big := x;
        goto 3;
4:      big := z;
3:      if not eof then goto 8
```

*(a)*

```
repeat
    read(x,y,z);
    if (x >= y) and (x >= z) then big := x
    else if (y >= x) and (y >= z) then big := y else big := z
until eof
```

*(b)*

**Figure 4.1.** Comparison of structured and unstructured code fragments.

Finally, an *iterative statement* repetitively executes a number of statements (called the *loop body*) until a specific condition is met. In general, either statement in Figure 4.2c is valid.

The Pascal **while** (Figure 1.9c), **repeat** (Figure 1.9d), and **for** (Figure 1.9e), are examples of iterative control statements.

**Figure 4.2a.** Sequence operation.

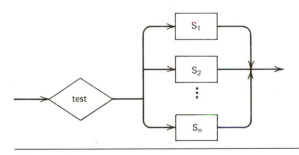

**Figure 4.2b.** Conditional operation.

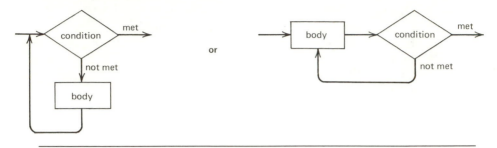

***Figure 4.2c.*** Iterative operation.

A well-structured program unit contains only those three statement types: sequence, conditional, and iterative. The flow chart of any well-structured code fragment can be written as a properly nested series of sequence, conditional, and iterative boxes of the type shown in Figures 4.2*a*, 4.2*b*, and 4.2*c*. For example, the flow chart of the code fragment from Figure 4.1*b* is shown in Figure 4.3. It can be written from repeated applications of the three control types.

In Pascal a well-structured program module is constructed using only the following high-level control structures.

**begin/end**
**if/then/else**
**case/end**
**while/do**
**repeat/until**
**for/do**

Specifically excluded are the **goto** construct and statement labels. The most fundamental characteristic of structured code is that it is composed of properly nested code segments that are entered only at the top and exited only from the bottom. These are called *single-entry, single-exit blocks*. The preceding statements and the flow charts in Figure 4.2 exhibit this characteristic. All computer programs can be coded this way, regardless of application, as long as the programming language has the appropriate control structures.

In 1966 C. Boehm and G. Jacopini formally demonstrated that any "well-behaved" computer program (i.e., one without an infinite loop) could be constructed using only the **if/then/else** and **while/do** control statements and the appropriate sequential operations (assignment, input, and output).[1] Therefore, in any programming language that contains those two control statements (e.g., Pascal, PL-1, Algol), the

---

[1]C. Boehm and G. Jacopini, "Flow Diagrams, Turing Machines, and Languages With Only Two Formation Rules," *Communications of the ACM*, May 1966.

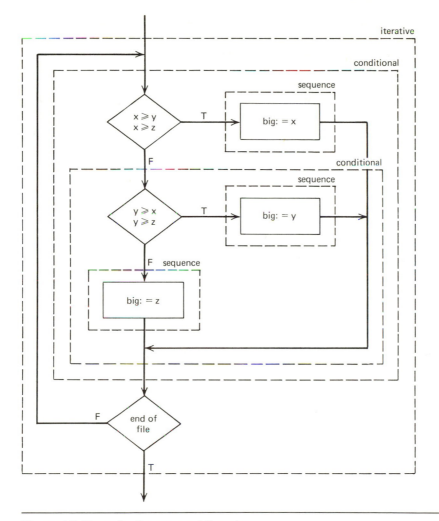

***Figure 4.3.*** Example of a structured flow chart.

**goto** statement is theoretically unnecessary. During the early work on structured coding, the **goto** statement became the center of a major debate called the "*goto* controversy." Misuse of the unconditional branch was blamed for a wide range of programming errors and crises, and the inclusion of even a single **goto** anywhere was considered to be the mark of a poor programmer.

Today, things have quieted down and most people recognize that the emphasis should not be on preaching against a particular construct such as the **goto** but on explaining why certain programming characteristics are desirable. Striving to attain those desirable characteristics will result naturally in programs that have few, if any, unconditional branches. However, we will also identify the situations in which the **goto** is actually a useful and appropriate control structure.

The basic problem with the **goto** is that is is too *primitive* and *low level*. (It is closely related to the unconditional branch in machine language that is executed directly by the hardware.) The **goto** and **if/then** are, in a sense, low-level "building blocks" to create more sophisticated and higher-level control statements. Pascal already provides these higher-level constructs as part of the language, and there is no reason to duplicate this work. The use of the **if/then** and **goto** to synthesize high-level control statements (as shown in Figure 4.4) defeats the purpose for including them in the language. It is similar to building a house from mud, clay, and water when finished bricks are readily available.

| Pascal | Synthesis using **if/then** *and* **goto** |
|---|---|
| a) **if** *be* **then** $s_1$ **else** $s_2$ | a)    **if** *be* **then goto** 1; |
| |     $s_2$; |
|     { *be is any boolean expression* } |     **goto** 2; |
|     { *s is any statement* } | 1: $s_1$; |
| | 2: |
| b) **while** *be* **do** *s* | b) 1: **if not** *be* **then goto** 2; |
| |     *s*; |
| |     **goto** 1; |
| | 2: |
| c) **repeat** | c) 1: *s*; |
|   *s* |     **if not** *be* **then goto** 1 |
|   **until** *be* | |
| d) **for** $v := n_1$ **to** $n_2$ **do** *s* | d)    $v := n_1$; |
| | 1: **if** $v > n_2$ **then goto** 2; |
| |     *s*; |
| |     $v := succ(v)$; |
| |     **goto** 1; |
| | 2: |
| e) **case** *exp* **of** | e) $v := exp$; |
|   $n_1 : s_1$; |   **if** $v = n_1$ **then** $s_1$; |
|   $n_2 : s_2$; |   **if** $v = n_2$ **then** $s_2$; |
|   . . . |   . . . |
|   **end** |   { *Note --this is not an exact translation since this works properly if v does not match any $n_i$. The **case** may not work.*} |

**Figure 4.4.** Synthesis of the Pascal control statements.

Another problem with the **goto** is that it has virtually no restrictions on its use. The programmer's freedom to transfer control anywhere at anytime becomes, like most totally unregulated and unrestricted freedoms, a license for *anarchy*. We may transfer control into or out of the middle of loops, clauses, or blocks. This seriously undermines any claims about the correctness or validity of our program. Some people feel that severely restricting the use of the **goto** would effectively solve this problem. For example, they would consider it proper to use the **goto** to transfer control to the first statement following a loop.

```
     i := 0;
     while i <= size do
     begin
         i := i + 1;
         if key = list[i] then goto 1
     end;
1:   if i <= size then { determine why we ended }
```

Their argument would be that the **while** loop is still a single-entry, single-exit block. We enter the loop only at the top and always exit to the first statement immediately following the loop. This may be true, but it still disregards the first point we made. The **goto** is being used to synthesize constructs already available within the language, sacrificing clarity and readability in the process. The following fragment uses the inherent capability of the **while** statement to achieve the same effect.

```
     i := 0;
     found := false;
     while (i <= size) and (not found) do
     begin
         i := i + 1;
         if key = list[i] then found := true
     end; { of while }
     if found then { determine why we ended }
```

The major advantage of the second example over the first is that the two conditions for loop termination—finding the desired item or coming to the end of the list—are clearly delineated by the conditions in the **while** loop. They are obvious to anyone reading the program. However, in the former example, you must look closely and carefully at the fragment to discover the second condition for loop exit. This code does not help readers to explain its purpose.

The unrestricted use of the **goto** can also result in a program that is difficult to read and understand and, consequently, difficult to work with and maintain. The fragment in Figure 4.1*a* should have made this point clear. If not, Figure 4.5 shows a program that is structured poorly and also violates almost all the stylistic guidelines introduced in Chapters 2 and 3. Study this program in order to experience firsthand

{ *assume that the arrays al,bl and the integer variables a,as,b,bs,c,i have been previously defined. This program will attempt to merge two sorted lists, al and bl, of length as and bs respectively, and produce a single merged list, cl. The merged list should contain every item in either list al or bl, and should be sorted into ascending order.* }

```
begin
            if (as < 0) or (bs < 0) or (as + bs > maxcsize) then goto 7;
            a := 1;
            b := 1;
            c := 1;
    4:      if (a > as) then goto 1;
            if (b > bs) then goto 1;
            if (al[a] < bl [b]) then goto 2;
            goto 3;
    2:      cl[c] := al[a];
            a := a + 1;
            goto 9;
    3:      cl[c] := bl[b];
            b := b + 1;
    9:      c := c + 1;
            goto 4;
    1:      if b > bs then goto 5;
            i := b;
    6:      cl[c] := bl[b];
            i := i + 1;
            c := c + 1;
            if i <= bs then goto 6;
            goto 7;
    5:      i := a;
    8:      cl[c] := al[a];
            i := i + 1;
            c := c + 1;
            if i <= as then goto 5
    7:
end.
```

---

*Figure 4.5*. Example of a poorly written unstructured program fragment.

the difficulty in attempting to work with and understand the logic of a poorly structured program. In addition, we have intentionally included some bugs (that you should try to find) to impress on you the difficulties involved with getting poorly written programs to work properly. By no standard whatsoever could we say that the program in Figure 4.5 "looks nice." We will rewrite this program according to our own stylistic guidelines in the following section. (The bugs in the program in Figure 4.5 are listed in the conclusion to this chapter.)

The problem with trying to read and understand a program with too many **goto**s is that it is necessary to follow two or more distinct paths that may be in totally separate parts of the program. This forking occurs every time we encounter the **goto**. For example, assume that on page 4 of a program listing we encounter the following.

> **if** $x = 5$ **then goto** *71;*
> **goto** *930*

We must search the listing for label 71 and, when we find it, say on page 9, start reading from that point. When we finish following that branch (which may include other branches), we must remember to come back to page 4 of the listing. This will send us out looking for the label 930 that may, for example, be on page 1. This jumping around continues with each additional use of the **goto** until we can no longer follow the flow of logic.

What we need is a program that flows naturally from beginning to end. The code should read like a book. We start reading at the first line of the program unit and keep reading, line by line, until we come to the end. This structure is obviously desirable and the way to achieve it is with the higher-level control structures provided automatically in Pascal.

To summarize, the major reasons for avoiding the unconditional branch (**goto**) are:

1. The **goto** is a primitive, low-level statement type that is really not needed by any programming language that has an **if/then/else** and a **while/do** construct (or any similar conditional and iterative statements).

2. The unrestricted use of the **goto** makes it much more difficult to debug, test, and verify a program for correctness.

3. The use of a **goto** to exit from a loop hides some of the conditions for loop termination.

4. The unrestricted use of the **goto** results in programs that are much more difficult for people to read and understand.

## 4.3  EXAMPLE OF STRUCTURED CODING

Let us rewrite the merge example in Figure 4.5 using the structured coding guidelines presented in the previous sections. The formal specifications of the merge problem are as follows.

You are given two lists of integer values — alist and blist of size m $>=0$, and n $>= 0$, respectively. The lists are already sorted into ascending order. We wish to merge alist and blist into a single list called clist of length (m + n), which is also sorted into ascending order. This new list should contain all items from alist and blist. For example:

| 5 | 3 | 3 |
|---|---|---|
| 9 | 5 | 5 |
| 10 | 5 | 5 |
| 13 | 11 | 5 |
| | 13 | 9 |
| | | 10 |
| | | 11 |
| | | 13 |
| | | 13 |
| alist | blist | clist |

Let us assume that alist and blist, and their lengths, now called asize and bsize for clarity, have been previously defined. First, we must verify that the values provided for asize and bsize are meaningful; that is, there must be something to merge. We must also check that the total length of the final merged list (asize + bsize) does not exceed the capacity of the clist array.

```
if (asize > = 0) and (bsize > = 0) and
      (asize + bsize <= maxcsize) then
begin
      { the data is valid so solve the problem }
end
else writeln('error in the size of the lists')
```

We can now concentrate on solving the problem for valid list sizes. Let us use the variables a, b, and c for indices pointing to the next item in alist, blist, and clist, respectively. To merge the two lists, we find the current top item of each list, determine which is smaller (we are sorting into ascending order), move it to clist, and adjust the pointers of the appropriate lists. This will work as long as there is at least one item in each list. The beginning of our solution is:

```
{ these initializations assume the arrays are indexed beginning at 1 }
a := 1;
b := 1;
c := 1;
while (a <= asize) and (b <= bsize) do
begin
      if alist[a] < blist[b] then
      begin
            "move an item from alist to clist"
      end
      else
      begin
            "move an item from blist to clist"
      end
end { of while }
```

Moving an item from one list to another involves copying that item and updating the index. For example, to move an item from alist to clist:

```
clist[c] := alist[a];
a := a + 1;
c := c + 1
```

A similar set of operations applies for blist.

This works until we run out of items from one of the two lists (i.e., until a > asize or b > bsize). Then the comparison alist[a] < blist[b] becomes meaningless. When this happens, we should copy the remaining items in the unfinished list over to clist directly, without performing a comparison.

```
if a > asize then
begin
     "move the remaining blist items to clist"
end
else
begin
     "move the remaining alist items to clist"
end { of if }
```

If we exhaust alist first, the remaining items in blist are those between the current position of the pointer (b) and the end of the list (bsize), inclusive.

```
for i := b to bsize do
begin
     clist[c] := blist[i];
     c := c + 1
end { of for }
```

The identical reasoning applies if blist is exhausted first.

The entire program fragment is shown in Figure 4.6. Comparing the programs in Figures 4.5 and 4.6 should clearly demonstrate the increase in clarity and readability that results from following the stylistic guidelines in Chapters 2 and 3 and the structured coding guidelines discussed in this chapter.

The purpose of structured coding is to achieve clarity and readability, but it is not the only reason for coding in this fashion. Many reasons will be discussed in later chapters, but we will briefly introduce them here.

One of the most important advantages of structured coding is that it brings *intellectual manageability* to the coding of program modules. Instead of coding randomly and haphazardly, we take advantage of the hierarchy of properly nested blocks. We code the statements in the outer block *first* and work our way inward. Each successive block allows us to narrow our focus and concentrate on finer and finer details of the program. Thus we work from the general aspects of the problem to the more specific.

```
if (asize > = 0) and (bsize > = 0) and
      (asize + bsize <= maxcsize) then
begin { the merge operation }
      a := 1;
      b := 1;
      c := 1;

      { this while loop merges alist and blist so long as there is at least one item
        in each list. }
      while (a <= asize) and (b <= bsize) do
      begin
            { see which list has the smaller item on top }
            if alist[a] < blist[b] then
            begin { select the next item from alist }
                  clist[c] := alist[a];
                  a := a + 1;
                  c := c + 1
            end
            else
            begin { select the next item from blist }
                  clist[c] := blist[b];
                  b := b + 1;
                  c := c + 1
            end
      end; { of while }

      { we arrive here when we have exhausted one of the two lists. first see
        which one is empty. }
      if a > asize then
      begin
            { we have copied all elements from alist to clist. copy the remaining
              items of blist into clist }
            for i := b to bsize do
            begin
                  clist[c] := blist[i];
                  c := c + 1
            end
      end
      else
      begin
            { we have copied all elements from blist to clist. copy the remaining
              items of alist into clist }
            for i := a to asize do
            begin
                  clist[c] := alist[i];
                  c := c + 1
            end { of for }
      end; { of else clause }
      csize := c - 1
end { of then clause testing for valid data }
else writeln (' error in list size -- cannot merge')
```

**Figure 4.6.** Structured code fragment for the merge operation.

For example, assume that we need to write a program to perform some arbitrary task on a large number of data sets. Our initial approach would be:

```
while not eof do
begin
     "read one data set"
     "solve the task for that one data set"
end
```

Instead of solving the problem for a number of data sets, we now narrow our focus and try to solve it for one data set. Our next refinement may be:

```
while not eof do
begin
     "read one data set"
     "validate the data"
     if valid then
     begin
          "solve the task for this one valid data set"
     end
     else
     begin
          "process error condition"
     end
end { of while }
```

Again, we have narrowed the task. Now we have to determine how to solve the problem for only a single *valid data set*. We might do that by saying:

```
while not eof do
begin
     "read one data set"
     "validate the data"
     if valid then
     begin
          "determine which case this belongs to"
          case datatype of
               case1 : "solve for case1 data set"
               case2 : "solve for case2 data set"
               case3 : "solve for case3 data set"
          end { of case }
     end
     else
     begin
          "process error condition"
     end { of if }
end { of while }
```

Now we can begin looking at how to solve and code just *one category* of *valid data set*. This process continues, each time focusing on more specific details of the problem. This approach, called *top-down design*, is a systematic method of writing programs that capitalizes on the characteristics of structured code. We will have much more to say about top-down design methods in Chapter 5 and will apply it to the design of individual program modules and to entire complex programming systems composed of dozens of modules.

The characteristics of structured code are also critically important during the debugging and testing of programs. If a program block has a single entry and a single exit, it is relatively easy to test whether that block is correct. We merely write out the values of all important variables upon entering and exiting the block.

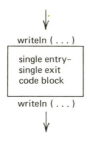

If the values were correct upon entering the block but incorrect coming out, the error must be in the block itself, since there is no other entry into the block. If, however, we allow arbitrary branching at any point in the program:

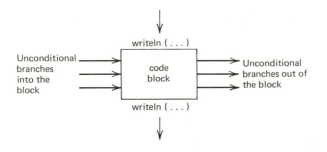

we cannot determine the exact site of the error. It may be in the block itself, or in any one of the other blocks linked to it by a transfer of control. This will significantly increase the complexity of the debugging task. We will say more about this in Chapter 10.

There is an important technique in programming called *program verification* that attempts to verify the correctness of programs in a radically new way. Instead of empirically testing a number of discrete data sets, we attempt to prove, using *formal mathematical* methods, that the program will produce correct values for all input. In program verification we think of the problem specification, p, as a function, with input, x, and output, y.

We now attempt to show that our program, P, is a correct realization of the formal specifications, p. That is, if for any input, x, the program produces a value y, y = P(x), the value y is correct with respect to the specifications of the problem—that is, y = p(x). Verification offers an exciting new alternative to the sometimes hit and miss approach to testing that presently exists. However, for verification to be realistic, the program must contain properly nested blocks, each of which exhibits well-known arithmetic and logical properties. Again, there is a demand for well-structured programs. (More will be said about program verification techniques in Chapter 10.)

*Style Review 4.1*

### Program Structure

Learn to use and master the six basic control structures of Pascal—**begin/ end, if/then/else, case/end, while/do, repeat/until**, and **for/do**, so that you can implement any program unit as a properly nested set of blocks created from the above six statements. Your program will be:

1. More readable and understandable
2. Easier to test and debug
3. Formally verifiable, if desired

## 4.4   SOME POINTS ABOUT STRUCTURED CODING

### 4.4.1   More About the Goto

Theoretically, the **goto** is never needed. All programs can be written using the higher-level control structures available in Pascal. However, the goal of structured code is not to eliminate the **goto** but to increase clarity and readability. The **goto** should be used whenever it improves the organization of a program.

Let us first describe some situations in which the **goto** is *not* needed. First, a **goto** is never needed to jump backward. By definition, a backward jump is a loop; it involves going back and repeating the execution of a group of statements. In Pascal a loop can always be implemented with one of the three iterative control constructs—**while, repeat,** or **for**. A program will always be clearer and more readable if loops are implemented in that fashion. For example, the following fragment sums by rows the elements of an n x n matrix called a. The summation stops either when we have summed up all $n^2$ elements or when the sum exceeds the value 1000.0.

```
sum := 0.0;
row := 0;
repeat
     col := 0;
     row := row + 1;
     repeat
          col := col + 1;
          sum := sum + a[row,col]
     until (col >= n) or (sum > 1000.0)
until (row >= n) or (sum > 1000.0)
```

(a)

```
         sum := 0.0;
         row := 0;
2:       col := 0;
         row := row + 1:
1:       col := col + 1;
         sum := sum + a[row,col];
         if col < n then goto 4;
         goto 3;
4:       if sum <= 1000.0 then goto 1;
3:       if row < n then goto 5;
         goto 6;
5:       if sum <= 1000.0 then goto 2;
6:
```

(b)

In the first fragment it is much easier to determine the scope of the loop as well as the criteria for termination. This is not true in the second fragment, which implements the same looping operations using two backward jumping **goto**s. This problem worsens as the loops become longer and more deeply nested. Avoid writing code such as that in the second fragment. Learn to feel comfortable using the high-level looping primitives provided by Pascal.

Programmers sometimes mistakenly believe that a **goto** is needed when it is necessary to exit from a loop before the normal loop condition is satisfied. For example, suppose we were counting characters in a line of text (n characters long) until the occurrence of the character '.'. We might initially be tempted to write:

```
    i := 1;
    while line[i] <> '.' do
    begin
        i := i + 1;
        if i > n then goto 1 { we came to the end of the line }
    end;
1:  { next statement }
```

We have used a **goto** to implement the abnormal branch that occurs when we have processed the entire line and not found a '.'. As mentioned earlier, this use of the **goto** is poor because it might lead us to infer, from looking at the **while** statement, a single criterion for loop termination: line[i] <> '.'. Only after more detailed investigation of the loop body itself would we discover the second condition—i > n.

Leaving a loop prior to normal termination is almost always handled better with a boolean variable that controls execution of the loop. By setting the variable to true or false, we can control continued execution or immediate termination of a loop. For example, we have rewritten the preceding fragment using a boolean variable endofline. The variable becomes true if we encounter the end-of-line condition i > n.

```
    count := 1;
    endofline := false;
    while (line[count] <> '.') and (not endofline) do
    begin
        count := count + 1;
        if count > n then endofline := true
    end
```

Now the criteria for loop termination are clearly reflected by the code, and clarity and readability are greatly enhanced. Using boolean variables to control loops is a very common and important programming technique that should be thoroughly understood and mastered.

The general structure of most loops that might terminate prematurely is as follows. (Assume done is a boolean variable.)

```
    done := false;
    "other loop initializations"
    repeat
        "process this pass of the loop"
        if "the premature exit conditions are not true" then
        begin
            "set up for the next iteration"
        end
        else done := true
    until ("normal termination") or done
```

Despite the many instances in which a **goto** is neither necessary nor desirable, there are times when prudent use can enhance the clarity and organization of a program. A **goto** can be helpful when an error or abnormal termination of an entire program unit causes a sudden break in the normal flow of logic. In such a case, the **goto** usually would jump to the end of the entire program unit.

For example, if execution is deeply nested inside a set of looping structures and a fatal error condition is encountered, it might be very inconvenient to exit from the loops one level at a time using boolean variables. It might be better to handle this multilevel exit with a single **goto**, as in the following fragment, which terminates whenever a negative element is encountered in a 3-dimensional array called matrix.

```
for i := 1 to depth do
begin
    planesum := 0;
    for j := 1 to row do
    begin
        rowsum := 0;
        for k := 1 to column do
        begin
            if matrix[i,j,k] < 0 then goto 99; { transfer to end of subprogram }
            rowsum := rowsum + matrix[i,j,k]
        end; { of for k loop }
        writeln('sum of row ', j, ' in plane', i, ' is ', rowsum);
        planesum := planesum + rowsum
    end; { of for j loop }
    writeln('sum of plane ', i, ' is ', planesum)
end { of for i loop }
       .
       .
       .
99: end; { of entire subprogram }
```

Notice that the **goto** is being used in a very specific and limited way: to terminate a program unit because of a fatal error. In this environment the **goto** is an effective way to handle the transfer of control. Notice also that we have added a comment to the **goto**. This is always a good idea, since the **goto** itself does not indicate the location to which you are transferring. (After all, the 99: label could be anywhere.) Situations in which a **goto** may be useful almost always have the following characteristics.

1.   A forward branch, usually through a number of levels.
2.   An abnormal condition, an error, or other break in the normal flow of logic.
3.   Usually a branch to the end of the program or program unit.

In those rare instances when a **goto** is needed, it should be used prudently and only to deal with errors or other abnormalities in the development of the logic. Most program units will have only one unconditional branch, or none at all. If you find yourself writing programs with many **goto**s, you have failed to understand the proper use of the Pascal control structures. You should reread Sections 4.1 to 4.3 and study additional examples of well-structured programs that properly utilize the **repeat, while, for, if/then/else,** and **case** constructs.

As another example of well-structured program units, Figure 4.7 presents a program that determines whether certain integer values are *prime*. A prime number is one that cannot be divided evenly by any values other than 1 and itself. Thus 4 is not prime (4/2 = 2) and 9 is not prime (9/3 = 3), but 3, 5, 7, 11, 13, and 17 are prime. The program, called "primefinder," is interactive and prompts the user for both numeric input values and a decision about whether or not to continue. The output (for legal numbers) will be either "nnn is not prime, mmm is a factor.", or "nnn is prime."

Users should study this and other programs presented so far as examples of good programming style and well-structured code. Mastery of the control structures of Pascal is critical to a proper understanding of many of the ideas that will be presented later in this text.

*Style Review 4.2* _____

> ### The Correct Use of the Goto
>
> *Do Not:*
>
>     Use the **goto** to jump backward.
>     Use the **goto** to implement one of a number of ways to exit a loop.
>     Use the **goto** too often.
>
> *Do:*
>
>     Consider using the **goto** to handle errors or other abnormalities in logic.
>     Consider using the **goto** to handle unusual circumstances such as long forward branches to the end of a program unit or multilevel forward branches out of a complex nested structure.

### 4.4.2  More About the While and Repeat Loop Constructs

Except for some minor syntactic differences and the fact that the **while** specifies a *continuation* criterion while the **repeat** specifies a *termination* criterion, the two loop constructs would seem to behave identically. This is not true, however, and a glance back at Figure 1.9 will illustrate one very important difference that can have a profound impact on program behavior. Specifically, the body of a **repeat** loop will

```pascal
program primefinder(input,output);
{ this program reads integer values and determines if they are prime. }
type
      responsetype = (yes, no, uncertain);

var
      factor       : integer;              { steps from 2 to upperlimit, checking
                                             for primes }
      firstmessage : boolean;              { true if user has not received
                                             message explaining yes and no
                                             response }
      found        : boolean;              { used to control search loop }
      number       : integer;              { the value entered }
      response     : responsetype;         { the user's response }
      upperlimit   : integer;              { the sqrt(number). the stopping point
                                             in searching for a prime }
```

{ getanswer

this procedure reads in the user's typed response and categorizes it into yes,
no, or uncertain.

entry conditions:
      program has asked user to type either yes or no

exit conditons:
      answer is set to yes, no, or uncertain
}

```pascal
procedure getanswer (var answer : responsetype);
const
      linelength    = 80;                  { maximum length of an input line }
var
      i             : integer;             { loop index }
      line          : packed array [1 .. linelength] of char;  { the input line }
begin { of procedure getanswer }
      { send a prompt to the user }
      readln;

      { read the user's response and place it one character per word in the array
        line }
      i := 0;
      while (not eoln) and (i <= linelength) do
      begin
          i := i + 1;
          read(line[i])
      end;
      { classify the response on the basis of the first 2 chars }
      answer := uncertain;
      if i >= 2 then
          if (line[1] = 'y') and (line[2] = 'e') and (line[3] = 'y) then answer :=
              yes
          else if (line[1] = 'n') and (line[2] = 'o') then answer := no
end; { of procedure getanswer }
```

```
begin { main program }
    writeln(' welcome to the prime finder program');
    firstmessage := true;
    repeat
        if firstmessage then
        begin
            writeln(' please indicate whether you wish to perform');
            writeln(' a calculation by entering ''yes'' or ''no'' followed');
            writeln(' by a carriage return');
            firstmessage := false
        end
        else
        begin
            writeln(' enter ''yes'' or ''no'' followed');
            writeln(' by a carriage return')
        end;
        getanswer(response);

        case response of
            no:
                writeln(' good bye');
            uncertain :
                begin
                    writeln(' your response was in error and could',
                        ' not be evaluated. please enter ''yes'' ');
                    writeln(' or ''no'' without imbedded blanks or',
                        ' other characters, immediately followed');
                    writeln(' by a carriage return')
                end;
            yes :
                begin
                    writeln(' please enter an integer value whose',
                        ' absolute value is >= 2');
                    readln; { prompt }
                    read(number);
                    if number < 0 then
                    begin
                        number := abs(number);
                        writeln(' you must mean ',number)
                    end;
                    if number = 1 then
                        writeln(' one is a prime. ')
                    else
                    begin
                        factor := 2;
                        upperlimit := trunc(sqrt(number));
                        found := false;
```

```
                              { now scan through all integers from 2 . . upperlimit
                                 seeing if any divide evenly into number }
                              while (factor <= upperlimit) and (not found) do
                                  if number mod factor <> 0 then
                                       factor := factor + 1
                                  else found := true;

                              { print the results }
                              if found
                              then writeln(number,' is not prime ',factor, ' is a
                                   factor')
                              else writeln(number, ' is prime')
                         end { number >= 2 }
                    end { response = yes }
               end { of case }
          until (response = no);
          writeln(' thank you')
     end. { of program primefinder }
```

*Figure 4.7.* Program to determine if numbers are prime.

always be executed *at least once*, but a **while** loop may *never* be executed if the condition for continuation is false initially.

The problem here is that situations often arise where we are given nothing whatsoever to do. It is imperative that our program behave properly and meaningfully when presented with this empty condition (the *null* case). This program characteristic is called *robustness* and was discussed at length in chapter 3.

For example, the fragment:

```
repeat
     readln(x,y,z);
     writeln(' x= ',x,' y= ',y,' z= ',z);
     processdata(x,y,z)
until eof
```

works properly only as long as there is at least one line of data in the text file input. If input were empty, we would attempt to perform an input operation (the readln on line 2) while eof was true and would terminate with a fatal error. Likewise:

```
{ search through table looking for key }
found := false;
place := 0;
repeat
     place := place + 1;
     if key = table[place] then found := true
until found or (place >= tablesize)
```

will work for any *nonzero* value tablesize. However, if the table were empty (tablesize = 0), we would incorrectly perform a comparison between key and table[1], possibly resulting again in a fatal run-time error (array index out of bounds) if table[1] were never defined. Both these examples would handle the null case properly if they were coded using the **while**.

```
while not eof do
begin
    readln(x,y,z);
    writeln(' x= ',x,' y= ',y,' z= ',z);
    processdata(x,y,z)
end { of while }

found := false;
place := 0;
while (not found) and not (place >= tablesize) do
begin
    place := place + 1;
    if key = table[place] then found := true
end { of while }
```

When selecting a loop construct, be particularly careful in determining whether the null case is possible and, if so, examining how the loop will behave when it is executed zero times as well as one or more times.

The difference in the time of actually performing the loop termination test has another important effect on writing a loop. We must remember that the **while** loop will evaluate its tests immediately; therefore all variables in the loop termination test must be defined on loop entry. For example, if we had a data file with a signal value of 9999 used to mark the end, we could *not* write the following.

```
while x <> 9999 do
begin
    readln(x);

        .       { process this value of x }
        .
        .
end
```

We cannot write that because x is not given a value until we are inside the loop, when it is already too late to do the test. If we wanted to use the **while** statement, we would have to write the loop as follows.

```
readln(x); { get the first value }
while x <> 9999 do
begin
    { process this value of x }
      .

      .

      .
    readln(x)     { get the next value }
end
```

### 4.4.3   More About the For Construct

In many programming languages a **for**-type loop is the only loop primitive available. Because the **for** loop in Pascal is very similar to these statements, programmers familiar with these other languages tend to rely too heavily on the **for** statement to implement loops in Pascal programs.

This is a bad habit to get into because, in Pascal, the **for** loop is the *least* flexible and *least* useful of the three major loop statements available. The statement allows for only a single criterion for loop termination—that the value of the control variable exceeds (or, in the case of the **downto** clause, falls below) the value of the terminating expression in the **for** statement. Restricting loop termination to a single condition is often impossible. A simple example is the ''table look up'' used in the previous section. If we were to code that fragment using a **for** loop, we would have the following.

```
for place := 1 to tablesize do
     if key = table[place] then ?
```

How do we specify that we no longer wish to remain in the loop? What should we put following the **then**? A **goto** is now the only way to exit this loop. The statement we selected limited us to a single termination condition—place > tablesize. The logic, however, has another termination condition—finding the desired item.

The solution to this problem is simple. Do not think of the **for** loop *first*; think of it *last*. Use the **for** loop only for the special case of simple loops that always proceed through to completion and that cannot unexpectedly terminate for any reason other than reaching the end. Zeroing out an array is a good example.

```
for i := 1 to n do a[i] := 0
```

So is summing a row of a matrix.

```
sum := 0.0;
for i := 1 to col do
     sum := sum + matrix[row,i]
```

In both these examples, the loop ends *only* after it has completed the specified number of executions.

Another reason for avoiding the **for** is that in Pascal the control variable is limited to *ordinal* data types (integer, boolean, character, user defined) and can be incremented or decremented only sequentially through its successors or predecessors (i.e., by $+1$ or $-1$). Some programmers, even when faced with a problem that does not conform to these restrictions, will attempt to use the **for** with temporary variables and confusing logic. For example:

```
for i := 1 to 50 do
begin
    theta := (i * 2.0) / 100.0;
    writeln(theta, sin(theta))
end
```

In this loop there is absolutely no need for the variable i or the confusing assignment statement. While it does no great harm, the programmer used a statement that was not really appropriate. How much more straightforward and clear it is to write:

```
{ generate a table of sin(0.02), sin(0.04),  ... , sin(1.0) }
theta := 0.0;
while theta < 1.0 do
begin
    theta := theta + 0.02;
    writeln(theta, sin(theta))
end { of while }
```

However, a word of caution is required about the robustness of this last example. It may result in an off-by-one error because of the round-off error inherent in representing the value 0.02 in binary. On the fiftieth iteration the variable theta may not have the exact value 1.0 but a value $1.0 - \epsilon$, where $\epsilon$ is some small positive quantity. This will cause the program to execute the loop improperly a fifty-first time. This point was discussed in Section 3.2.3. To correct this error we could set up an ordinal counter to count from 1 to 50 or else write our loop termination condition as follows.

```
while theta < 0.99 do
```

Limit your use of the **for** to the special cases first discussed. The majority of loops should be coded using the more general and more powerful **while** and **repeat** constructs. Learn to feel comfortable using them.

## 4.5  CONCLUSION

We have now presented ways to write programs that both look and act **nice**. By following proper stylistic guidelines and using only the higher-level control statements of Pascal, we can develop programs that are easy to read, follow, and understand. This should reduce errors and facilitate testing, maintenance, and modification.

Chapters 5 and 6 discuss how to design and implement large complex programs composed of many units. The method will be virtually identical to the top-down coding approach introduced in this chapter. In Chapter 5 we will apply it not just to a single module but to the overall problem.

The errors in Figure 4.5 are:

1. The statement labeled 6: should read 6: $cl[c] := bl[i];$.
2. The statement labeled 8: should read 8: $cl[c] := al[i]$.
3. The penultimate statement should read **if** $i <=$ as **then goto** 8.

## EXERCISES FOR CHAPTER 4

1. Rewrite the following fragment so it achieves the same results without using **goto** statements.

```
var
         hand   :   array [1 .. 13] of cards;

         cards  :   record
                         rank : 1 .. 13; { j = 11, q = 12, k = 13 }
                         suit : (spade, heart, diamond, club)
                    end;

     points,i  :   integer;
                .
                .
                .

{ count the points in the hand using the following point values: ace = 4
  pts, k = 3, q = 2, j = 1 }
         points: = 0;
         i: = 1;
3:       if hand[i].rank = 1 then
         begin
                 points : = points + 4;
                 goto 2
         end;
         if hand[i].rank = 13 then
         begin
                 points : = points + 3;
                 goto 2
         end;
         if hand[i].rank = 12 then
         begin
                 points : = points + 2;
                 goto 2
         end;
         if hand[i].rank = 11 then points : = points + 1;
2:       i : = i + 1;
         if i <= 13 then goto 3;
         writeln('total points in the hand = ', points)
```

Discuss how the fragment could also be improved through the use of symbolic constants.

2. Extend your rewritten fragment from Exercise 1 so that it also awards points for distribution according to the following rules.

Three points if you have no cards of a suit (*void*).

Two points if you have 1 card in a suit (*singleton*).

One point if you have 2 cards in a suit (*doubleton*).

3.   Rewrite the following **repeat** loop so that all conditions for loop termination are specified directly in the **repeat** statement.

```
{ find the vowel pair 'ae' in the current line. the characters of the current
  line are stored in line[1], . . . , line[max]. a period encountered in the line
  should also stop the scan }

i := 0;
repeat
      i := i + 1;
      if line[i] = '.' then goto 2;
      if i >= max then goto 2
until (line[i] = 'a') and (line[i + 1] = 'e');
2:      .

        .

        .
```

In addition, discuss the programming style of this fragment with respect to overall generality and robustness in handling the null case.

4.   Rewrite the following **for** loop, using either a **repeat** or a **while** statement, so that it achieves the same effect but without the **goto**.

```
{ generate a table of function values, f(x), for x = 0,1, . . . , 500. however,
  stop if the value of f(x) ever exceeds some upper bound, called limit, for
  any x. }

for x := 0 to 500 do
begin
      value := f(x);
      if value > limit then
      begin
            writeln('value of function unacceptable at the point ', x);
            goto 1
      end { of then clause }
      else table(x) := value
end; { of for loop }
1:
```

5.   Try modifying the unstructured program fragment from Figure 4.5 so that it eliminates *duplicates* between lists al and bl. That is, if a value occurs in both lists, move just one value to the merged list cl and discard all other identical values. Keep track of the time and effort involved. Now try making the identical

change to the fragment in Figure 4.6. Discuss how the program's structure and style facilitated this modification.

6. Modify the program in Figure 4.7 so that, if the number being tested is *not* prime, the program will print out a list of *all* the factors of that number, not just any one factor.

7. In Section 4.4.1 we discussed places where the **goto** might be a convenient and appropriate control structure. We then showed an example of triply nested **for** loops with a **goto** used as an error exit. It is important to remember that the **goto** was used out of convenience, *not* necessity. Rewrite that fragment so it achieves the same results but without a **goto**. Compare the ease of reading and understanding the two different fragments.

8. Code the following problems as a well-structured, "elegant" Pascal procedure called path.

(a) You will be given, as input, a two-dimensional matrix called connection that is n x n. Connection(i,j) = 1 if there is a direct physical path (i.e., a road, a bridge, a connection) from node i to node j. Otherwise, connection(i,j) = 0. For example:

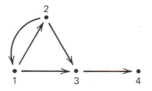

Your procedure should take as input the matrix connection along with the index of any two nodes i,j; i,j = 1, . . . , n. The output of the procedure should be a boolean value that is true if there is a path that eventually leads from node i to node j and false otherwise. For example, referring to the preceding chart:

path(1,4) is true (1 → 2 → 3 → 4 or, alternatively, 1 → 3 → 4)
path(4,3) is false
path(3,1) is false

(b) Modify the procedure of part a so that, in addition to telling you that a path exists from node i to node j, it also returns what that path is. That is, it returns the exact sequence of nodes $a_1, a_2, . . . , a_k$ such that there is a connection between $a_i$ and $a_{i+1}$ for i = 1, . . . , k − 1, and such that $a_1$ = i and $a_k$ = j.

9. Write a well-structured Pascal program to simulate the operation of a grocery store checkout system. The program should initially build a data structure containing information on all the products for sale in the store. For each item, keep the following information:

**type**
    *productrecord* =
        **record**
               *number*       : *integer*; { *the 6 digit universal product code* }
               *description*  : **array** [1 .. 10] **of** *char*; { *item name or description* }
               *unitprice*    : *real*;
               *taxable*      : *boolean* { *does sales tax apply to this item* }
           **end**; { *of record productrecord* }

Overall inventory can now be described as an array of product-records.

**type**
    *inventory* = **array** [1 .. *max*] **of** *productrecords*;

After building the inventory, your program should input a series of product numbers. (This is equivalent to running the bar code on the product label under a bar code reader.) The program should be printing out the cash register ticket as well as acccumulating totals. At the end of the transaction (indicated by a product number of 0), the program should print the subtotal, the tax (4% of the value of all taxable items), and the grand total.

For example, the following input:

101240
391662
557330
    0

might lead to the following output:

| 1 can soup | $0.85 |
|---|---|
| 1 loaf bread | $0.65 |
| 1 steak | $3.50 |
| subtotal | $5.00 |
| tax | .20 |
| grand total | $5.20 |

If your program cannot find a product number in the inventory, write out a message that there is no such product and continue with the next item. After finishing with one customer, continue on to the next. The end of all customers is signified by an end-of-file condition.

*Chapter 5*

# PROGRAM DESIGN METHODS

## 5.1 INTRODUCTION

Chapters 1 to 4 concentrated almost exclusively on the characteristics (style, structure) of individual program units or modules. In this chapter we begin our discussion of *program design methods*—techniques for managing, both intellectually and administratively, the overall development of large, complex computer programs composed of many modules.

As indicated in Table 3.8, the length of real-world programs differs markedly from typical student assignments. An implementation technique that works well for a 100-line program may do poorly with a 1000-line program and may be totally inappropriate for a 10,000-line one.

We will be using an extended analogy—a typical small grocery store where "Mom and Pop" do everything—to clarify our discussion. In our grocery, Pop loads the shelves, orders goods, sweeps the floor, and makes deliveries. Mom works the cash register, writes out the checks, and goes to the bank. Together, they directly perform every task, from sweeping floors to setting long-term inventory policies. Their company has no formal management structure or organization. As the store grows, they may add one or two part-time helpers to perform specific tasks, but the business structure is still minimal, and Mom and Pop are still involved with all operational details of the entire company on a day-to-day basis.

This type of informal management structure is appropriate only as long as the business remains small; as the business expands, Mom and Pop must eventually select one of two totally different management policies (Figure 5.1).

If they do not wish to change the way they manage their business, they will essentially set an upper bound on the size of their store based on how much detail they can handle personally and how much work they can squeeze into one day (path

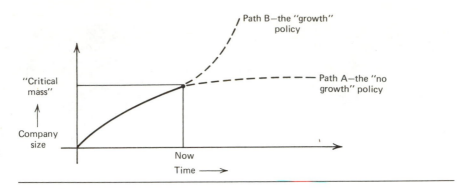

*Figure 5.1.* Patterns of growth.

A). If they decide to continue growing (path B), they must drastically change their way of doing business. They must set up a hierarchically structured *management organization* that recognizes the differences between low-level day-to-day details and high-level policy decisions. Mom and Pop must realize that now not everyone in the company will know every operational detail. Each individual is responsible for certain aspects of the business, and a superior monitors and coordinates these individual efforts. With this structure, their corner grocery store can grow substantially and handle a significantly increased level of business complexity.

The identical trends can be seen in the handling of programming projects. The small Mom and Pop grocery store is analogous to the small student programming assignment usually seen in an introductory programming course. Students have no "management" organization or "grand plan." There is really no need because, like the small grocery store, the task is so simple that everything can be kept in your head or on scratch paper. One person handles all algorithm selection, coding, data entry, debugging, testing, and documentation. The same person keeps track of all current program listings and the current working/not working state of the code. Finally, he or she manages a personal work schedule in order to complete the assignment by the deadline.

This approach works well for small programs but quickly collapses when applied to longer efforts. Confusion usually occurs when beginning students attempt, without any planning or organization, their first large programming project (e.g., on the order of 1000 lines of code). These attempts are rarely successful.

Specifically, the following problems are encountered when attempting a large programming project.

1.   It is difficult to predict all the ramifications of an early decision. The programs are so large that the effect of a decision may not become evident for days, weeks, or even months. At that time it may be difficult to undo its effects, and we are stuck with our early choice, even though it may not be the best one.

2. Debugging and testing become extremely complex. The programs are so large that looking for an error can be like searching for the proverbial needle in a haystack.

3. We find ourselves engulfed in detail. There is so much to do and remember that we forget to perform certain operations, miss key completion dates, or omit critical sections from the code. We cannot keep details in our head or on scratch paper and hope to remember them. We easily lose track of what has already been done and what still needs to be done.

Like Mom and Pop, we have reached our "critical mass." However, unlike them, we do not have the luxury of the two choices shown in Figure 5.1. We cannot opt for path A and say to an employer, "Sorry, I choose to work on small specialty problems whose solutions will consist of programs no longer than 100 lines!" We must select path B and be prepared to work on problems that can grow larger and larger; consequently, we must be willing to change how we implement these programs. We must develop and use an organizational structure that facilitates the correct development of large, complex computer programs. The remainder of this chapter will discuss just such an organizational tool—*structured programming*.

*Style Review 5.1* _____

> **Large Programs**
>
> Length is not the only difference between small and large programs. Large programs are also conceptually and intellectually different and require different mental operations, such as generalization and abstraction. You cannot predict how you will perform on a large program from how you handle small programs. Being able to write a 100-line program in x units of time does *not* mean that you will automatically be able to complete a 10,000-line program correctly in 100x units of time.

## 5.2 STRUCTURED PROGRAMMING

*Structured programming* is a program implementation technique that systematizes and organizes the entire cycle of program design, coding, and testing. The ultimate goal of this technique is to develop correct, reliable programs by preventing errors and facilitating debugging. Structured programming also attempts to develop software that minimizes personnel costs and increases productivity.

The structured programming methodology is composed of three components.

1.  Top-down design philosophy.
2.  Independent program modules.
3.  Structured coding principles.

We will now discuss the top-down design philosophy. The next two points (independent modules and structured coding) will be discussed in Chapter 6.

### 5.2.1  Top-Down Program Design

The techniques of top-down design are quite old, and you have probably used them in other areas of study. For example, assume that you were assigned to write an historical essay about World War II. After some thought you decide to structure your paper as follows.

I.   The Simmering Pot: 1930–1936
II.  The Testing Period: 1936–1939
III. All-Out War: 1939–1945
IV.  Reconstruction: 1945–1950

Then, piece by piece, you flesh out the details of each section.

I.   The Simmering Pot: 1930–1936
    A.  The Great Depression and its Effect on German Employment
    B.  Worsening Inflation and Disillusionment
    C.  The Rise of Anti-Semitism
    D.  Birth of the Nazi Party

This outlining process is an excellent example of the top-down design method. It is currently being applied to essays, book reports, critical reviews, and now to *computer programs* as well.

The top-down design method begins with the overall goals of the program—*what* we wish to achieve instead of *how* we will achieve it. Initially we specify a very broad solution that achieves these goals by invoking a series of generalized tasks. In a sense, we can say that we are specifying the "top-level" structure of our overall program, much like our first attempt at an outline of our World War II historical essay.

The design process continues with a series of *stepwise refinements*. Each generalized high-level task is elaborated on with more specific and detailed subtasks that, when properly executed, will accomplish the high-level goal. As the subtasks are described, we will also develop the essential data structure details needed by that subtask.

This refinement process continues, each time describing in greater detail the lower-level steps needed to accomplish a high-level operation. Development proceeds in a "treelike" fashion, as shown in Figure 5.2.

In Figure 5.2 subtasks $t_1$, $t_2$, and $t_3$ are needed to solve the original problem. Subtasks, $t_{11}$, $t_{12}$, and $t_{13}$ are needed to accomplish subtask $t_1$, and so forth. The refinements stop when a subtask (e.g., task $t_{121}$ of Figure 5.2) is simple enough to be directly implemented without having to define any additional lower-level subtasks.

Each subtask is *validated* (certified to be correct) before it is further refined. Thus task $t_1$ would be thoroughly tested or formally proved correct *prior* to designing and coding subtasks $t_{11}$, $t_{12}$, or $t_{13}$. This will eliminate many of the problems involved with trying to debug a program as a single, large, complex entity.

As an example, we will apply a top-down design philosophy to the coding of a *text editor*—a program that accepts and manipulates character strings. Users will give the text editor commands that describe exactly how to manipulate the character string (e.g., insert new characters, locate a particular substring). The text editor will immediately carry out that operation and wait for the next command. (Note that the refinement steps in this example are unrealistically fine to highlight the concepts we have been discussing. A larger and more realistic programming example will be presented in Chapter 6.)

An interactive text editor is a large and complex program (usually 1000 to 5000 lines of code, depending on what it does). Any attempt to enter directly into the code will fail. However, at the highest and most generalized level, even a complex task such as that of a text editor can be simply and straightforwardly described, as shown in Figure 5.3a on the following page.

This is a very broad but nonetheless accurate description of our text editor. The only major algorithmic decision we have made so far is that the text editor will be *interactive* in nature. We will get one command from the user and process it immediately. We will repeat this get/process cycle until there is nothing more to do. We have made only one "data structure" decision:

**var**
    *working*    : *boolean;*              { *working is true while there is still something to do* }

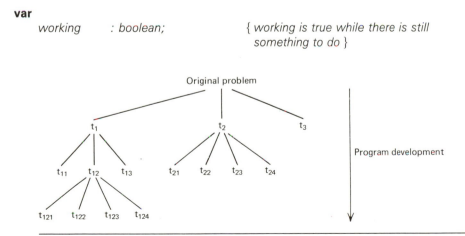

*Figure 5.2.* A program development tree.

Finally, we have defined three lower-level subtasks.

1. *Initialize.* To perform all initialization operations necessary for the proper execution of the text editor.
2. *Getacommand.* To request a command from the user, input the command, and classify it according to command type.
3. *Processcommand.* To process the command just entered by the user.

If these three subtasks existed, our initial top-level design would be a proper solution to the problem.

Before we begin to design and code these subtasks, we must attempt to validate the correctness of the design so far. The current refinement as shown in Figure 5.3*a* is so trivial and so short that manual verification methods should be sufficient. A *hand simulation* of the program involves working through each executable statement with paper and pencil and doing exactly what the program says to do. In a sense, we are "playing computer." A more formalized type of manual test method is called a *structured walkthrough.* During a formal review session with technical colleagues, we present our program and explain the rationale behind our design decisions. We may also "walk through" or simulate program operation for some typical cases. The colleagues act as troubleshooters—probing, questioning, and trying to ferret out errors or logical inconsistencies. The goal of the session is to help spot errors that the original authors overlooked, perhaps because they were too close to and too familiar with the project.

Let us assume that the top-level design has been validated and that we now want to refine the "processcommand" subtask. First, we must make a decision about the types of text editing commands users will be allowed to enter from the console. Say that we decide to allow the following five commands.

*Insert.* Insert a new character string into the current file.

*Delete.* Remove a character string from the current file.

*Find.* Locate a character string in the current file.

```
begin
    working := true;
    initialize;
    while working do
    begin
        getacommand;
        if "valid command" then processcommand
    end { of while }
end
```

**Figure 5.3a.** Top-level description of a text editor.

*Change.* Replace every occurrence of one character string with another character
string.

*Stop.* Stop editing and save the current file.

If we specify a new user-defined data type of the following form:

```
type
      commands        = (insert, delete, find, change, stop, error);
var
      commandtype     : commands;
```

the job of getacommand can be defined as accepting whatever the users type, determining what type of command it was, and setting the variable commandtype accordingly. If getacommand cannot classify the input into one of these five commands, the
procedure should set commandtype to the value error.

With these additional decisions made,we are now ready to add a little more detail
to the program (See Figure 5.3*b* on the next page).

We have now specified six additional subtasks that are needed to implement
processcommand—one routine for each of the five command types and an errorhandler
to process all erroneous input.

Again, we should test our refinement before proceeding. We can do a hand
simulation as well as a structured walkthrough, as discussed earlier. However, at some
point we should *run* the program on the computer to double-check our manual methods. Initially, that may seem impossible. The top-down method we are using always
describes a solution in terms of lower-level procedures that are as yet *unwritten*. How
is it possible to test our interim solution when initialize, getacommmand, insertroutine,
deleteroutine, findroutine, changeroutine, stoproutine, and errorhandler do not yet
exist?

The answer is that we will test our program using *stubs*. A stub is a dummy
procedure included in a program to allow compilation and execution to proceed meaningfully. Stubs do as *little* as possible. They will usually carry out only the following
two operations.

1.  Write out a message that you have reached this procedure. This guarantees
    that we are correctly linking to the procedure.
2.  If the calling program references a value that is defined by this procedure,
    the stub *must* also define that value in order to prevent a run-time error caused
    by an undefined variable.

A reasonable stub for insertroutine might be:

```
procedure insertroutine;
begin
      writeln('in insertroutine, not yet coded')
end; { of procedure insertroutine }
```

```
begin
    working := true;
    initialize;
    while working do
    begin
        getcommand(commandtype);
        { now we process the command just entered }
        case commandtype of
            insert   : insertroutine;
            delete   : deleteroutine;
            find     : findroutine;
            change   : changeroutine;
            stop     :
                begin
                    working := false;
                    stoproutine
                end;
            error    : errorhandler
        end { of the case statement }
    end { of while }
end
```

*Figure 5.3b.* First refinement of the text editor.

Similar stubs would suffice for initialize, errorhandler, and the four other command routines. For getcommand, however, we must somehow assign a value to the variable commandtype, since that value is immediately referenced by the main program. (Refer back to lines 6 to 8 of Figure 5.3b). We could do this initially by using the integer values 1 to 5 to represent the five possible commands, with a 0 representing all errors. Now, by simply typing in a one-digit integer, we can simulate the inputting of a command to the text editor.

```
procedure getcommand (var class:commands);
{ temporary stub for the procedure getcommand }
var
    x :     0 .. 5;
begin
    readln(x);
    case    x   of
        0   :   class := error;
        1   :   class := insert;
        2   :   class := delete;
        3   :   class := find;
        4   :   class := change;
        5   :   class := stop
    end { of case }
end; { of procedure getcommand }
```

With the stubs in place, we can now test our design, at least as far as we have progressed. The result of running the current program will be:

in initialize routine, not yet coded
? 0
in errorhandler, not yet coded
? 1
in insert routine, not yet coded
? 2
in delete routine, not yet coded
? 3
in find routine, not yet coded
? 4
in change routine, not yet coded
? 5
in stop routine, not yet coded
END OF PROGRAM

and everything seems to be operating properly. We can be relatively confident that this level is correctly specified and proceed to refine the other tasks.

Let us assume that we now want to describe the getacommand operation in more detail. Specifically, we are concerned with command interpretation—how to translate the characters as they are entered by the users (e.g., "insert") into the internal representation used by the program (the constant *insert* of the data type *commands.*)

At this point in our design we are *not* concerned with any of the following details.

1. How to input the command from the input file.
2. The exact format of the command (fixed versus free format, separator characters, etc.).
3. The number and type of parameters required by each command. All we care about now is that the parameters, if any, come *after* the command name. For example:

*change/abc/xyz/*

   might mean change every occurrence of the string "abc" into the string "xyz."

These three details can be postponed until a subsequent refinement. To handle the command interpretation, we will create a table, called *comtable,* that carries the following information about each text-editing command.

1. The exact string of characters that denotes this command (e.g., "insert").
2. The number of parameters required by this command (initially 0 for all commands, since we are not currently interested in parameters).

3. The command type.

By setting up our table in this way, we can map more than one input string (e.g., "change," "ch," "c") into the same command, thus allowing synonyms to be used. The Pascal declaration to create this table is:

```
const
    maxcommand              = 25; { maximum number of unique text editing
                                    commands }
type
    commandrecord           =
        record
            name            : nametype; { the input string corresponding to
                                          this command }
                paramcount     : integer;
                classification : commands
            end; { of command record }
var
    comtable                : array [1 .. maxcommand] of commandrecord;
    tablesize               : integer;    { actual length of comtable }
```

Now we can see that the procedure getacommand will accept a character string entered by the users, look that string up in comtable, classify it according to the value in the "classification" field and, finally, read in the parameters specified for that command (although we have not yet specified exactly how to do this).

Before we code getacommand we must decide which algorithm to use for searching the command table (e.g., sequential, binary search, or one of various hashing methods). The decision will be based on the size of the command table, the number of times the table is searched, and the amount of extra time and memory space available. If we select a sequential search technique, getacommand might appear as in Figure 5.3c.

We have described getacommand in terms of three lower-level subtasks.

1. *Getinput (string, size).* A procedure to read in and return the exact character "string" entered by users as the command name along with its "size" in characters.

2. *Equal(a,b,size).* A boolean function that tests for exact equality between two character strings, a and b, each containing "size" characters. Equal returns the value true if the strings a and b are identical, else it returns the value false.

3. *Getparameters(k).* A procedure that reads in and stores the next "k" parameters entered by users following the command name.

Note how we have effectively "hidden" the details that we have not as yet chosen to address by pushing them into lower-level procedures. Getacommand, as written,

```
procedure getacommand (var commandtype : commands);
var
      found    : boolean;                 { switch to indicate if command was
                                            found }
      i        : 0 . . maxcommand;        { loop index }
      k        : integer;                 { parameter count }
      size     : integer;                 { the number of characters in the
                                            command name }
      string   : nametype;                { the string entered by the user }

begin
      writeln('now entering getacommand');
      getinput(string,size);
      commandtype : = error;
      found : = false;
      i : = 0;
      { now search sequentially through the table of all legitimate commands
        looking for the character string typed by the user }
      while (i <= tablesize) and (not found) do
      begin
            i : = i + 1;
            if equal(comtable[i].name,string,size) then
            begin
                  writeln('command found at position', i, 'of the command table');
                  found : = true;
                  commandtype : = comtable[i].classification;
                  k : = comtable[i].paramcount;
                  getparameters(k) { will fetch the next k parameters }
            end { of if }
      end { of while }
      if not found then writeln('command not found')
end; { of procedure getacommand }
```

*Figure 5.3c.* Refinement of procedure getacommand.

does not require a knowledge of the exact format of the commands. That detail is hidden in the procedure called ''getinput'' and in the unspecified data type ''name-type.'' Likewise, the details of command parameters are hidden in the procedure ''getparameters.'' In this refinement we have concentrated on only two problems: the data structure comtable (its format, contents, and maximum size) and the algorithm for searching it.

Again we should stop and test our current refinement. (Remember that the refinements have intentionally been kept small to illustrate the concepts.) Testing will require development of the following stubs.

**type**
      *nametype*                   *: char;*

For simplicity, all text-editing commands will be one character long—adequate for test purposes. Later refinements can add the full naming capacity. The procedure initialize will have the following code.

```
procedure initialize; { this is a testing stub for initialize }
begin
      writeln('in the initialize procedure');
      comtable[1].name := 'I'; { this is the insert command }.
      comtable[1].paramcount := 0;
      comtable[1].classification := insert;
              .

              .
      comtable[5].name := 'S'; { for stop }
      comtable[5].paramcount := 0;
      comtable[5].classification := stop;
      tablesize := 5 { we will test with only five unique commands }
end; { of procedure initialize }
```

The stub for procedure getinput can (for now at least) limit itself to reading one-character commands.

```
procedure getinput(var ch : nametype; var length : integer);
begin
      writeln('please enter the next text editing command');
      readln(ch);
      length:=1
end; { of procedure getinput }
```

The stub for the function equal checks for equality between two one-character variables.

```
function equal(a , b : nametype; length : integer) : boolean;
begin
      equal := (a = b)
end; { of function equal }
```

Finally, getparameters need not do anything, since we have not as yet decided on what, if any, parameters will go with each command.

```
procedure getparameters(i : integer);
begin
    writeln('now we would retrieve any parameters ',
            ' associated with this command')
end; { of procedure getparameters }
```

With all these stubs in place, we can begin testing the current design as specified in Figures 5.3*b* and 5.3*c*. The execution of the program will produce results similar to these.

```
in the initialize procedure
now entering getacommand
please enter the next text editing command
? I
command found at position 1 of the command table
now we would retrieve any parameters associated with this command
in insert routine, not yet coded

now entering getacommand
please enter the next text editing command
? D
command found at position 2 of the command table
now we would retrieve any parameters associated with this command
in delete routine, not yet coded

now entering getacommand
please enter the next text editing command
? Z
command not found
we are now inside the error handler

now entering getacommand
please enter the next text editing command
? S
command found at position 5 of the command table
now we would retrieve any parameters associated with this command
in stop routine, not yet coded
END OF PROGRAM
```

Once again, the program at this stage of development seems to be working properly.

As the final refinement to be discussed here, we will write procedure getinput. This procedure should take the individual characters entered by users, scan them to find the character string that forms the command name, and pack them into an array. Then we will be able to work with the name as a single indivisible unit called a *token*. This step requires us to specify exactly how a command name will be typed by users at the console. A reasonable formatting decision might be:

1. A command name must be the first item on a line, but it may be preceded by one or more blanks.

2. A command name may be 1 to 10 alphabetic characters in length and is terminated by the first occurrence of any nonalphabetic character. The string "insert" is a valid command, but "insertaletter" is not because it is too long and "insert2" is not because it contains a nonalphabetic character ("2").

These decisions will allow us to begin to code the procedure getinput. As with our previous refinements, we have postponed certain lower-level decisions. In this refinement we will not worry about the specific input terminal to be used or its particular hardware characteristics (e.g., line terminator). These concerns will be delayed yet another level.

The additional data structure declarations that are needed for this refinement are:

```
const
    linelimit      = 80; { maximum length of an input line }
    namelimit      = 10; { maximum length of a command }

type
    nametype       = packed array [1 . . namelimit] of char;

var
    inputline      : packed array [1 . . linelimit] of char;
    linesize       : 0 . . linelimit; { actual size of the input line in characters }
```

Getinput will now appear as in Figure 5.3*d*.

We have again used a lower-level procedure, called readstring, to define the procedure getinput. Readstring allows us to hide, for now, the messy low-level details of reading in characters from an input device, worrying about code sets, fiddling with end-of-line characters, and checking for end-of-file conditions. These details can be pushed down one level to the coding of readstring. All that we need to know is that readstring will input a character string from some input device and return the string and its length.

Development of the text editor will continue downward through a tree of tasks from the most general to the most detailed. Figure 5.4 shows the task tree at this point. Likewise, our data structures are evolving and growing as critical decisions are made and refinements implemented.

In reality, a program as complex as a text editor may contain as many as 50 to 100 subtasks, each with up to 100 lines of code. However, the approach to the problem and our management of its implementation would be the same, regardless of length.

```
procedure getinput(var command : nametype; var length : integer);

{ procedure to get a line typed by the user, and find and isolate the command
  name which should be the first item on the line. The procedure will return the
  'command' and its length in characters. It will return length=0 if no command
  was found. }

const
    blank    = ' ';
var
    i          : integer; { loop index }

begin
    writeln('please enter next text editing command');
    readstring(inputline,linesize);
    if linesize = 0 then { user typed an empty line }
        length := 0 { couldn't locate any command }
    else
    begin
        i := 1;
        length := 0;
        { first let's discard leading blanks on the line }
        while (inputline[i] = blank) and (i < linesize) do
            i := i + 1;
        { now put the command name together }
        while (inputline[i] in ['a' .. 'z']) and
                (length < namelimit) and
                (i < linesize) do
        begin
            length := length + 1;
            command[length] := line[i];
            i := i + 1
        end { of while clause on putting command together }
    end { of else clause for non-empty line }
end; { of procedure getinput }
```

*Figure 5.3d.* Refinement of procedure getinput.

To prove this point, in Chapter 9 we will develop and present a complete text editor that finishes the work we have begun here. This editor contains over 1200 lines of code and 43 subprograms. Without the implementation scheme we are discussing (or something quite similar), this task would be totally unmanageable.

## 5.2.2   Advantages of the Top-Down Design Method

Probably the single most important advantage of using top-down design is the contribution it makes to the *intellectual manageability* of a program. By developing

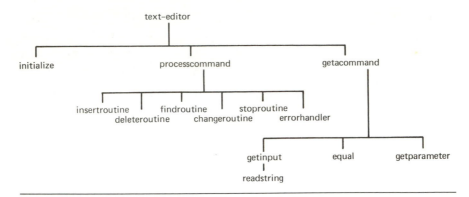

*Figure 5.4.* Development tree for the text editor.

a problem from the most generalized goals to the specific details, we can maintain intellectual control. We know where we are and where we are going. By considering the program in its entirety, we are not mired in detail in the early stages of development.

The popular alternative (called *bottom-up design*) involves going immediately into the coding of some of the elementary modules, such as initialize or readstring in the previous example. With this approach we would have to make early decisions on low-level details such as input formats or table entries well before we realized the implications of such decisions. We would spend more time worrying about these low-level details than about what is happening and where we are going.

We have achieved this intellectual manageability through the process called *abstraction*. That is, we initially deal with an operation only from a general viewpoint, totally disregarding its detailed substructure. It is abstraction that allows us, for example, to understand a large, complex map of the world. Initially, we ignore the many low-level details and concentrate on higher-level structures such as continents and countries. Only later, when we have understood these higher-level concepts, do we narrow our focus to the details of any specific country. In our development of computer programs, our approach is the same. We can manage the development of large, complex, multimodule programs by disregarding at first all insignificant lower-level details and concentrating on a few higher-level constructs. Only when we feel that we have a thorough understanding of these points do we begin to investigate and develop the lower-level details that were originally ''abstracted away.''

Another important point is that it is the *number* of tasks that we must keep track of, not the sophistication of each one, that is most critical to the proper intellectual management of large programs. That is why it is easy to understand the refinements of the text-editing problem in Section 5.2.1. These refinements are expressed in terms of a relatively small number of new tasks. And, even though the tasks may themselves become quite complex when expanded (e.g., processcommand or insertroutine), the *number* of things we must comprehend at once will always be manageable. You should always limit the length of a procedure and the number of new subtasks defined during refinement of a higher-level task. If either parameter (module length and new procedures) becomes excessive, you may have inadvertently skipped a level and defined a much lower-level task than intended. Referring to Figure 5.2, if the first refinement of the original program had had many new subtasks, $t_{11}$, $t_{12}$, $t_{21}$, $t_{22}$, $t_{23}$, $t_{24}$, $t_{31}$, . . . , we might have considered adding an intermediate level of abstraction: tasks $t_1$, $t_2$, and $t_3$. We will say more about the length of subtasks in the next chapter.

The text-editing example in Section 5.2.1 clearly illustrates another important advantage of top-down design: *delayed decisions*. Decisions do not have to be made until we code the modules they affect. For lower-level decisions, this will occur later in the project development, when we have a better idea of what to do.

As an example of a delayed data structure decision, let us refer to our development of the command table data structure, *comtable*. In the early development stages shown in Figures 5.3a and 5.3b, we had not yet decided to create such a table. With the refinement step depicted in Figure 5.3c, we decided on a 25-element table containing one record per command, with each record containing three distinct fields: (1) name; (2) parameter count; and (3) command type. However, only command type had been totally defined. The ''name'' field existed only as an unspecified data type called nametype (temporarily set to char). The ''parameter count'' field, while known to be integer, had all its entries arbitrarily set to 0. By the time we reached the refinement of Figure 5.3d, we had declared ''nametype'' to be a packed array of 1 to 10 characters but had not yet decided on the number of parameters required by each command of our text editor.

The data structures (like the algorithm) are developing in a top-down fashion. Initial decisions are directed at the *need* for such data structures and the generalized and high-level *organization* of the data structures. Lower-level decisions on the exact format of individual fields and the initial values within the fields can be postponed for later refinements. This is a crucial point: top-down design, abstraction, and delayed decisions apply to *both* the algorithms and the data structures of the developing solution.

We can also see in the text-editing example a progression of decisions about editing commands. The highest-level decision concerned which commands to support. The next decisions were directed at how these commands would be entered into the system. Subsequent lower-level decisions (not done in that example) would address points such as the parameters required by each command, the options supported by each command, and the error condition that can occur during the processing of each command. Again, the key point is that the decisions were not all made at once but in a distinct order—from the most general to the most detailed. Initial decisions will be directed at the fundamental structure of the developing program. Our decisions become more specialized, more direct, and more detailed only when we are in a better position to make the correct decision.

Another advantage of the top-down design method is that we *validate* each program unit for correctness as it is developed, not after the entire program is completely coded. The number of individual statements is not the only factor that affects debugging time. The effects of *interactions* between statements also contribute to the problem. This means that the time, t, needed to debug a program consisting of n lines will not increase at the same rate as program length but will do so *faster*. The relationship between t and n is:

$$t \approx n^k \qquad k > 1$$

As programs get longer, debugging time increases dramatically, eventually becoming the dominant step in the entire programming project. Reports show that on large projects with inadequate programming management and organization, debugging consumed about 50 to 70% of the overall project hours and, even then, the programs were not completely free of bugs.

It is definitely to our advantage to debug a program as a series of smaller units rather than as one large "lump." The top-down design method, which develops a program as a hierarchical set of tasks, defines a natural set of small subunits that can be individually tested, verified, and integrated into the overall solution, as in our previous example. We will discuss debugging, testing, and formal verification further in Chapter 10.

*Style Review 5.2* _____

### Resist That Urge to Start Coding!

Henry Ledgard, in *Programming Proverbs* (Hayden Book Co., 1975), introduced a corollary to Murphy's law. "Murphy's Law of Programming" states that "The sooner you start coding your program, the longer it is going to take." Trying to write anything but the simplest program without a well-organized outline and an overall design plan is like building a house without a blueprint.

Students are used to solving problems by coding them directly from the specification document. However, real problems cannot be handled this way. The first part of any programming project must include adequate high-level planning, designing, organizing, and goal setting.

Do not worry about the amount of debugged code during the early stages of the project. Resist that inner urge to pick up a coding form immediately.

### 5.2.3  Programming Teams

Another important advantage of the top-down design method is our increased ability to *delegate* and *manage* the overall work load. The development sequence of program modules in a programming project typically looks like this.

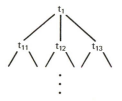

This is identical to the employer-employee relationship in any business organization chart. So it is quite natural for one individual (e.g., the project supervisor) to have overall responsibility for the description of high-level task $t_1$ and its three subtasks $t_{11}$, $t_{12}$, and $t_{13}$. That person may then choose to delegate the actual implementation of some or all of these subtasks to programmers under his or her supervision. The project supervisor's main responsibilities are to approve subsequent design decisions made by staff, coordinate the efforts of the separate programmers, and guarantee that the individually developed pieces fit together into a cohesive unit. This last point is called the *interface responsibility*. Without it, you may face the same problem encountered by two railroad teams laying track between points A and B.

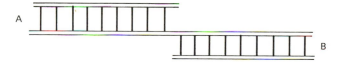

Each team did its job properly, but the finished product is useless.

Recently there has been a great deal of academic and industrial concern about the optimal organization and management of *programming teams*. As mentioned in Table 3.8, large production programs are rarely the work of a single individual. Gone are the days of the "lone wolf" programmer who disappears for 6 months and emerges with a working program. Now programs are the result of a coordinated team effort involving customers, managers, programmers, coders, and clerical staff. The reasons for this are:

1.  Most current software projects are too complex to be handled by a single programmer.
2.  More than one person will have technical knowledge of a piece of software.

Users will not be at the mercy of the employment whims of a single individual.

3.   Responsibilities can be separated and specialized. One person would no longer have to handle diverse tasks such as program design, coding, data entry, writing, and typing.

In the early days of software development, the basic approach to programming teams was the "army of ants" philosophy; keep throwing bodies at a problem and you will eventually solve it. Teams might include 50, 100, or even 200 programmers. At its peak, in 1965, the programming team involved with designing IBM's 360 Operating System numbered about 1000. However, more recent research into team structure and programmer productivity has shown that more is not always better. At some point, the time and effort spent attempting to coordinate and manage such a large group can cause an actual *decrease* in overall group productivity.

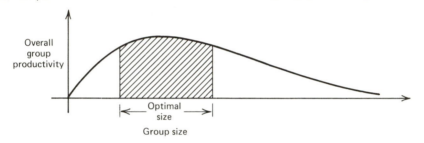

During the early 1970s, case studies in team organization began to develop models, or *paradigms*, for the structure of a programming team. One of the most famous of these models is called the *Chief Programmer Team* (CPT).[1] The CPT philosophy is based on the program design concepts discussed in this book and includes:

1.   Top-down design.
2.   Adherence to the principles of structured coding.
3.   Validation of individual modules through stubs and structured walkthroughs.

However, the CPT also has a very specific *functional organization* based on a rigidly assigned set of responsibilities. Like a crack surgical team (which was its model), all members of the CPT must know their exact responsibility and carry it out flawlessly. The CPT is made up of 3 to 10 members who have the following titles and responsibilities.

1.   *Chief Programmer*. The technical manager of both the project and the team who must have extensive programming experience. He or she has responsibility for all high-level design decisions and for coding and testing critical

[1]F. T. Baker, "Chief Programmer Team Management of Production Programs," *IBM System Journal*, Vol. 11, No. 1, 1972.

program units. The chief programmer assigns the work to other members of the team and coordinates and integrates their work into the overall solution.

2. *Assistant Chief Programmer*. A senior-level programmer who is totally familiar with all phases of the developing project. He or she will be able to move into the role of chief programmer if it becomes necessary.

3. *Programmers*. Depending on the complexity of the project, there may be none to five additional junior- or senior-level programmers. They implement and code the algorithms and modules as designed by both the chief and assistant chief programmers.

4. *Librarian*. A nontechnical clerical person responsible for maintaining the *Program Production Library* (PPL). The PPL consists of both *on-line*, machine-readable information (source code, test data files, object code) and *external information* (binders containing up-to-date program listings and output). The librarian allows the team to clearly separate the *technical* and *clerical* duties of program development. In addition, there may be none to two lower-level clerical assistants who report to the librarian and handle other nontechnical details such as:
   (a) Entering and editing data.
   (b) Filing results in chronological order and keeping all external documents current.
   (c) Preparing progress reports and memoranda.
   (d) Delivering and picking up output and tapes.

The organizational structure of a typical CPT is summarized in Figure 5.5 on the next page.

Case studies of software development using the CPT organization are quite impressive. The New York Times Bibliographic Project[2] involved the development of a data base system with 83,000 lines of code. It was completed in 11 man-years. More recently, the mission simulation system for the Skylab project was completed by a CPT in 2 years, even though it included about 400,000 lines of source code![3]

There are a number of other organizational models similar to the CPT model shown in Figure 5.5. They all have the same objective—the effective management of human resources to solve a programming project. Students should realize that real-world programming is not an individual effort, but a cooperative effort usually involving 3 to 10 specialists.

At the end of this chapter we have provided a number of team programming assignments. We hope that every student will have the experience of working with such a team at least once. However, students must be warned in advance of the pitfalls that inevitably plague initial team efforts. Multistudent projects often fail miserably, not because of any lack of enthusiasm or technical expertise, but because of *group*

[2]F. T. Baker, op. cit.

[3]C. McGowan, and J. Kelly, "Top Down Structured Programming Techniques," Petrocelli-Charter, New York, 1975, p. 147.

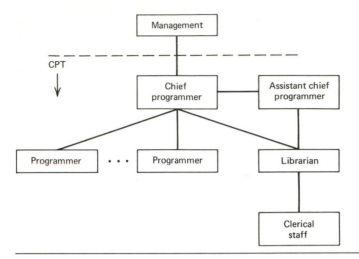

***Figure 5.5.*** Organization of a typical chief programmer team.

*dynamics* — the ups and downs of the personal, intellectual, and emotional relation-ships between group members — and the lack of a specific management structure (e.g., Figure 5.5).

The following complaints are common.

Mr. V is never available for our meetings.

Ms. W is not carrying her share of the workload.

Mr. X did not finish his module as promised.

Ms. Y wants to do it all herself and doesn't let us contribute.

Mr. Z's work is poor and below the level of the other team members.

These problems, which are difficult for other student team members to resolve, de-crease team productivity and can jeopardize the success of the project. This point illustrates the critical importance of organizational structure in creating an effective programming team. Without such an organization, chaos reigns. When attempting a team project, even within a classroom environment, you must impose some minimal management structure. Someone must be responsible for monitoring individual student progress and insuring success. Otherwise the "team" is nothing more than a collection of individuals.

Despite the inevitable logistic and grading problems, we still feel that team as-signments are essential and well worth any trouble they may cause. There is no better way to introduce students to the *human* aspects of programming problems.

1.  The need for unambiguous verbal and written communication.
2.  The ability to compromise and cooperate with other technical specialists.

3. Handling the differences in the ability and temperament of different team members.

4. Being able to sublimate one's own ego for overall team rewards and success.

There are no easy solutions to these problems. What is important is the exposure to working *with* and *for* people and to the human and technical aspects of program development.

## 5.3  SUMMARY

In this chapter we have addressed top-down design, the first of the three aspects of structured programming listed in Section 5.2. Top-down program design is a way to organize and manage mentally the development of large, complex programming systems, thus increasing the likelihood of developing correct, reliable software. In addition, the top-down programming method is a natural framework for managing the *personnel* involved in developing software. The hierarchical structure of the developing solution lends itself well to the allocation of portions of the overall programming task to different programmers, with one person managing the work and integrating the separate program units. The CPT example discussed in this chapter relies heavily on a top-down design approach.

In Chapter 6 we will discuss the second aspect of structured programming: the characteristics of program modules.

*Style Review 5.3* _____

### *Working with Other People*

Many of the differences between academic and production environments are summarized in Figure 3.8. One of the greatest challenges students face is to work effectively *with* people as well as *for* them. Students are used to accepting direction from a superior, whether it is a manager or a professor, but they are often not used to cooperating with others to complete the work. Students are rarely required to adjust their thinking, sublimate their ego, accept the solution methods of others, or even modify their work habits and schedules to accommodate others. In any group effort this can lead to personnel problems and decreased productivity.

An in-depth discussion of the psychological and emotional aspects of programming can be found in the following sources.

Weinberg, Gerald. *The Psychology of Computer Programming*. New York: Van Nostrand Reinhold, 1971.

Shneiderman, Ben. *Software Psychology*. Englewood Cliffs, N.J.: Winthrop, 1980.

## EXERCISES FOR CHAPTER 5

1. Write the function "equal" as described in Section 5.2.1 and used in Figure 5.3c.

2. Show a stub for the routine "readstring" that could be used to test the procedure getinput in Figure 5.3d. After you have written the stub, show the exact output of the entire text editor as it has developed to this point, including the actual coded modules and all stubs.

3. Write the routine called readstring assuming that the following operations are to be carried out.

   (a) The input line is terminated by either an eoln or by reading 72 characters.
   (b) The lowercase alphabetic characters are to be converted to uppercase alphabetic characters.
   (c) The characters "$\equiv$," "$\neq$," "$\vec{\Gamma}$," "$\vee$," "$\wedge$," "$\downarrow$," and "$\rceil$"are to be converted to blanks.

* * *

The following programming assignments are all intended to be done as *team assignments* in groups of three to four students. After a project has been completed, write up a report that discusses the following points.

I. How much time was spent by the group on each phase of the programming project: design, organization, coding, debugging, testing? How does this division of time on a large programming project compare with the way you utilized your time on the small programs written in introductory classes (50 to 200 lines)?

II. What difficulties were encountered in working on a problem with a group of two to three other people? Discuss the problems (if any) that occurred in organizing and utilizing the available manpower. Was any time lost or wasted in simply trying to coordinate the individual efforts? How might you do things differently in your next group project?

* * *

4. Develop a program to produce a *cross-reference table* of a Pascal program. A cross-reference table is a list of every user-defined identifier (type name, variable, constant, module name, etc.) appearing in the program along with every line on which it occurred. The list should be alphabetized and, if a name appeared more than once on a line (e.g., n:=n+1), there should only be one reference to it.

Finally, the letter "d" (for defined) should be appended to those line numbers in which a name is being assigned a value (e.g., read in, **const**, left side of an assignment operator, **for** loop control variable). Two other rules to follow in producing the cross-reference table are:

(a)  Do not include reserved words such as **begin, end,** and **if**. However, do include standard identifiers.

(b)  Do not include names appearing within comments.

The output of your program should look something like this.

| *Name* | *References* |
|---|---|
| amount | 15, 31, 32, 50d, 160 |
| bill | 22, 23d |
| bundle | 1, 2, 11d, 15d, 308 |
| . | . |
| . | . |
| . | . |

For input data, try running the cross-referencing program on itself!

5.  Develop a system that will allow you to do *plotting* on a line printer. The plotting should be done *vertically* on the page so that the plot can be as long as desired. For example, y = sin(x) might look like this.

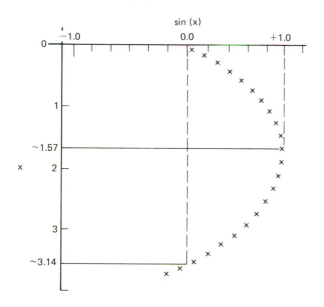

The input to your system will be a function f(x) and a set of limits a, b (b > a). Your system should plot the value of f(x) from a to b. You should scale the horizontal axis so that all function values will properly fit on the page. You should also scale the vertical axis so the overall plot takes no more than 200 lines or about four pages. That is, the step size on x should be no more than (b − a)/200. Other features you may want to include are:

(a)  The ability to plot more than one curve on a single graph.
(b)  The ability to let users set certain system values, such as the plotting character ("*" in the preceding example) and the width of the output device in columns.

Be sure to handle the special case of a function that becomes *undefined* at a point x, a <= x <= b. The plot should be able to be continued past the point x.

6.   Write a program that simulates the behavior of a *hand-held calculator*. The program will interactively accept commands in two forms.

expression
variable ← expression

In the first case the program will simply print out the value of the expression. In the second case it will assign the variable the value of the expression. The syntax of the expressions will follow the same rules as in Pascal, although you may change the precedence rules for determining the order of evaluation of operators if desired. For example, the following might be a typical dialogue with such a program. (The output from the program is underlined.)

    3 + 3
6
    a ← 1
    a + 3
4
    b ← 2
    a + (b * 2)
5
    50 div (a + 14)
3
    1 div 0
error
    1 + (2 + (3 + (4 + (5 + (6)))))
21
    b ← 3
    b
3

Initially begin with just the integer data type and the operators +, −, *, **div**, **mod**, and ( ). Then, if time permits, allow some of the following to be included in your expressions.

(a) Standard functions.

(b) Real data type and the / operator.

(c) The boolean data type and the **and, or,** and **not** operators.

(d) The character data type.

(e) Some operators that do not exist in Pascal, such as ↑ (exponentiation) and **xor** (exclusive-**or**).

7. Write a program to simulate the playing of the game called *Blackjack*, also called *Twenty-One*. (If you do not know the rules, look in any good book of games.) The game should be played between a dealer and from 1 to n players. Each player should be able to bet from 1 to m chips against the dealer on each hand. The game should be played using a single deck of cards.

```
type
    cards =    record
                  rank : 1 . . 13;
                  suit : (spades, hearts, diamonds, clubs)
               end;
    deck =     array [1 . . 52] of cards;
```

You should shuffle the deck, place your bets, and deal out the cards to the dealer and all players according to the rules. Continue to play until you determine whether each player has won or lost. This constitutes a round. Have your program play some number k of rounds and then print out how much each player has won or lost and how much the house (the dealer) has won or lost. Determine what the advantage to the house is in the game of Blackjack.

Also try different types of betting and playing strategy to see if you can reduce the house advantage or even turn it to an advantage for the players against the house.

8. Write a set of procedures to do "infinite precision arithmetic." In infinite precision computations each digit or character is individually stored in a separate position of an array. Thus the quantities 53 and 12.56 would be stored as:

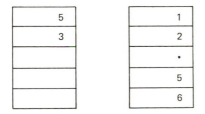

and the result of adding them together would be:

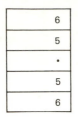

Therefore the only limit to the size of our numbers is the size of our arrays. This will usually allow numbers with thousands of digits and virtually "infinite" accuracy.

Write a set of procedures to do the following operations.

(a) Read a real or integer number of arbitrary length into the preceding array representation. The value may occupy one or more lines of output and is terminated by the first blank.

(b) Validate that the number just read in is syntactically valid in that it contains only the characters $+$, $-$, $.$, and $0 \ldots 9$ and follows the syntax rules for integers and reals presented in Chapter 1.

(c) Write procedures to add and subtract two infinite precision values.

(d) Output a value in infinite precision representation. The value may occupy one or more lines in the output file.

Try out your procedures with the following values:

$\pi = 3.14159272834652184709118435726559 11003842749$
$e = 2.71828273984006158$

and print out the values of $2\pi$, $\pi + e$, $\pi - e$, and $e - \pi$.

# Chapter 6

# PROGRAM MODULARITY AND CASE STUDY

## 6.1 INTRODUCTION

In Chapter 5 we introduced top-down programming. At each step in the design process, we add to the details of one program module and may, in the process, define the specifications for a number of new lower-level modules. The program develops in the classic downward-growing tree pattern shown in Figures 5.2 and 5.4.

However, Chapter 5 did not discuss the *characteristics* of those modules. For example, in terms of a top-down programming approach, it is immaterial whether we expand a module by adding to an existing program (as we did in going from Figure 5.3*a* to Figure 5.3*b,* or by developing a totally separate procedure (as in Figures 5.3*c* and 5.3*d*). However, these two approaches could have very different effects on the resulting program.

In this section we will discuss the programming characteristics of the individual modules that comprise the overall solution. (Note that we will use *module* to denote either a procedure or a function. Usually which of the two program units you choose is only a question of style.)

## 6.2 MODULE CHARACTERISTICS

### 6.2.1 Logical Coherence

Probably the most important requirement for any program unit is that it address only a single, *logically coherent* task. It should do one thing and do it well. We must resist the temptation to incorporate more than one task into a module because the result is usually an overly large, complex, and confusing program unit.

When designing logically coherent program modules, we must be careful to avoid including other tasks, either from the *same level* or from *different levels*. For example, see Figure 6.1. When coding procedure $t_{11}$, we could err in two ways—by including aspects of tasks $t_{12}$ or $t_{13}$, or by including lower-level details from task $t_{111}$ or $t_{112}$.

It is an easy mistake to include operations from related but distinct tasks in a single module. For example, in Figure 6.1, tasks $t_{11}$ and $t_{12}$ are both required to perform task $t_1$; therefore they will probably be closely related. However, it is incorrect to conclude that two related tasks are actually one operation that can be handled by a single program module.

The following pairs of tasks are obviously related to one another.

1.  Read in the input.
    Validate the input for legality.
2.  Read in a string.
    Search a string for a specific pattern.
3.  Compute the mean.
    Compute the coefficient of variation.
4.  Sort a list.
    Merge two sorted lists.

However, in almost all situations, these pairs of tasks would represent two distinct operations that should be implemented as separate procedures.

There are a number of important reasons for keeping modules logically coherent; the most important has to do with *modification*. Problem specifications and users' needs change frequently. Supporting these changes usually requires program modification. To minimize errors during the modification process, we should change as little of the code as possible. The sections of the code that are unaffected by the new specification should not be affected by the modification process. One way to insure this is to isolate each indivisible operation.

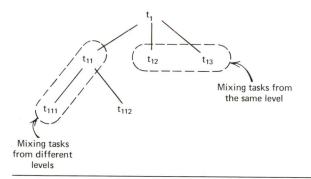

**Figure 6.1.** Overlapping of task responsibility.

For example, assume that A and B are separate tasks (e.g., sort and merge) but are coded as a single procedure. If we wish to modify task A to develop a more efficient sorting method, there is always the danger of inducing an error in task B, even though it has not changed and should be unaffected by the modification. For example, the reassignment of a value to a local variable i within task A might affect task B if B also uses a variable with the same name. If the tasks had been isolated into distinct program units with separate name spaces, there would be virtually no chance of inducing errors in tasks other than the one we are changing. Isolation of responsibility enhances security.

Aside from the advantages to be gained during modification, we can also benefit significantly during the testing phase. Assume that we have task A with m distinct control paths and task B with n distinct paths. If we write a procedure that incorporates both tasks into a single unit, we may need at least (m*n) unique data sets to test exhaustively all possible paths through this program unit (see Figure 6.2a). But, if we coded it as two separate tasks, it would require only m cases to test A and n cases to test B—a total of (m+n) data sets. For any m,n $>= 2$, $(m+n) < (m*n)$. It is obviously useful to develop and test a program using short, simple procedures.

We have established the importance of maintaining logical coherence between tasks at the same level. Of equal importance is logical coherence between tasks at *different levels* (see Figure 6.1). Including unnecessary and inappropriate lower-level details will obscure the purpose and function of a procedure.

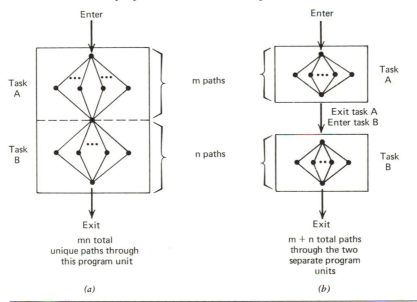

*Figure 6.2.* Testing of program modules. (*a*) mn total unique paths through this program unit. (*b*) m + n total paths through the two separate progrm units.

For example, look at the following program fragment that reads in a string of characters.

```
i := 0;
while (i < arraymax) and (not eoln) do
begin
    i := i + 1;
    read(ch);
    if (ch >= 'a') and (ch <= 'z') then
        ch := chr(ord(ch) − 32);
    if (ch = '[') or (ch = '{') then ch := '(';
    if (ch = ']') or (ch = '}') then ch := ')';
    if ((ord (ch) >= 0) and
        (ord (ch) <= 31)) or
        (ord (ch) > 127) then ch := '?';
    text[i] := ch;
    write(ch)
end { of the while loop }
```

The loop is actually quite simple. It reads characters coded in the ASCII code set (see Appendix B), stores them in an array called text, and then echo prints them. What makes this simple loop so difficult to interpret is the presence of several "messy" lower-level details about how to handle certain special characters. The first **if** statement converts all lowercase characters to uppercase characters, the next two **if** statements map the braces and brackets into parentheses, and the last **if** statement takes all nonprinting characters (which in ASCII are 0-31 and 128-255) and sets them to the character "?". However, these details are certainly *not* essential to an understanding of the overall input operation, and the entire fragment would be much more lucid if these items were relegated to a logically separate, lower-level procedure, as in the following.

```
i := 0;
while (i < arraymax) and (not eoln) do
begin
    i := i + 1;
    read(ch);
    fixupchar(ch); { correct certain special characters }
    text[i] := ch;
    write(ch)
end { of the while loop}
```

Naturally, deciding which operations are "logically coherent" and therefore deserving of a distinct module will depend on the specific problem we are trying to solve. However, the general rule still applies; each procedure should do *one* thing and do it well. If you are unsure about how to decompose a task, err on the side of defining a greater number of shorter procedures. It is much better stylistically to have many short procedures than to have a few large, complex, highly interrelated units. The former will usually be easier to work with, understand, test, and modify.

### 6.2.2 Independence

A program module (procedure or function) should be self-contained and independent of other modules in the system. Specifically, it should be totally independent of:

1.  The source of the input.
2.  The destination of the output.
3.  The past history of activation of this module.

If we design program units to be independent, we can treat the program unit much like a circuit board. We can remove one procedure and ''plug in'' a different one without affecting the remainder of the program, as long as the input/output specifications for the two procedures are identical. We use this approach with Pascal's library routines when we code the following assignment.

  $y := sin(theta) - 1.0$

We do not care at all what algorithm is used to evaluate the sine function. Whether it uses a Taylor Series Expansion or the Modified Gliebowitz Technique is immaterial (and it is even immaterial to us if the method *changes* during the life of our program) as long as the function being used has these characteristics.

1.  It accepts one real parameter, x, whose units are radians.
2.  It returns a real value, v, such that v = sin(x) to a sufficient level of accuracy.
3.  It does it in a ''reasonable'' amount of time.

Failure to develop independent modules results in what is called the *ripple effect*. This occurs when a change to one program unit causes unexpected changes to a number of other units throughout the program. This severely complicates the program modification process, turning minor changes into major undertakings.

The best way to achieve module independence is by:

1.  Avoiding the unnecessary modification of global variables, thus making the procedure free of undesirable *side effects*.
2.  Declaring all temporary variables local to the procedure in which they occur.
3.  Avoiding changes to input parameters that were passed by reference because of memory space limitations. Whenever possible, input parameters should be passed by value so they cannot be modified.
4.  Not doing anything other than what you were supposed to do as defined by the specifications of the problem.

These rules clearly indicate that, as viewed by the outside world, a procedure will affect only its call-by-reference parameters. All other values outside the scope of the procedure should be unaffected. (This is not to imply that one must never modify a global variable. There are times when this may be necessary, but it should always be done *cautiously*, and the modification should be a documented and understood effect of the procedure invocation.)

For example, the following sort procedure is given a list with n numbers in the range 0 to 10. Some entries in list have the value $-1$ which represents missing data. The procedure sort should sort the values in list into ascending order, with the $-1$'s at the end of the list, as in the following.

| | |
|---|---|
| 10 | 3 |
| 3 | 5 |
| 9 | 8 |
| $-1$ | 9 |
| 5 | 10 |
| $-1$ | $-1$ |
| 8 | $-1$ |
| *Input* | *Output* |

A "too-clever" programmer might reason that if the $-1$'s were instead set to $+11$'s, the ascending sort would naturally place them at the end of the list. The result might be the sort procedure shown in Figure 6.3.

This procedure does sort a list into ascending order. However, the procedure has a disastrous side effect. Any other procedure using the output of procedure sort will work improperly if it still assumes that the missing data symbol is a $-1$. Now we must either change this procedure or search through every module activated subsequent to sort to see if there are any occurrences of the missing data constant $-1$. Even worse, if at some time in the future we change procedure sort to utilize a faster algorithm, we must insure that the new procedure changes the missing data indicator in precisely the same way as before. Otherwise, we will again be forced to go through every module, making changes to procedures that should be totally unaffected by a change in the sorting method. This is a classic example of the ripple effect in action.

Related to this is the process of *localizing a data structure*. A data object (integer, array, pointer, etc.) should be declared in the innermost (i.e., most local) module possible. Assume that we have the following program structure.

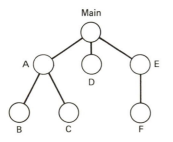

```
procedure sort(var list : arraytype; n : integer;
        var m : integer);

{ procedure to sort a "list" of n values in the range 1-10 into ascending order
    using a bubble sort. The procedure returns the sorted list and "m," the number
    of missing data items. }

var
    i                : integer;               { for loop index }
    sorted           : boolean;               { flag to test if list is sorted }
    temp             : integer;               { temporary used for interchanging
                                                items }
begin
    m := 0;
    for i := 1 to n do
        if list[i] = -1 then
        begin
            m := m + 1;
            list[i] := +11 { bad, bad, bad! }
        end; { of if }

    { Now do the bubble sort. The 11's will naturally percolate to the end }
    repeat
        sorted := true;
        for i := 1 to n - 1 do
            if list[i] > list[i + 1] then
            begin
                sorted := false;
                { exchange the items that are out of place }
                temp := list[i];
                list[i] := list[i + 1];
                list[i + 1] := temp
            end { of if statement }
    until sorted
end; { of procedure sort }
```

**Figure 6.3.** Bubble sort procedure with improper side effect.

Furthermore, assume that procedure A creates and manipulates a data structure called DS. B builds DS; C searches DS. Even though no other module in the program references DS, we could (as discussed in Chapter 1) legally declare DS in either main or procedure A. However, declaring DS in main also makes it available (through the scope rules of Pascal) to modules D, E, and F (unless they declare a structure of the same name), which could accidentally access and/or modify DS. Placing the declaration in module A makes this impossible, since modules D, E, and F are outside the scope of the declaration of DS. In addition, if we should decide to modify DS, we can be certain that only modules A, B, and C will be affected. The data structure modification cannot possibly affect either "main" D, E, or F, since they are independent of DS.

In summary, *independence* requires a set of modules constructed in such a way that a change to any one module does not cause an unexpected change to any other module in the program.

### 6.2.3  Module Size

The "proper size" for a module is not an isolated characteristic; it is the result of adhering to the guidelines already discussed. If we design each module to address only one task, the modules will naturally be relatively small.

While there is no standard maximum, one rule is that a module should never exceed one page of code—about 50 to 60 lines. This way we never have to flip pages back and forth to look at and understand an individual module (Figure 6.4*a*). Other programmers set a limit of two pages—about 100 to 120 lines—which allows them to keep the listing in a book binder and still see the entire unit (Figure 6.4*b*).

These limits define a valid and workable range for maximum module size that we will adhere to throughout the text. Remember, however, that the critical characteristic we are striving for is not proper size *per se* but *logical coherence*. If we find ourselves writing modules that are 200 to 300 or more lines long, it is very likely that we are doing too much within a single unit or including too many extraneous and lower-level details.

### 6.2.4  Structured Coding

The third and final principle of structured programming is *structured coding*—the implementation of each individual module using only the three basic coding forms of sequence, iteration, and conditional.

Structured coding was discussed in detail in Chapter 4; it should now be reviewed in light of the discussion on programming methodology in Chapters 5 and 6.

The three phases of structured programming that we have now discussed cover all aspects of the programming process. *Top-down design* addresses the problem of how to approach a large, complex task by developing a hierarchy of smaller subtasks.

about 50 lines                    about 100 lines

*(a)*                                    *(b)*

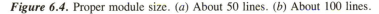

**Figure 6.4.** Proper module size. (*a*) About 50 lines. (*b*) About 100 lines.

*Program modularity* involves the implementation of these subtasks as distinct, logically coherent, independent modules with certain desirable programming characteristics. Finally, the techniques of *structured coding* insure proper coding of the modules we have defined. These three principles define a programming management technique for the efficient implementation of correct, reliable programs.

## 6.3  CASE STUDY

In this section we will follow the development of a large, complex computer program. It is not as large as the real-world program sizes quoted in Figure 3.8, but it will certainly be large enough to illustrate the program design concepts presented in Chapters 4 to 6.

This case study will involve the construction of a *simulation model*—specifically, a terminal room for computer science students. Computer simulation is a well-known and popular technique for studying the behavior of complex systems. There are other ways to study system behavior, but they are often impractical or impossible.

1.  *Study the Actual System.* With existing systems, this approach is often either unsafe or prohibitively expensive. It obviously does not apply in cases where we want to determine if the system is unstable *before* we build it.

2.  *Study a Physical Model of the System.* This is done with wind tunnel tests of airplane and rocket models. But how does one build a miniature student?

3.  *Study a Mathematical Model of the System.* If the system can be described in terms of explicit mathematical formulas, solving the formulas will explain the behavior of the system. For large, complex systems this is often difficult or impossible.

Simulation is a useful tool that frequently provides good results when the preceding techniques cannot.

The system that we will be simulating is shown in Figure 6.5. Students enter the terminal room; they either go directly to one of the k identical terminals, if one is available, or they enter a waiting line (called a *queue*) to wait for an available terminal. There are no special priorities in the queue; students are given a terminal according to when they arrived. (This is called a *first-come-first-served* (FCFS) queueing discipline.)

After working on the terminal, the students will either finish their work and leave the room or they will encounter some difficulty. In the latter case, they stand in another FCFS queue waiting for assistance from a programming consultant. After getting the help they need, they get back in line and wait for another terminal.

The technique we will use to simulate this system is called *discrete event simulation*. With this method we do not simulate the continuous behavior of a system, only a very specific class of *events* in the life of the system. We will observe only the

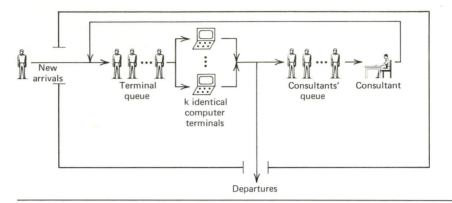

Departures

*Figure 6.5.* The terminal room system to be simulated.

events that *change* the state of the system. A special data structure, called a *calendar*, contains a list of all the events to be processed and keeps them sorted by the time they occur. For example:

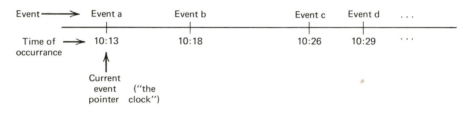

The highest level of our simulation program can now be explained in terms of a calendar and its events. We move the current event pointer (sometimes called the simulation "clock") to the next event on the calendar and process that event by changing the system according to the specifications of that event. We continue this until either there are no more events on the calendar or the time on the clock exceeds some preset upper bound. For example, in the preceding diagram, after completing event a we would move the current event pointer to 10:18 and begin processing event b.

The three events that will change the state of our terminal room system are:

1. *Arrival Event.* A new student walks in and either goes to a terminal or is placed in a waiting line for one.

2. *Terminal Completion Event.* A student finishes using a terminal and either leaves the room or goes to the consultant for assistance.

3. *Consultant Completion Event.* A student finishes talking to the consultant and reenters the terminal queue; the consultant becomes free to help the next person in line.

There is another detail about simulation that will affect the top level of our program. Simulation is a design tool. This means that we do not know beforehand how to set certain parameters so the system will behave properly. First, we try a set of parameters, note how the system behaves, and adjust the parameters accordingly. This is done repeatedly, always with the goal of improving performance.

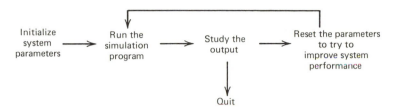

Therefore our simulation program must be able to be run frequently, with a different set of parameters each time. Initially, let us assume that the only design parameters we wish to change are *nterms,* the number of terminals in the room, and *maxtime,* the total time (in minutes) for which we will run the simulation.

In order to avoid having to pass around dozens of parameters from procedure to procedure or, even worse, refer to them globally (and possibly have disastrous side effects throughout our program), we will define two important **record** structures. The **record** variable simstate will include all variables related to the status of the simulation. By simply passing around simstate, we can make the entire status of the simulation available to a procedure. The definition of simstate will, for now, look like this.

```
type
    staterec              =
        record
            calendar      : ?          { not sure yet what it will look
                                         like }
            clock         : real;      { the running time, from 0.0 to
                                         maxtime }
            maxtime       : real;      { total time to run the simulation }
            nterms        : integer    { number of terminals in the
                                         room }
        end; { of record staterec }

var
    simstate              : staterec;  { state of the simulation }
```

The second basic **record** structure will be stats, which will carry around all of the statistics that we are accumulating about the system. Again, by passing only one **record** variable, we can make all the necessary information available to a procedure. However, since we have not decided anything at all about statistics, we can say nothing more about this data structure than simply:

```
type
    statsrec          =
          record
                  { this will be filled in later in the development }
          end; { of record statsrec }

var
    stats                 : statsrec; { the accumulated statistics }
```

Finally, we need a user-defined data type to define the possible types of events that can occur.

```
type
    eventclass          = (arrival, consult, term, error);
var
    eventtype           : eventclass;
```

We are now in a position to code the top level of our simulation program. The result is shown in Figure 6.6 on pages 201-202.

Before proceeding, we should point out a few important characteristics of the program in Figure 6.6.

1.  The program unit defines a total of eight lower-level procedures.
    (a)  Initialize.
    (b)  Getnextevent.
    (c)  Arrivalevent.
    (d)  Consultevent.
    (e)  Errorhandler.
    (f)  Termevent.
    (g)  Everybodyout.
    (h)  Reportwriter.
2.  The program has about 45 lines of code, which is within our guidelines for proper module length.
3.  The program hides most of the details of event processing and the representation for the calendar data structure. The details will come later, when we refine the eight lower-level procedures.
4.  Our program adheres to the stylistic guidelines presented in Chapters 2 to 4 with respect to:
    (a)  Mnemonic names.                (e)  Input validation.
    (b)  Commenting standards.          (f)  Robustness.
    (c)  Indentation.                   (g)  Output presentation.
    (d)  Control structure.

```
begin { program simulator }
    writeln(' welcome to the terminal room simulator ');

    repeat
        repeat
            writeln(' please enter values for nterms -- the number of',
                ' terminals,');
            writeln(' and maxtime -- the total simulation time in minutes');
            readln; { get user response }
            read(simstate.nterms, simstate.maxtime);

            if (simstate.nterms >= 1) and (simstate.maxtime > 0.0) then
            begin
                gooddata := true
            end
            else
            begin
                gooddata := false;
                writeln(' *** you made an error in the input. nterms must be',
                    ' >= 1');
                writeln(' *** and maxtime must be real and non-negative.');
                writeln(' *** please try again')
            end { of else }
        until gooddata;

        { this will put the first event on the calendar and set clock to 0 }
        initialize(simstate);

        { now we run the simulator for as long as the user specified }
        while (simstate.clock <= simstate.maxtime) do
        begin
            { getnextevent will pick off the next event on the calendar, and
              return the type and the time of occurrence }
            getnextevent(simstate.calendar, eventtype, eventtime);
            simstate.clock := eventtime;

            { now simply call the appropriate routine for this event type }
            case eventtype of
                arrival : arrivalevent(simstate, stats); { process arrival event }
                consult: consultevent(simstate, stats); { consultant
                                                   completion event }
                error : errorhandler; { something bad happened }
                term : termevent(simstate, stats) { terminal completion event }
            end { of case }
        end; { of while loop to handle one complete simulation }
```

```
                { we have finished one run. Throw everyone out of the room, produce
                   the results and see if the user wants to try again }
                everybodyout;
                reportwriter(stats);
                writeln(' do you wish to run the simulation again for a different');
                writeln(' number of terminals or time value? y(es) or n(o)');
                readln;
                read(response); { we will read only the first character }
                done := (response = 'n')
        until done;

            { we will go through the above repeat loop as many times as the user
               wants. we will exit here only when he/she is completely finished }
            writeln(' exiting from the simulator program. thank you')
    end. { of program simulation }
```

*Figure 6.6.* Top level of the simulation program.

Before continuing with the development of the program, we should validate what we have written using either structured walkthroughs or execution through stubs, which were discussed earlier. We will omit this set of operations in our case study in order to concentrate on the program development process. However, we will say more about testing and verification methods in Chapter 10.

Now let us look at the three event processing routines—arrivalevent, termevent, and consultevent. Before we can begin to code these routines, we must make an important decision. Exactly what *output results* do we want our simulation model to produce? What *statistics* do we need to understand the behavior of our system? There are many possible values we could produce, but we will limit ourselves to the following five.

1.   Average waiting time in the terminal queue.
2.   Percent utilization of all nterms terminals.
3.   Average waiting time in the consultant queue.
4.   Percent utilization of the one consultant.
5.   Total average time in the room for each student.

These five measures should allow us to characterize the behavior of the terminal room in terms of usage, bottlenecks, and wasted resources. We will incorporate these values (as well as the values needed to determine them) as fields in the record structure stats.

This decision will also allow us to begin designing other necessary data structures. Each student who enters the terminal room will be represented by a single **record** containing the following two pieces of information.

($t_1$) *Time of Arrival*. Time first entered the terminal room.

($t_2$) *Time in Queue*. Time first placed in the current queue (either terminal or consultant).

A *queue* is simply an ordered list of these individual student records. We cannot use an array to implement a queue because we have no idea what the maximum number of people in line will be at any one time. For example, if we declare our array to contain 100 students, what would happen if there were 101 people waiting for a terminal? (A fatal error). A better structure for this situation is a *linked list*. (We will discuss this type of data structure more fully in Chapter 8. You may wish to read Section 8.2 now.) A *queue* can be implemented as a linked list of individual student records.

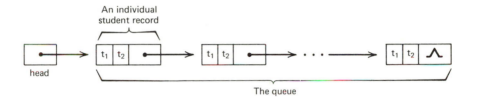

These data structures can be created with the following declarations.

```
type
      studentptr                        = ↑ student;
      student                           =
          record
              timeofarrival             : real;
              timeinqueue               : real;
              next                      : studentptr
          end; { of record student }
var
      termqueue                         : studentptr; { pointer to head of terminal
                                                        queue }
      consultqueue                      : studentptr; { pointer to head of
                                                        consultant queue }
```

We will also need to add the following fields to the simstate record.

ntermsinuse: number of terminals being used.

helperbusy: true if the consultant is currently helping a student and false otherwise.

Finally, we must decide how new events will be placed on the calendar. Notice that so far only a single event has been put on the calendar, corresponding to the very first student who walks into the room. (It will be placed there by the procedure initialize.) How are the other events placed on the calendar? Each event we have defined will spawn one or more additional events so that new events are constantly being created. In fact, one of the major purposes of the event procedures will be to schedule new events and place them on the calendar. Thus the simulation will terminate *not* by running out of events but by running past the preset time limit called "maxtime."

In our system the creation of new events will proceed as follows.

1. Each arrival event will determine the interval until the next person enters the room and then schedule an arrival event for that time.

2. If an arriving student can get on a terminal immediately, we will determine the length of the terminal session and schedule a terminal completion event for that time.

3. The terminal completion event procedure will check if a student is waiting and, if so, give the terminal to that student and schedule another terminal completion event.

4. In addition, if the terminal completion event procedure determines that the current student will go to the consultant and if the consultant is idle, it will determine the length of the consulting session and schedule a consultant completion event.

5. Finally, the consultant completion event will send that student back to a terminal (if one is available) and schedule a terminal completion event. It will also check if anybody is in line and, if so, schedule another consultant completion event.

The summary of the event scheduling process is shown in the following diagram.

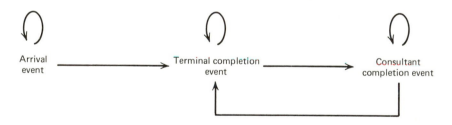

Placing the first arrival event on the calendar is like knocking over the first domino in that it indirectly creates all other events that will be processed by our system.

Now that we have decided on data structures for our student records and queues and have specified the event generation sequence, we are ready to encode the three second-level event processing routines defined earlier. The code for these routines is shown in Figure 6.7 on pages 205-208.

{ *arrivalevent*

*handle a student who just arrived, giving him a terminal if one is available or putting him in line if one is not.*

*entry conditions:*
      *simstate*          *: current state of the simulation*
      *stats*             *: current statistics*

*exit conditions:*
      *simstate*          *: updated appropriately*
      *stats*             *: also updated appropriately*
}

**procedure** *arrivalevent* (**var** *simstate : staterec;* **var** *stats : statsrec);*

**var**
    *newp*                *: studentptr;*    { *temporary student pointer* }
    *t*                  *: real;*        { *temporary variable used for scheduling* }

**begin** { *procedure arrivalevent* }
    { *schedule the next arrival event* }
    *nextarrival(t);* { *procedure which determines interarrival time* }

    { *schedule next arrival in t minutes* }
    *schedule(simstate.calendar,arrival,***nil**,*simstate.clock + t);*

    { *now create that new student and set initial parameters* }
    *new(newp);*
    *newp↑.timeofarrival := simstate.clock;* { *when student first walked in* }

    { *see if there are any terminals free* }
    **if** *simstate.ntermsinuse = simstate.nterms* **then** { *all busy* }
    **begin** { *put student in queue* }
        *putinqueue(newp,simstate.termqueue);*
        *newp↑.timeinqueue := simstate.clock* { *when student started waiting* }

    **end** { *of if busy* }
    **else**
    **begin** { *give student a terminal* }
        { *determine length of terminal session* }
        *uselength(t);*

        { *schedule terminal completion event* }
        *schedule(simstate.calendar,term,newp,simstate.clock + t);*
        *simstate.ntermsinuse := simstate.ntermsinuse + 1;*
        *stats.termcount := stats.termcount + 1;* { *this is the number of terminal sessions* }
        *stats.totaltermtime := stats.totaltermtime + t* { *total time a terminal has been in use* }
    **end**; {*of else clause* }

    *stats.totalstudents := stats.totalstudents + 1*
**end**; { *of procedure arrivalevent* }

```
{ consultevent

    take student p away from the consultant and give him a terminal if one is open,
    putting him in line if there are none. if anyone is waiting, he will be taken to the
    consultant.

    entry conditions:
            p                       : points to student who is done being helped
            simstate                : current state of the simulation
            stats                   : accumulated statistics

    exit conditions:
            simstate                : updated appropriately
            stats                   : updated appropriately
}

procedure consultevent (p : studentptr; var simstate : staterec;
        var stats : statsrec);

var
        newp                    : studentptr;   { temporary student pointer }
        t                       : real;         { temporary used for scheduling }

begin { procedure consultevent }
        { first put this student back in line for a terminal }
        if simstate.ntermsinuse = simstate.nterms then { must wait }
        begin
                putinqueue(p, simstate.termqueue);
                p↑.timeinqueue := simstate.clock
        end { of if terminals full }
        else
        begin { give student a terminal }
                uselength(t);
                schedule(simstate.calendar, term, p, simstate.clock + t);
                simstate.ntermsinuse := simstate.ntermsinuse + 1;
                stats.termcount := stats.termcount + 1;
                stats.totaltermtime := stats.totaltermtime + t
        end; { of else get an open terminal }

        { now see if anybody else wants to talk to the consultant }
        if queueempty(simstate.consultqueue) then { nobody is waiting }
                simstate.helperbusy := false
        else
        begin { somebody wants to see the consultant }
                takeoffqueue(newp, simstate.consultqueue);
                { collect statistics about this person's wait in line }
                stats.consultwait := stats.consultwait
                        + (simstate.clock − newp↑.timeinqueue);
                stats.consultcount := stats.consultcount + 1;

                { schedule completion time of consultant }
                helplength(t);
                schedule(simstate.calendar, consult, newp, simstate.clock + t);
                stats.totalconsulttime := stats.totalconsulttime + t
        end
end; { of procedure consultevent }
```

{ termevent

remove student p from his terminal. if he needs help, send him to the consultant. if anyone is waiting for a terminal, it will be given to the first person in line.

entry conditions:
        p                           : points to the student who is done
        simstate                    : current state of the simulation
        stats                       : accumulated statistics

exit conditions:
        p                           : will be shown the door or go to the consultant
        simstate                    : will be updated appropriately
        stats                       : will be updated appropriately
}

**procedure** termevent(p : studentptr; **var** simstate : staterec;
    **var** stats : statsrec);

**var**
    newp                    : studentptr;   { temporary student pointer }
    t                       : real;         { temporary used for scheduling }
    leaving                 : boolean;      { true if a student is going to leave the
                                            room, false if he/she is going to the
                                            consultant }

**begin** { procedure termevent }
    { determine if this student needs help or will leave the room }
    needhelp(leaving);
    **if** leaving **then** { collect statistics about this student }
    **begin**
        stats.finished := stats.finished + 1;
        stats.totaltime := stats.totaltime + (simstate.clock −
            p ↑ .timeofarrival);
        dispose(p) { say goodbye to student }
    **end** { of if leaving }
    **else**
    **begin** { he/she is going to the consultant }
        **if** simstate.helperbusy **then** { wait in line }
        **begin**
            putinqueue(p, simstate.consultqueue);
            p ↑ .timeinqueue := simstate.clock
        **end** { of if helperbusy }
        **else**
        **begin**
            simstate.helperbusy := true;
            helplength(t); { determine length of consultation }
            schedule(simstate.calendar, consult, p, simstate.clock + t);
            stats.consultcount := stats.consultcount + 1;
            stats.totalconsulttime := stats.totalconsulttime + t
        **end** { of else helper not busy }
    **end**; { of else person goes to consultant }

```
        { now see if anybody else wants the terminal just freed up }
        if queueempty(simstate.termqueue) then
                simstate.ntermsinuse := simstate.ntermsinuse − 1
        else
        begin
                takeoffqueue(newp,simstate.termqueue); { get next person from queue }
                { determine how long this person waited, adding it to total }
                stats.termwait := stats.termwait + (simstate.clock −
                        newp↑.timeinqueue);
                stats.termcount := stats.termcount + 1;
                uselength(t); { determine length of terminal session }
                { schedule terminal completion }
                schedule(simstate.calendar,term,newp,simstate.clock + t);
                stats.totaltermtime := stats.totaltermtime + t
        end
end; { of procedure termevent }
```

---

*Figure 6.7.* Second-level event processing procedures.

Again, let us make a few observations about this code.

1.  The three modules are about 32, 36, and 46 lines long, respectively; they are nice and small and readily understood.

2.  These three second-level modules have defined a total of eight new third-level modules. These new modules fall into three classes.

    (a)  *Statistical Procedures that Characterize the System*
        *Nextarrival.* Computes student interarrival times.
        *Uselength.* Computes how long students use the terminal.
        *Needhelp.* Computes what percentage of students use the consultant.
        *Helplength.* Computes how long students talk to the consultant.

    (b)  *Procedures that Manage the Two Waiting Lines*
        *Putinqueue.* Adds a new student to the end of a queue.
        *Takeoffqueue.* Takes the first person from the front of a queue.
        *Queueempty.* Boolean function that is true if a queue is empty and false otherwise.

    (c)  *Procedures that Manage the Calendar*
        *Schedule.* Adds a new event to the calendar at a designated time.

    The overall development tree to this point is summarized in Figure 6.8. The underlined procedures are those that have already been developed.

3.  Even though we have come a long way in the development of our program, there is still a good deal of very low-level detail that is "hidden" in lower-level procedures. For example, we have not even begun to select a data representation for the calendar. That level of detail is currently hidden in the modules called getnextevent and schedule. Also, we have not yet specified

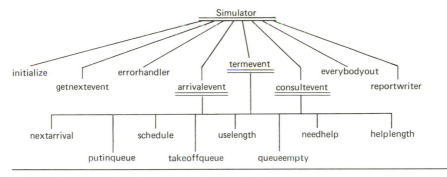

***Figure 6.8.*** Current development tree for the simulator.

the algorithms for searching or sorting the calendar. These and other details have been saved for lower-level routines.

4.  As a final point, notice that some of the *calling sequences* for our event processing routines have changed. In our higher-level module of Figure 6.6 we activated the routines as follows.

    *arrivalevent(simstate,stats);*
    *termevent(simstate,stats);*
    *consultevent(simstate,stats);*

    But after defining some data structures and adding to the specifications of these routines, our declarations in Figure 6.7 now look like this.

    *arrivalevent(simstate,stats);*
    *termevent(p,simstate,stats);*
    *consultevent(p,simstate,stats);*

    This is not uncommon. The development of a lower-level module should not change the *logic* of higher-level routines, but it may change the *interface* between the two (i.e., the calling sequence and the procedure declaration). This difference may be caused by newly defined data objects being passed back to the calling routine, or data that were needed by the lower-level routine but were unanticipated earlier. In our case it was due to the definition of the student record data structure. We will now need to go back to the top-level module of Figure 6.6 and adjust the calling sequences to match our new declaration.

As mentioned earlier, before we develop any of the unwritten modules listed in Figure 6.8, we must again test our three new modules for correctness. We will again omit this step to concentrate on program development.

We will now develop the two calendar handling routines: getnextevent, which takes the next event from the calendar, and schedule, which puts a new event on the calendar in the proper (i.e., time-ordered) sequence. This will require us to select a data structure for the calendar.

An array is probably an inappropriate data structure for a calendar, since we have no prior knowledge of how many events may be on the calendar at any one time. If we declared the calendar array [1. .500], what would we do if there were 501 scheduled events? (A fatal error) Therefore a *linked list structure* is again the best way to represent a calendar. It will consist of a linked list of events sorted by increasing value in the time field.

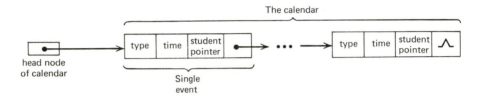

This structure can be created through the following declarations.

```
type
    eventptr                = ↑ event;
    event                   =
        record
            eventtype       : eventclass;     { the type of event }
            eventtime       : real;           { time the event will occur }
            studptr         : studentptr;     { pointer to the student who
                                                caused the event. See p. 203}
            link            : eventptr
        end; { of record event }
var
    calendar                : eventptr;       { head of the calendar }
```

With out data structure decision made, we can better understand the purpose of the two new routines we are about to write.

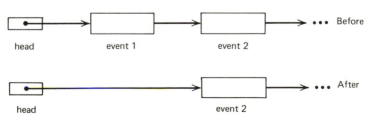

schedule:

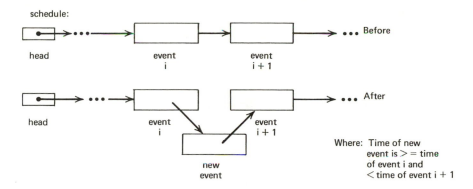

We can now write these routines, which are shown in Figure 6.9 on pages 209-212.

Neither of these routines defines any lower-level routines (both new and dispose are procedures that are part of the proposed standard Pascal language), so we can immediately begin coding some of the other lower-level routines already specified. (See Sections 8.2.1 and 8.2.2 for an explanation of what the new and dispose procedures accomplish.)

The queue handling routines—putinqueue, takeoffqueue, and queueempty—are somewhat similar to the calendar routines just developed. They both manage (add, delete, search) linked list data structures. However, queue management is simple by comparison, since we never insert items in the middle of the data structure. Because we are using a first-come-first-served queueing discipline, we will always add items to the end of the list and take items from the beginning. However, if we look at our original definition of the queue data structure (pp. 202-203), we will see that it is a most inefficient selection based on the types of operations we will be doing. The queue structure we set up looked like this.

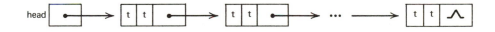

This is fine for taking items from the front of the list, but what about adding items to the end? Given the current data structure definition, we would have to walk through the entire queue, following the links, until we came to the end—a very inefficient procedure.

We need pointers to both the *head* and *tail* of the queue.

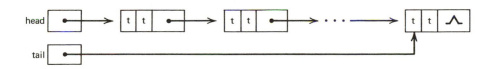

*{ schedule*

*search through the calendar to determine where a new event should be placed. then add this new event to the calendar, keeping the calendar ordered by the time field.*

*entry conditions:*
        *calendar*          *: pointer to a list of 0 or more events*
        *evtype*            *: type of the event to be scheduled*
        *sptr*              *: pointer to the student involved (may be nil)*
        *evtime*            *: time that event is to be scheduled*

 *exit conditions:*
        *the event is inserted into the calendar list, preserving ordering by the time field.*

*}*

**procedure** *schedule (***var** *calendar : eventptr; evtype : eventclass;*
    *sptr : studentptr; evtime : real);*
**var**
    *done*               *: boolean;*     *{ true when searching should stop }*
    *eventp*           *: eventptr;*    *{ temporary for node creation }*
    *oldp*              *: eventptr;*    *{ temporary used for calendar search }*
    *p*                  *: eventptr;*    *{ temporary used for calendar search }*

**begin** *{ procedure schedule }*
    *{ first build a new node with the proper values }*
    *new(eventp);*
    **with** *eventp ↑* **do** *{ the with statement is described in Chapter 7 }*
    **begin**
        *eventtype : = evtype;*
        *eventtime : = evtime;*
        *studptr : = sptr*
    **end***; { of with }*

    *{ now see where this new node should be placed and reset the links }*
    *p : = calendar;*
    **if** *p <>* **nil then**
    **begin**
        **if** *p ↑ .eventtime <= evtime* **then** *{ we must search through the list }*
        **begin**
            **repeat**
                *oldp : = p;*
                *p : = p ↑ .link;*

                *{ see if we are done without doing a nil pointer reference }*
                **if** *p <>* **nil then** *done : = (p ↑ .eventtime > evtime)*
                **else** *done : = true*
            **until** *done;*

```
            { now insert the new item }
            eventp ↑ .link : = p;
            oldp ↑ .link : = eventp
        end { of search and insert new item }
        else
        begin { it goes at the head of the list }
            eventp ↑ .link : = calendar;
            calendar : = eventp
        end { of else }
    end { of if p <> nil }
    else
    begin { calendar is empty }
        eventp ↑ .link : = nil;
        calendar : = eventp
    end { of else empty calendar }
end; { of procedure schedule }
```

{ getnextevent

  get the next item from the calendar, removing it from the head of the list.

  entry conditions:
        calendar           : pointer to the list of 1 or more events

  exit conditions:
        calendar           : has had it's first event removed
        evtype             : returns the type of the event.
                              if the calendar list was empty, error is returned.
        sptr               : returns a pointer to the student involved
        evtime             : returns the time this event is to take place
}

```
procedure getnextevent (var calendar : eventptr; var evtype : eventclass;
    var sptr : studentptr; var evtime : real);
var
    newlink                : eventptr;     { temporary used for updating links }
begin { procedure getnextevent }
    if calendar <> nil then
    begin
        with calendar ↑ do
        begin
            evtype : = eventtype;
            evtime : = eventtime;
            sptr    : = studptr
        end; { of with }
```

```
                    { now adjust the links and discard that node }
                    newlink := calendar↑.link;
                    dispose (calendar); { dispose is a built-in Pascal procedure }
                    calendar := newlink
                end { of if calendar <> nil }
                else evtype := error { calendar empty }
            end; { of procedure getnextevent }
```

***Figure 6.9.*** Calendar handling routines.


We can set up this structure in the following way.


```
type
        queue                              =
            record
                head                       : studentptr;
                tail                       : studentptr
            end; { of record queue }

var
        termqueue                          : queue;
        consultqueue                       : queue;
```


This change in the queue data structure will not change the procedures we have written
so far because all details of queue searching were hidden in the lower-level queueing
procedures we are about to write. This is a perfect example of the advantage of
modularization of code and isolation of responsibility. A major data structure change
has caused us no problems at all.

The responsibilities of the queueing procedures can be represented as follows.

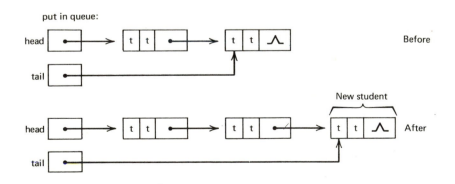

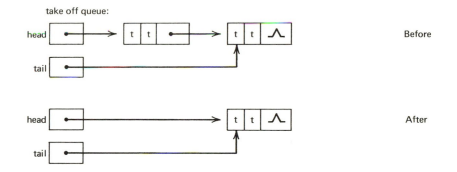

take off queue:

The three queue management routines are coded and shown in Figure 6.10.

```
{ putinqueue

        put student p at the end of queue q.

        entry conditions:
            p                   : points to student being placed in line
            q                   : points to a queue of 0 or more students

        exit conditions:
            p                   : the next field is set to nil
            q                   : has had student p put at the end of the line
}
procedure putinqueue (p : studentptr; var q : queue);
begin { procedure putinqueue }
    if q.tail <> nil then { queue not empty }
    begin
        q.tail ↑ .next := p;
        p ↑ .next := nil;
        q.tail := p
    end { of add to nonempty queue }
    else
    begin { queue is empty }
        q.head := p;
        q.tail := p;
        p.next := nil
    end { of empty queue }
end; { of procedure putinqueue }
```

{ *takeoffqueue*

*take the first student p from queue q. abort if queue is empty.*

*entry conditions:*

        *q*                      *: points to a queue of 1 or more students*

*exit conditions:*

         *p*                   *: points to the student who was first in line*
         *q*                   *: has had its first entry removed*
}
**procedure** *takeoffqueue (***var** *p : studentptr;* **var** *q : queue);*
**begin** { *procedure takeoffqueue* }
    **if** *q.head* <> **nil then** { *there is at least one item in the queue* }
    **begin**
        *p* := *q.head;*
        { *now remove that item and reset pointers* }
        **if** *p↑.next* <> **nil then**
            *q.head* := *p .next* { *remove from list* }
        **else**
        **begin** { *the queue had only one item and is now empty* }
            *q.head* := **nil***;*
            *q.tail* := **nil**
        **end** { *of else* }
    **end** { *of queue was not empty* }
    **else** *halt(' takeoffqueue called with empty queue')*
        { *this will abort if the programmer is careless enough not to check if the queue is empty before calling this routine. This is a questionable stylistic practice and a future modification may wish to change this operation.* }
**end**; { *of procedure takeoffqueue* }

{ *queueempty*

*determine if queue q has any students left in it*

*entry conditions:*

        *q*                     *: points to a queue of 0 or more students*

*exit conditions:*
        *function returns true if the queue is empty, false otherwise.*
}
**function** *queueempty (q : queue) : boolean;*
**begin** { *function queueempty* }
    *queueempty* := *(q.head* = **nil***)*
**end**; { *of function queueempty* }

---

**Figure 6.10.** Queue handling routines.

Again, these routines did not define any new low-level routines to be added to the development tree. Looking back at what we have done, we see that we have coded 9 of the 17 modules so far defined and written about 250 lines of code (about 350 lines, including those with comments). Despite the size of the program we are developing, it is clear and understandable (except possibly for some problems understanding the basics of simulation). We have a good idea of where we are going and what tasks still need to be done. It might not be a bad idea to review what we have developed and imagine trying to write the same program without any planning or organization! (Remember that this program is still very small compared to most real-world applications programs.)

We will now start coding the four statistical routines that characterize system behavior—nextarrival, uselength, needhelp, and helplength. More than any other routine we will write, these four modules will determine the ultimate overall accuracy of our simulator. It is imperative that our statistical assumptions about arrival rates, terminal session lengths, and the like, closely match the actual values in the real system (the terminal room). If they do not, our entire simulator is an exercise in futility, and our results will be meaningless. (Remember: garbage in, garbage out.) If the terminal room actually existed, we might try to sit by the door and time the arrivals or measure terminal sessions with a stopwatch in order to get accurate empirical observations. We would then try to fit our observations to one of a number of well-known statistical distributions (e.g., exponential, normal, and uniform) that could then be coded into our procedures.

However, since this is only a case study and since we cannot assume an understanding of statistics, we will use some very simple and unrealistic numerical values to describe the behavior of our system. These statistics are summarized in Figure 6.11 on the next page.

Using the tables from Figure 6.11, we can code the four statistical routines. We will use a random number generator, which produces uniformly distributed random numbers between 0.0 and 1.0. Then, using the frequencies in the second column of Figure 6.11, we can determine the appropriate category and, with a second call to the random number generator, can find the exact value.

For example, to determine the length of a terminal session (Table B of Figure 6.11), we can generate a random number in the range 0.0 to 1.0. If the value were between 0.000 and 0.250, we would say the session lasted between 0 and 15 minutes. We could then generate a random number between 0.0 and 15.0 to get the exact length in minutes. If the original random number were 0.250 to 0.800, we would say the session lasted 15 to 30 minutes and generate a second random value from 15.00 to 30.00 to determine the length. If the original random number were between 0.800 and 0.980, we would generate a random value between 30.0 and 60.0. Finally, if the original random number exceeded 0.980, we would generate a second random value between 60.0 and some reasonable upper bound.

Notice that the four routines in Figure 6.12 create the need for yet another lower-level routine called random(x,y). This function will generate a random real number,

A.   *Interarrival Distribution*

| Interval (min) | Percent of Students |
|---|---|
| 0-1 | 12 |
| 1-2 | 24 |
| 2-4 | 35 |
| 4-8 | 20 |
| 8-15 | 7 |
| Over 15 | 2 |

B.   *Length of Terminal Session*

| Time (min) | Percent of Sessions |
|---|---|
| 0-15 | 25 |
| 15-30 | 55 |
| 30-60 | 18 |
| Over 60 | 2 |

C.   *Length of Consulting Sessions*

| Time (min) | Percent of Sessions |
|---|---|
| 0-5 | 50 |
| 5-10 | 33 |
| Over 10 | 17 |

D.   *Percentage of Students Who Use the Consultant = 30%*

***Figure 6.11.*** Statistical tables describing the system.

```
{ helplength

  compute length of consulting session based on the distribution in figure 6.11

  entry conditions:
        none

  exit conditions:
        time                    : set to the length of a consulting session
}
procedure helplength (var time : real);
var
    r                       : real;        { a pseudorandom number }
begin { procedure helplength }
    r := random(0.0,1.0);
    if r <= 0.50 then time := random(0.0,5.0)
    else
    if r <= 0.83 then time := random(5.0,10.0)
    else time := random(10.0,15.0)
    { assume maximum consulting time is 15 minutes }
end; { of procedure helplength }
```

{ *uselength*

*determine the length of a terminal session based on the distribution in figure 6.11*

*entry conditions:*
      *none*

*exit conditions:*
      *time*               *: set to the length of a terminal session*
}

**procedure** *uselength (***var** *time : real);*
**var**
    *r*                       *: real;*         { *a pseudorandom number* }
**begin** { *procedure uselength* }
    *r := random(0.0, 1.0);*
    **if** *r <= 0.25* **then** *time := random(0.0, 15.0)*
    **else**
    **if** *r <= 0.80* **then** *time := random(15.0, 30.0)*
    **else**
    **if** *r <= 0.98* **then** *time := random(30.0, 60.0)*
    **else** *time := random(60.0, 180.0)*
    { *assume nobody sits at a terminal for more than three hours* }
**end;** { *of procedure uselength* }

{ *needhelp*

*determine if a student does or does not go to the consultant based on the distribution shown in figure 6.11*

*entry conditions:*
      *none*

*exit conditions:*
      *exit*               *: true if the consultant is needed, false otherwise.*
}

**procedure** *needhelp (***var** *exit : boolean);*
**var**
    *r*                       *: real;*         { *a pseudorandom number* }
**begin** { *procedure needhelp* }
    *r := random(0.0, 1.0);*
    *exit := (r <= 0.70)*
**end;** { *of procedure needhelp* }

```
{ nextarrival

  determine how long until the next student arrives based on the distribution
  shown in figure 6.11

  entry conditions:
      none

  exit conditions:
      time                  : set to the length of time before the next student
                              arrives.
}
procedure nextarrival (var time : real);
var
    r                       : real;          { pseudo-random number }
begin { procedure nextarrival }
    r := random(0.0,1.0); { pseudorandom number in the range [0..1] }
    if r <= 0.12 then time := random(0.0,1.0)
    else
    if r <= 0.36 then time := random(1.0,2.0)
    else
    if r <= 0.71 then time := random(2.0,4.0)
    else
    if r <= 0.91 then time := random(4.0,8.0)
    else
    if r <= 0.98 then time := random(8.0,15.0)
    else time := random(15.0,30.0)
    { assume that the maximum interarrival time is one half hour }
end; { of procedure nextarrival }
```

**Figure 6.12.** Statistical procedures.

r, in the range x <= r <= y. As with all our other procedures, we must add this one to the tree of procedures and eventually code it. However, as we mentioned in Style Review 3.3, almost all large computer installations have extensive *program libraries* with useful functions and procedures. They perform common operations such as root finding, averaging, sorting, merging, and searching. It is very likely that the function random as we have defined it (or one very similar) already exists in some library. Since we never want to duplicate useful work, we would always opt to use this library routine if it were available. We could do this through the following *external declaration*.

```
function random(x , y : real) : real; extern;
```

The preceding declaration states that the function random will not be contained within the body of the program but will be found in an external library. If the routine were not available in some library, we would merely replace the preceding external declaration with the actual code for the function. (Since our library did not include a random number generator in exactly the form we needed, we wrote our own. It is shown in the listing of the entire program at the end of this chapter.)

The last four routines left to code are ones that appear at the beginning and end of the program: the initialize procedure, the errorhandler, the cleanup routine everybodyout, and the output procedure called reportwriter. These routines are usually done last because only when we are finished with all *other* routines do we know for certain exactly what values to initialize and what values should be presented as output.

For initialize, we must go through all the modules and determine which data values should be initialized and to what values. Also, at this late stage we cannot do a sloppy job of printing the results in the procedure reportwriter. All the time and hard work will be wasted (from the end-user's standpoint) if the output of our program is difficult to interpret or incomplete. We must put as much care and effort into the final output procedure as we have into any of the other modules. Our errorhandler need only handle one case—an empty calendar. This should never happen and would indicate a significant flaw in the system. We will inform the user of this fact and then halt. Finally, the routine everybodyout will clean up all the memory space used for student records and queues and return it to the main memory pool. This is necessary in Pascal to insure that we do not run out of memory when we execute the simulator over and over again. This cleanup is done using the standard Pascal procedure dispose, which frees up memory that had previously been allocated to a particular data structure. (The procedure dispose is described in Section 8.2.2.)

The final four procedures of our simulator are shown in Figure 6.13.

This completes our case study. By textbook standards, this is a monumental program—spanning 17 pages and containing 20 different procedures with approximately 450 lines of code (with comments, about 650 lines). But, more than anything else in this text, this case study graphically illustrates the fundamentals of programming. It demonstrates the critical need for a management policy to handle large, complex programs. It exemplifies the top-down design of programs and data structures. It illustrates procedure modularity and logical coherence. Finally, it again demonstrates the importance of clarity of expression. A program of this size would have been much more difficult to follow if we had not adhered to the style and expression guidelines presented in earlier chapters.

After spending all that time designing and coding, we would be remiss if we did not present the results of our simulation program. The program just developed was coded and run on a CDC/Cyber 74; some results are presented in Figure 6.14. Before we can analyze the data and say what the "proper" number of terminals is, we must decide on acceptable limits for waiting times, waiting lines, and utilizations in our

{ *errorhandler*

*handle an error by halting the program. in future versions of the program we might try to recover from the error.*

*entry conditions:*
        *none*

*exit conditions:*
        *the program will be halted*
}

**procedure** *errorhandler;*
**begin** { *procedure errorhandler* }
        *writeln(' *** fatal error -- we encountered an empty calendar');*
        *writeln(' *** dump followed by a system halt');*
        *halt(' empty calendar')*
**end**; { *of procedure errorhandler* }

{ *initialize*

*initialize the state of the simulation and zero out statistics.*

*entry conditions:*
        *none*

*exit conditions:*
        *simstate*          : *all fields are initialized except nterms and maxtime*
                               *which have been read from input*
        *stats*             : *all fields are set to zero*
}

**procedure** *initialize (***var** *simstate : staterec;* **var** *stats : statsrec);*
**begin** { *procedure initialize* }
        **with** *simstate* **do**
        **begin**
                *clock := 0.0;*
                { *put first arrival event on the calendar for time 0.0* }
                *calendar :=* **nil**;
                *schedule(calendar, arrival,* **nil**, *0.0);*
                { *set the queues to be empty* }
                *termqueue.head :=* **nil**;
                *termqueue.tail :=* **nil**;
                *consultqueue.head :=* **nil**;
                *consultqueue.tail :=* **nil**;
                { *initially all terminals are free, and consultant is idle* }
                *ntermsinuse := 0;*
                *helperbusy := false*
        **end**; { *of with simstate* }

```
    with stats do
    begin { zero all totals used for gathering data }
        consultcount := 0;
        consultwait := 0.0;
        finished := 0;
        termcount := 0;
        termwait := 0.0;
        totalconsulttime := 0.0;
        totalstudents := 0;
        totaltermtime := 0.0;
        totaltime := 0.0
    end { of with stats }
end; { of procedure initialize }

{ reportwriter

 produce the finished report.

 entry conditions:
        nterms            : the number of terminals used in the simulation
        duration          : the time in minutes that the simulation ran

 exit conditions:
        the report will be written to output
}

procedure reportwriter (nterms : integer; duration : real; stats : statsrec);
var
    consultpercent        : real;          { consultant utilization level }
    termpercent           : real;          { terminal utilization level }
begin { procedure reportwriter }
    writeln;
    writeln;
    writeln(' results of the simulation run');
    writeln(nterms:8,' terminals were available');
    writeln(duration:8:2,' minutes simulation time');
    with stats do
    begin
        writeln(totalstudents:8,' students entered the terminal room');
        writeln(finished:8,' students exited during the simulation');
        if termcount > 0 then
        writeln(termwait / termcount :8:2,' average minutes (per student)',
        ' spent waiting for a terminal')
    else writeln(' ':8,' no terminals were used');
    if consultcount > 0 then
        writeln(consultwait / consultcount :8:2,' average minutes',
            ' (per student) spent waiting for the consultant')
    else writeln(' ':8,' nobody saw the consultant');
```

```
        if finished > 0 then
                writeln(totaltime / finished :8:2,' average minutes (per student)',
                        ' spent in the terminal room');

        { compute the fraction of usage of the terminals and consultant
          recall that zero length simulations cannot be requested }
        termpercent : = (totaltermtime * 100.0) / (nterms * duration);
        consultpercent : = (totalconsulttime * 100.0) / duration
    end; { of with stats }

    { it is possible to get terminal or consultant utilizations beyond 100 percent
      because we tally the last few sessions even though they may extend past
      closing time. to avoid confusing the reader, these values will be reset to
      100. }
    if termpercent > 100.0 then termpercent : = 100.0;
    if consultpercent > 100.0 then consultpercent : = 100.0;

    writeln(termpercent:8:1,' percent usage made of all terminals');
    writeln(consultpercent:8:1,' percent usage made of the consultant')
end; { of procedure reportwriter }

{ everybodyout

  it is closing time so all students will be asked to leave, freeing space to be
  used for the next simulation.

  entry conditions:
            calendar              : points to a list of 0 or more events
            consultqueue          : points to a queue of 0 or more students
            terminalqueue         : points to a queue of 0 or more students
  exit conditions:
                    all nodes remaining in the above lists are returned to available
                    storage by disposing them, and the pointers are set to nil.
}

procedure everybodyout (var calendar : eventptr; var consultqueue : queue;
        var termqueue : queue);
var
        evptr                   : eventptr; { used to scan calendar }
```

{ *escortq*

 *quietly escort those still waiting in line to the nearest exit*

 *entry conditions:*
  *q         : points to a queue of 0 or more students*
 *exit conditions:*
  *all nodes in the queue are disposed, and the pointers set to nil*
}

**procedure** *escortq (***var** *q : queue);*
**var**
  *studentp      : studentptr; { the current deadbeat }*
**begin** { *procedure escortq* }
   **while** *q.head* <> **nil do**
   **begin**
     *studentp := q.head;*
     *q.head := studentp↑.next;*
     *dispose(studentp)*
   **end**; { *of while loop to remove a student* }
   *q.tail :=* **nil**
 **end**; { *of procedure escortq* }

**begin** { *procedure everybodyout* }
   **while** *calendar* <> **nil do**
   **begin**
     **if** *calendar↑.studptr* <> **nil then** *dispose(calendar↑.studptr);*
     *evptr := calendar .link;*
     *dispose(calendar);*
     *calendar := evptr*
   **end**; { *of while loop emptying calendar* }

   *escortq(consultqueue);*
   *escortq(termqueue)*
**end**; { *of procedure everybodyout* }

---

*Figure 6.13.* Initialize, errorhandler, reportwriter, and everybodyout procedures.

| Number of Terminals | Average Waiting Time for a Terminal (min) | Utilization of Terminals (%) | Utilization of Consultant (%) |
|---|---|---|---|
| 1 | 617 | 100.0 | 6.6 |
| 2 | 544 | 100.0 | 12.0 |
| 3 | 425 | 100.0 | 25.7 |
| 4 | 321 | 100.0 | 27.1 |
| 5 | 204 | 100.0 | 37.4 |
| 6 | 191 | 99.9 | 40.4 |
| 8 | 64 | 97.0 | 52.8 |
| 10 | 5 | 86.6 | 57.3 |
| 15 | 0.03 | 54.4 | 55.9 |
| 20 | 0 | 43.3 | 56.8 |
| 50 | 0 | 16.8 | 58.7 |

**Figure 6.14.** Results of the simulation model.

terminal room. For example, using our data, if we do not want students to wait more than 5 minutes on the average for a terminal, we would need at least 10 terminals in the room. If we wanted the waiting time to be just a few seconds, we would need about 15 terminals. The data in Figure 6.14 show clearly how important it is to study the behavior of this system thoroughly *before* we buy terminals. For example, if we had only purchased 3 terminals for the room, the average waiting time would be about 7 hours!

We will leave it to the interested student (drawing on his/her own frustration with trying to find a free terminal the day before homework is due!) to analyze the data in Figure 6.14 further and decide how many terminals we should purchase.

A complete listing of the simulator program developed in this chapter appears in Figure 6.15.

**program** *simulator(input,output);*
*{ simulator - simulate computer terminal room; g. m. schneider; jan 1981*

  *this is a program to simulate the behavior of a computer terminal room using
  discrete event simulation. the system is described in chapter 6 of advanced
  programming and problem solving by g. m. schneider, s. c. bruell.*
*}*

**const**
     *version*                          = *' 2.1, last mod 11 feb 81';*
**type**
     *eventclass*                       = *(arrival, term, consult, error);*

     *studentptr*                       = *↑ student;*
     *student*                          =
          **record**
               *next*                   : *studentptr;*   *{ points to next person in
                                                             queue }*
                    *timeinqueue*       : *real;*          *{ time when inserted }*
                    *timeofarrival*     : *real*           *{ to terminal room }*
          **end***; { of student record }*

     *eventptr*                         = *↑ event;*
     *event*                            =
          **record**
               *eventtype*              : *eventclass;*
               *eventtime*              : *real;*          *{ time of occurrence }*
               *studptr*                : *studentptr;*
               *link*                   : *eventptr*
          **end***; { of event record }*

     *queue*                            =
          **record**
               *head*                   : *studentptr;*
               *tail*                   : *studentptr*
          **end***; { of queue record }*

```
statsrec                            =
    record { all statistical sums }
        consultcount        : integer;     { counts consultations }
        consultwait         : real;        { accumulates time
                                             students waited for a
                                             consultation }
        finished            : integer;     { counts students leaving
                                             terminal room }
        termcount           : integer;     { counts terminal events
        termwait            : real;        { accumulates student
                                             waiting time }
        totalconsulttime    : real;        { accumulates consultation
                                             times }
        totalstudents       : integer;     { counts student arrivals }
        totaltermtime       : real;        { accumulates useful
                                             terminal time }
        totaltime           : real         { accumulates time spent in
                                             room }
    end; { of statsrec record }

staterec                            =
    record { the current state of the simulation }
        calendar            : eventptr;    { head of calendar list }
        clock               : real;        { the running time. starts at
                                             0.0, up to maxtime, and is
                                             reset to eventtime for
                                             each new event }
        consultqueue        : queue;       { points to ends of
                                             consultant queue }
        helperbusy          : boolean;     { true if the consultant is
                                             busy, false otherwise }
        maxtime             : real;        { total time to run the
                                             simulation }
        nterms              : integer;     { number of terminals in
                                             room }
        ntermsinuse         : integer;     { number of terminals being
                                             used }
        termqueue           : queue        { points to ends of terminal
                                             queue }
    end; { of staterec record }
```

**var**

| | | |
|---|---|---|
| done | : boolean; | { boolean flag to signal completion } |
| eventtime | : real; | { the time of the next calendar event } |
| eventtype | : eventclass; | { the type of the next calendar event } |
| gooddata | : boolean; | { boolean flag used to validate input } |
| p | : studentptr; | { student involved in current event } |
| response | : char; | { user response about continuation } |
| simstate | : staterec; | { state of the simulation } |
| stats | : statsrec; | { accumulates statistics } |

{ schedule

    search through the calendar to determine where a new event should be placed. then add this new event to the calendar, keeping the event list ordered by the time field.

    entry conditions:

| | |
|---|---|
| calendar | : pointer to a list of 0 or more events |
| evtype | : type of the event to be scheduled |
| sptr | : pointer to the student involved (may be nil) |
| evtime | : time that event is to be scheduled |

  exit conditions:

    the event is inserted into the calendar list, preserving ordering by the time field.

}

**procedure** schedule(**var** calendar : eventptr; evtype : eventclass;
    sptr : studentptr; evtime : real);
**var**

| | | |
|---|---|---|
| done | : boolean; | { true when searching should stop } |
| eventp | : eventptr; | { temporary for node creation } |
| oldp | : eventptr; | { temporary used for calendar search } |
| p | : eventptr; | { temporary used for calendar search } |

**begin**
    { first build a new node with the proper values }
    new(eventp);
    **with** eventp ↑ **do**
    **begin**
        eventtype : = evtype;
        eventtime : = evtime;
        studptr    : = sptr
    **end**; { of with }

```
{ now see where this new node should be placed and reset the links }
p := calendar;
if p <> nil then
begin
    if p ↑ .eventtime <= evtime then { we must search through
                                         the list}
    begin
        repeat
            oldp := p;
            p := p ↑ .link;

            { see if we are done without doing a nil pointer reference }
            if p <> nil then done := (p ↑ .eventtime > evtime)
            else done := true
        until done;

        {now insert the new item }
        eventp ↑ .link := p;
        oldp ↑ .link := eventp
    end { of search and insert new item }
    else
    begin { it goes at the head of the list }
        eventp ↑ .link := calendar;
        calendar := eventp
    end { of else }
end { of if p <> nil }
else
begin { calendar is empty }
    eventp ↑ .link := nil;
    calendar := eventp
end { of else empty calendar }
end; { of procedure schedule }

{ getnextevent

get the next item from the calendar, removing it from the head of the list.

entry conditions:
    calendar            : pointer to the list of 1 or more events

exit conditions:
    calendar            : has had it's first event removed
    evtype              : returns the type of the event.
                          if the calendar list was empty, error is returned.
    sptr                : returns a pointer to the student involved
    evtime              : returns the time this event is to take place

}
```

```
procedure getnextevent(var calendar : eventptr; var evtype : eventclass;
     var sptr : studentptr; var evtime : real);
var
     newlink                  : eventptr; { temporary used for updating links }
begin
     if calendar <> nil then
     begin
          with calendar ↑ do
          begin
               evtype : = eventtype;
               evtime : = eventtime;
               sptr    : = studptr
          end; { of with }

          { now adjust the links and discard that node }
          newlink : = calendar ↑ .link;
          dispose(calendar); { dispose is a built-in Pascal procedure }
          calendar : = newlink
     end { of if calendar <> nil }
     else evtype : = error { calendar empty }
end; { of procedure getnextevent }
```

```
{ random

   get a random number in the specified range from a uniform distribution. this is
   a system dependent routine

   entry conditions:
          low                       : lower bound of range
          high                      : upper bound of range

   exit conditions:
          the function returns a number on the open interval (low,high).
}
```

```
function random (low, high : real) : real;

     function ran : real; extern; { a Pascal 6000 release 3 library routine }

begin
     random : = ran * (high − low) + low
end; { of function random }
```

{ *helplength*

*compute length of consulting session based on the distribution in figure 6.11*

*entry conditions:*
        *none*

*exit conditions:*
        *time*                                  *: set to the length of a consulting session*
}

**procedure** *helplength(***var** *time : real);*
**var**
        *r*                              *: real;*          { *a pseudorandom number* }
**begin**
        *r := random(0.0,1.0);*
        **if** *r <= 0.50* **then** *time := random(0.0,5.0)*
        **else**
        **if** *r <= 0.83* **then** *time := random(5.0,10.0)*
        **else** *time := random(10.0,15.0)*
        { *assume maximum consulting time is 15 minutes* }
**end***;* { *of procedure helplength* }

{ *uselength*

*determine the length of a terminal session based on the distribution in figure*
*6.11*

*entry conditions:*
        *none*

*exit conditions:*
        *time*                                  *: set to the length of a terminal session*

}

**procedure** *uselength(***var** *time : real);*
**var**
        *r*                              *: real;*          { *a pseudorandom number* }
**begin**
        *r := random(0.0,1.0);*
        **if** *r <= 0.25* **then** *time := random(0.0,15.0)*
        **else**
        **if** *r <= 0.80* **then** *time := random(15.0,30.0)*
        **else**
        **if** *r <= 0.98* **then** *time := random(30.0,60.0)*
        **else** *time:= random(60.0,180.0)*
        { *assume nobody sits at a terminal for more than three hours* }
**end***;* { *of procedure uselength* }

{ *needhelp*

*determine if a student does or does not go to the consultant based on the distribution shown in figure 6.11*

*entry conditions:*
    *none*

*exit conditions:*
    *exit*                      *: true if the consultant is needed, false otherwise.*

}

**procedure** *needhelp(***var** *exit : boolean);*
**var**
    *r*                               *: real;*        { *a pseudorandom number* }
**begin**
    *r := random(0.0,1.0);*
    *exit := (r <= 0.70)*
**end;** { *of procedure needhelp* }

{ *nextarrival*

*determine how long until the next student arrives based on the distribution shown in figure 6.11*

*entry conditions:*
    *none*

*exit conditions:*
    *time*                      *: set to the length of time before the next student arrives.*

}

**procedure** *nextarrival(***var** *time : real);*
**var**
    *r*                               *: real;*        { *pseudorandom number* }
**begin**
    *r := random(0.0,1.0);* { *pseudorandom number in the range [0. .1]* }
    **if** *r <= 0.12* **then** *time := random(0.0,1.0)*
    **else**
    **if** *r <= 0.36* **then** *time := random(1.0,2.0)*
    **else**
    **if** *r <= 0.71* **then** *time := random(2.0,4.0)*
    **else**
    **if** *r <= 0.91* **then** *time := random(4.0,8.0)*
    **else**
    **if** *r <= 0.98* **then** *time := random(8.0,15.0)*
    **else** *time := random(15.0,30.0)*
    { *assume that the maximum interarrival time is one half hour* }
**end;** { *of procedure nextarrival* }

{ *putinqueue*

*put student p at the end of queue q.*

*entry conditions:*

| | |
|---|---|
| *p* | *: points to student being placed in line* |
| *q* | *: points to a queue of 0 or more students* |

*exit conditions:*

| | |
|---|---|
| *p* | *: the next field is set to nil.* |
| *q* | *: has had student p put at the end of the line* |

}

**procedure** *putinqueue (p : studentptr;* **var** *q : queue);*
**begin**
    **if** *q.tail* <> **nil then** { *queue not empty* }
    **begin**
        *q.tail* ↑ *.next := p;*
        *p* ↑ *.next :=* **nil**;
        *q.tail := p*
    **end** { *of add to nonempty queue*}
    **else**
    **begin** { *queue is empty* }
        *q.head := p;*
        *q.tail := p;*
        *p* ↑ *.next :=* **nil**
    **end** { *of start queue* }
**end**; { *of procedure putinqueue* }

{ *takeoffqueue*

*take the first student p from queue q. abort if queue is empty.*

*entry conditions:*

| | |
|---|---|
| *q* | *: points to a queue of 1 or more students* |

*exit conditions:*

| | |
|---|---|
| *p* | *: points to the student who was first in line* |
| *q* | *: has had its first entry removed* |

}

**procedure** *takeoffqueue(***var** *p : studentptr;* **var** *q : queue);*
**begin**
    **if** *q.head* <> **nil then** { *there is at least one item in the queue* }
    **begin**
        *p := q.head;*

```
        { now remove that item and reset pointers }
        if p↑.next <> nil then
            q.head := p↑.next { remove from list }
        else
        begin { the queue had only one item and is now empty }
            q.head := nil;
            q.tail := nil
        end { of else }
    end { of queue was not empty }
    else halt(' takeoffqueue called with empty queue')
        { this will abort if the programmer is careless enough not to check if the
            queue is empty before calling this routine }
end; { procedure takeoffqueue }


{ queueempty

    determine if queue q has any students left in it

    entry conditions:
        q                                  : points to a queue of 0 or more students

    exit conditions:
            function returns true if the queue is empty, false otherwise.

function queueempty(q : queue) : boolean;
begin
    queueempty := (q.head = nil)
end; { of procedure queueempty }


{ arrivalevent

    handle a student who just arrived, giving him/her a terminal if one is available or
    putting him/her in line if one is not.

    entry conditions:
        simstate                    : current state of the simulation
        stats                       : current statistics

    exit conditions:
        simstate                    : updated appropriately
        stats                       : also updated appropriately
}
procedure arrivalevent(var simstate : staterec; var stats : statsrec);
var
    newp                            : studentptr;  { temporary student pointer }
    t                               : real;        { temporary time for
                                                        scheduling }
begin
    { just schedule the next arrival event }
    nextarrival(t); { determines inter-arrival time }
```

```
    { schedule next arrival in t minutes }
    schedule(simstate.calendar, arrival, nil, simstate.clock + t);

    { now create that new student and set initial parameters }
    new(newp);
    newp↑.timeofarrival := simstate.clock; { when student first walked in }

    { see if there are any terminals free }
    if simstate.ntermsinuse = simstate.nterms then { all busy }
    begin { put student in queue }
        putinqueue(newp, simstate.termqueue);
        newp↑.timeinqueue := simstate.clock
    end { of if busy }
    else
    begin { give student a terminal }
        { determine length of terminal session }
        uselength(t);

        { schedule terminal completion event }
        schedule(simstate.calendar, term, newp, simstate.clock + t);
        simstate.ntermsinuse := simstate.ntermsinuse + 1;
        stats.termcount := stats.termcount + 1;
        stats.totaltermtime := stats.totaltermtime + t
    end; { of else clause }

    stats.totalstudents := stats.totalstudents + 1
end; { of procedure arrivalevent }

{ consultevent
```

take student p away from the consultant and give him a terminal if one is open, putting him in line if there are none. if anyone is waiting, he will be taken to the consultant.

entry conditions:

| | |
|---|---|
| p | : points to student who is done being helped |
| simstate | : current state of the simulation |
| stats | : accumulated statistics |

exit conditions:

| | |
|---|---|
| simstate | : updated appropriately |
| stats | : updated appropriately |

```
}
```

**procedure** *consultevent(p : studentptr;* **var** *simstate : staterec;*
    **var** *stats : statsrec);*
**var**
    *newp*                             *: studentptr;*   *{ temporary student pointer }*
    *t*                                 *: real;*         *{ temporary time for scheduling }*
**begin**
    *{ first put this student back in line for a terminal }*
    **if** *simstate.ntermsinuse = simstate.nterms* **then** *{ must wait }*
    **begin**
        *putinqueue(p,simstate.termqueue);*
        *p ↑ .timeinqueue : = simstate.clock*
    **end** *{ of if terminals full }*
    **else**
    **begin** *{ give student a terminal }*
        *uselength(t);*
        *schedule(simstate.calendar,term,p,simstate.clock + t);*
        *simstate.ntermsinuse : = simstate.ntermsinuse + 1;*
        *stats.termcount : = stats.termcount + 1;*
        *stats.totaltermtime : = stats.totaltermtime + t*
    **end***; { of else get an open terminal }*

    *{ now see if anybody else wants to talk to the consultant }*
    **if** *queueempty(simstate.consultqueue)* **then** *{ nobody is waiting }*
        *simstate.helperbusy : = false*
    **else**
    **begin** *{ somebody wants to see the consultant }*
        *takeoffqueue(newp,simstate.consultqueue);*

        *{ collect statistics about this person's wait in line }*
        *stats.consultwait : = stats.consultwait + (simstate.clock −*
            *newp ↑ .timeinqueue);*
        *stats.consultcount : = stats.consultcount + 1;*

        *{ schedule completion time of consultant }*
        *helplength(t);*
        *schedule(simstate.calendar,consult,newp,simstate.clock + t);*
        *stats.totalconsulttime : = stats.totalconsulttime + t*
    **end**
**end***; { of procedure consultevent }*

*{ errorhandler*

    *handle an error by halting the program. in future versions of the program we might try to recover from the error.*

    *entry conditions:*
        *none*

    *exit conditions:*
        *the program will be halted*

*}*

```
procedure errorhandler;
begin
      writeln(' *** fatal error-- we encountered an empty calendar');
      writeln(' *** dump followed by a system halt');
      halt(' empty calendar')
end; { of procedure errorhandler }
```

{ termevent

remove student p from his terminal. if he needs help, send him to the consultant. if anyone is waiting for a terminal, it will be given to the first person in line.

entry conditions:

| | |
|---|---|
| p | : points to the student who is done |
| simstate | : current state of the simulation |
| stats | : accumulated statistics |

exit conditions:

| | |
|---|---|
| p | : will be shown the door or go to the consultant |
| simstate | : will be updated appropriately |
| stats | : will be updated appropriately |

}

```
procedure termevent(p : studentptr; var simstate : staterec;
      var stats : statsrec);
var
      leaving                          : boolean;    { true if a student is going to
                                                        leave the room, false if he/
                                                        she is going to the
                                                        consultant }
      newp                             : studentptr; { temporary student pointer }
      t                                : real;       { temporary time for
                                                        scheduling }
begin
      { determine if this student needs help or will leave the room }
      needhelp(leaving);
      if leaving then { collect statistics about this student }
      begin
            stats.finished := stats.finished + 1;
            stats.totaltime := stats totaltime + (simstate.clock −
                  p ↑ timeofarrival);
            dispose(p) { say goodbye to student }
      end { of if leaving }
```

```
      else
      begin { he/she is going to the consultant }
            if simstate.helperbusy then { wait in line }
            begin
                  putinqueue(p,simstate.consultqueue);
                  p↑.timeinqueue := simstate.clock
            end { of if helperbusy }
            else
            begin
                  simstate.helperbusy := true;
                  helplength(t); { determine length of consultation }
                  schedule(simstate.calendar,consult,p,simstate.clock + t);
                  stats.consultcount := stats.consultcount + 1;
                  stats.totalconsulttime := stats.totalconsulttime + t
            end { of else helper not busy }
      end; { of else person goes to consultant }

      { now see if anybody else wants the terminal just freed up }
      if queueempty(simstate.termqueue) then
            simstate.ntermsinuse := simstate.ntermsinuse - 1
      else
      begin
            takeoffqueue(newp,simstate.termqueue); { get next person from queue }
            { determine how long this person waited, adding it to total }
            stats.termwait := stats.termwait + (simstate.clock -
                  newp↑.timeinqueue);
            stats.termcount := stats.termcount + 1;
            uselength(t); { determine length of terminal session }

            { schedule terminal completion }
            schedule(simstate.calendar,term,newp,simstate.clock + t);
            stats.totaltermtime := stats.totaltermtime + t
      end
end; { of procedure termevent }
{ initialize

   initialize the state of the simulation and zero out statistics.

   entry conditions:
         none

   exit conditions:
         simstate                    : all fields are initialized except nterms and
                                       maxtime which have been read from
                                       input
         stats                       : all fields are set to zero
}
```

```
procedure initialize(var simstate : staterec; var stats : statsrec);
begin
    with simstate do
    begin
        clock := 0.0;
        { put first arrival event on the calendar for time 0.0 }
        calendar := nil;
        schedule(calendar, arrival, nil, 0.0);
        { set the queues to be empty }
        termqueue.head := nil;
        termqueue.tail := nil;
        consultqueue.head := nil;
        consultqueue.tail := nil;
        { initially all terminals are free, and consultant is idle }
        ntermsinuse := 0;
        helperbusy := false
    end; { of with simstate }

    with stats do
    begin { zero all totals used for gathering data }
        consultcount := 0;
        consultwait := 0.0;
        finished := 0;
        termcount := 0;
        termwait := 0.0;
        totalconsulttime := 0.0;
        totalstudents := 0;
        totaltermtime := 0.0;
        totaltime := 0.0
    end { of with stats }
end; { of procedure initialize }

{ reportwriter

produce the finished report.

entry conditions:
        nterms                          : the number of terminals used in the
                                          simulation
        duration                        : the time in minutes that the simulation
                                          ran

exit conditions:
        the report will be written to output
}
```

```
procedure reportwriter(nterms : integer; duration : real; stats : statsrec );
var
    consultpercent          : real;        { consultant utilization level }
    termpercent             : real;        { terminal utilization level }
begin
    writeln;
    writeln;
    writeln(' results of the simulation run');
    writeln(nterms:8,' terminals were available');
    writeln(duration:8:2,' minutes simulation time');
    with stats do
    begin
        writeln(totalstudents:8,' students entered the terminal room');
        writeln(finished:8,' students exited during the simulation');
        if termcount > 0 then
            writeln(termwait / termcount :8:2,' average minutes (per student)',
                ' spent waiting for a terminal')
        else writeln(' ':8,' no terminals were used');

        if consultcount > 0 then
            writeln(consultwait / consultcount :8:2,' average minutes',
                ' (per student) spent waiting for the consultant')
        else writeln(' ':8,' nobody saw the consultant');

        if finished > 0 then
            writeln(totaltime / finished :8:2,' average minutes (per student)',
                ' spent in the terminal room');

        { compute the fraction of usage of the terminals and consultant
          recall that zero length simulations cannot be requested }
        termpercent := (totaltermtime * 100.0) / (nterms * duration);
        consultpercent := (totalconsulttime * 100.0) / duration
    end; { of with stats }

    { it is possible to get terminal or consultant utilizations beyond 100 percent
      because we tally the last few sessions even though they may extend past
      closing time. to avoid confusing the reader, these values will be reset to
      100. }
    if termpercent > 100.0 then termpercent := 100.0;
    if consultpercent > 100.0 then consultpercent := 100.0;

    writeln(termpercent:8:1,' percent usage made of all terminals');
    writeln(consultpercent:8:1,' percent usage made of the consultant')
end; { of procedure reportwriter }
```

{ *everybodyout*

*it is closing time so all students will be asked to leave, freeing space to be used for the next simulation.*

*entry conditions:*

| | |
|---|---|
| *calendar* | *: points to a list of 0 or more events* |
| *consultqueue* | *: points to a queue of 0 or more students* |
| *terminalqueue* | *: points to a queue of 0 or more students* |

*exit conditions:*

> *all nodes remaining in the above lists are returned to available storage by disposing them, and the pointers are set to nil.*

}

**procedure** *everybodyout(***var** *calendar : eventptr;* **var** *consultqueue : queue;*
    **var** *termqueue : queue);*
**var**

> *evptr                                  : eventptr;     { used to scan calendar }*

> { *escortq*

> *quietly escort those still waiting in line to the nearest exit*

> *entry conditions:*

> > *q                                  : points to a queue of 0 or more students*

> **procedure** *escortq(***var** *q : queue);*
> **var**

> > *studentp                      : studentptr; { the current deadbeat }*

> **begin**
>     **while** *q.head <>* **nil do**
>     **begin**
>         *studentp := q.head;*
>         *q.head := studentp↑.next;*
>         *dispose(studentp)*
>     **end**; *{ of while loop to remove a student }*
>     *q.tail :=* **nil**
> **end**; *{ of procedure escortq }*

**begin** *{ procedure everybodyout }*
    **while** *calendar <>* **nil do**
    **begin**
        **if** *calendar↑.studptr <>* **nil then** *dispose(calendar↑.studptr);*
        *evptr := calendar↑.link;*
        *dispose(calendar);*
        *calendar := evptr*
    **end**; *{ of while loop emptying calendar }*

    *escortq(consultqueue);*
    *escortq(termqueue)*
**end**; *{ of procedure everybodyout }*

```
begin { program simulator }
    writeln(' welcome to the terminal room simulator version ',version);

    repeat
        repeat
            writeln(' please enter values for nterms -- the number of ',
                ' terminals,');
            writeln(' and maxtime -- the total simulation time in minutes');
            readln; { get user response }
            read(simstate.nterms,simstate.maxtime);

            if (simstate.nterms >= 1) and (simstate.maxtime > 0.0) then
            begin
                gooddata := true
            end
            else
            begin
                gooddata := false;
                writeln(' *** you made an error in the input. nterms must be ',
                    ' >= 1');
                writeln(' *** and maxtime must be real and non-negative.');
                writeln(' *** please try again')
            end { of else }
        until gooddata;

        initialize(simstate,stats); { put the first event on the calendar }

        { now we run the simulator for as long as the user specified }
        while (simstate.clock <= simstate.maxtime) do
        begin
            getnextevent(simstate.calendar,eventtype,p,eventtime);
            simstate.clock := eventtime;

            { now simply call the appropriate routine for this event type }
            case eventtype of
                arrival   : arrivalevent(simstate,stats); { process arrival event}
                consult : consultevent(p,simstate,stats); { consultant completion }
                error   : errorhandler;       { something bad happened }
                term    : termevent(p,simstate,stats) { terminal completion}
            end { of case }
        end; { of while loop to handle one complete simulation }
```

> { *we have finished one run. produce the results and see if the user*
>    *wants to try again* }
> **with** *simstate* **do**
>        *everybodyout(calendar,consultqueue,termqueue);*
>    *reportwriter(simstate.nterms,simstate.clock,stats);*
>
>    *writeln(' do you wish to run the simulation again for a different');*
>    *writeln(' number of terminals or time value ? y(es) or n(o)');*
>    *readln;*
>    *read(response); { we will read only the first character }*
>    *done := (response = 'n')*
> **until** *done;*
>
> { *we will go through the above repeat loop as many times as the user*
>    *wants. we will exit here only when he/she is completely finished* }
> *writeln(' exiting from the simulator program. thank you')*
> **end**. { *of program simulation* }

---

***Figure 6.15.*** The completed simulation program.

## EXERCISES FOR CHAPTER 6

1. Rewrite the program in Figure 6.3 so that it accomplishes the same task but without the disastrous side effect.

2. Review and discuss the modularity of the terminal room simulator developed in this chapter. Do you think any of the existing modules do too much or too little? Would you have decomposed the problem in a different way? If so, how?

3. The following program converts a character string to an integer value. It uses global variables to accomplish its task. Rewrite the procedure so it does the same thing using only:

   (a) Reference parameters.
   (b) Value parameters.
   (c) Local variables.

   **procedure** chartsint;

   { *convert an unsigned string of characters stored in text*[1] *to text*[20] *into a decimal integer. The conversion terminates when we*

*1)   encounter a character that cannot syntactically be part of a valid unsigned integer*

*2)   have converted all 20 characters*

*3)   exceed the value of maxint }*

```
begin
    i      := 1;
    done : = false;
    sum  := 0;
    while (i <= numofcharacters) and (not done) do
    begin
        if (text[i] < ord('0')) or (text[i] > ord('9')) then
            done : = true
        else
        begin
            sum : = (sum * 10) + (text[i] − ord('0'));
            if sum > maxint then
                done : = true
            else
                i : = i + 1
    end { of while loop }
end; { of procedure chartsint }
```

In addition, what assumption does this procedure make about the collating sequence '0', . . . ,'9' that might make this procedure not portable to other computer systems? Rewrite the procedure so it will work for any character set.

4.  Rewrite the procedure chartsint from Exercise 3 to increase its generality in the following way. Assume that the character string in the text represents the digits of an integer represented in *base B* ($B >= 2$). Convert that character string to the correct decimal integer value. For example, if the text contained '1,' '4,' '2,' and it was interpreted as a base 6 value, the procedure would compute $142_6 = (1 * 6^2) + (4 * 6^1) + (2 * 6^0) = 62_{10}$

5.  A student was asked to write a very simple Pascal function to compute tan($\theta$) for values of $\theta$ in the range $-2\pi <= \theta <= 2\pi$. For values outside this range the procedure should indicate an error. Here is what was produced.

```
function tangent (var theta : real) : real;
const pi = 3.1415927;
begin
    if abs(theta) > 2.0 * pi then
    begin
        writeln(' error in the argument to the tangent function ');
        theta := 0.0;
        tangent := 0.0
    end { of then clause }
    else
    begin
        { tangent is not defined at either ± π/2 or ± 3π/2 }
        if (abs (theta) = pi/2.0) or (abs (theta) = 3.0*pi/2.0) then halt
        else tangent := sin (theta) / cos (theta)
    end { of else clause }
end; { of function tangent }
```

Comment on the following programming characteristics of the function.

(a)  Its general applicability.

(b)  Its robustness against bad data.

(c)  Side effects.

(d)  Its use of halt and writeln.

(e)  Its treatment of real values.

Rewrite the function to overcome what you feel are its shortcomings.

6.  Rewrite the terminal room simulator developed in this chapter so it correctly handles the following changes.

   (a)  Instead of allowing only one consultant, users may specify a value for n, the number of consultants to be used in the simulation.

   (b)  Instead of having a single queue of people waiting for terminals, set up three separate queues with the following priority.

   | Faculty | (highest priority) |
   |---|---|
   | Graduate students | ↓ |
   | Undergraduates | (lowest priority) |

   The queueing discipline is always to take a person from the highest-priority queue before taking one from a lower-priority queue. Within a queue the priority is still FCFS. The simulator should produce the average waiting time for each of the preceding three classes.

   (c)  The average arrival intervals, average service time, and average consulting times should not be explicitly given by a table such as those in Figure 6.11

but should be assumed to be exponentially distributed with values $\lambda_1$, $\lambda_2$, and $\lambda_3$, respectively. The values of $\lambda$ are input to the simulator. You may assume that there is a function to generate exponentially distributed random numbers of the form:

**function** *exprandom (lambda : real) : real; extern;*
*{ generate exponentially distributed random numbers with expected value*
  *lambda }*

7. Write a program that uses discrete event simulation to determine the optimum number, n, of elevators in a building of k stories. The input should consist of the average arrival rate of people to the elevator for each of the k floors, their destination (a value 1, . . . , k), and the capacity, c, of each elevator. Generate a table of average waiting time for an elevator as a function of n. (Assume k and c are fixed quantities that cannot be changed by the user.) Determine the value of n so that the average waiting time for an elevator is less than 1 minute.

When coding the program, follow the implementation guidelines presented in this chapter. Develop the program as a series of independent, logically coherent modules without side effects.

8. Develop a program to do *text formatting*. The input will consist of text lines, intermixed with formatting comands.

.l n        set the left margin to column n
.r n        set the right margin to column n
.p          terminate the current line and begin a new paragraph
.s n   .    skip n blank lines

The output of the program should be the input text printed with left and right margins aligned according to the formatting commands. For example, the following input:

.l 1
.r 10
Now is the time
for all good men.
.s 1
Anonymous

should produce this output:

Now is    the
time       for
all       good
men.

Anonymous

Set up your program so new commands can be easily added later. After your text formatter is working properly for the preceding four commands, think about what other commands might be useful in the preparation of pleasing text. Add these new commands to the next formatter.

9. Write a program that finds its way through a maze (**program** rat?). The program should read in an nxn matrix containing the values 0, 1, and 2. 1 represents a wall that cannot be penetrated, 0 represents open space, and 2 represents the goal to be reached (a pellet of food, perhaps). The program should then read in the x, y coordinates of the starting position and attempt to find the *shortest* path to the goal. (Assume that there is only a single 2 located in the maze.)

For example, in the following maze:

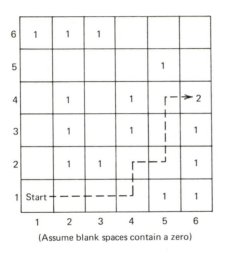

(Assume blank spaces contain a zero)

if the starting point were (1,1) then the output would be the path indicated by the dotted line:

$(1,1){\mapsto}(1,2){\mapsto}(1,3){\mapsto}(1,4){\mapsto}(2,4){\mapsto}(2,5){\mapsto}(3,5){\mapsto}(4,5){\mapsto}(4,6)$

Your program should be careful about the following situations.

(a) Going off the edge of the maze.
(b) Going around in an infinite circle [i.e., $(1,1){\rightarrow}(1,2){\mapsto}(2,2){\mapsto}(2,1){\mapsto}$ (1,1)${\mapsto}$etc.]
(c) Trying hopelessly to find a path where none exists.

# BIBLIOGRAPHY FOR PART III

Aron, J. *The Program Development Process*. Reading, Mass.: Addison-Wesley, 1974.

Baker, F. T. "Chief Programmer Teams." *Datamation,* Vol. 19, No. 12, December 1973.

Dahl, O., E. Dijkstra, and C. A. R. Hoare. *Structured Programming*. New York: Academic Press, 1972.

Dijkstra, E. *A Discipline of Programming*. Englewood Cliffs, N.J.: Prentice-Hall, 1976.

Dijkstra, E. "*Goto* Statement Considered Harmful." *Communications of the ACM,* Vol. 11, No. 3, March 1968. (One of the original papers on structured coding.)

Hughes, R., and M. Michtom. *A Structured Approach to Programming*. Englewood Cliffs, N.J.: Prentice-Hall, 1977.

Knuth, D. "Structured Programming with *Goto* Statements." *ACM Computing Surveys,* Vol. 6, 1974.

Lecarme, O. "Structured Programming, Programming Teaching and the Language Pascal." *SIGPLAN Notices,* Vol. 9, No. 7, July 1974.

Ledgard H. "The Case for Structured Programming." *BIT,* Vol. 13, 1973.

McGowan, C., and J. Kelly. *Top-Down Structured Programming Techniques*. New York: Petrocelli, 1975.

Weinberg, V. *Structured Program Analysis*. Englewood Cliffs, N.J.: Prentice-Hall, 1979.

Wirth, N. "On the Construction of Well-Structured Programs." *ACM Computing Surveys,* Vol. 6, No. 4, 1974.

Wirth, N. *Algorithms + Data Structures = Programs*. Englewood Cliffs, N.J.: Prentice-Hall, 1976.

Yourdon, E. *Techniques of Program Structure and Design*. Englewood Cliffs, N.J.: Prentice-Hall, 1977.

# Part IV

# DATA
# STRUCTURES

# Chapter 7

# LINEAR DATA STRUCTURES

## 7.1 INTRODUCTION

Most computer applications manipulate data in one form or another. Usually the data involved are not random assortments of unrelated objects; on the contrary, the objects are grouped together precisely because they share certain properties. This chapter demonstrates how to store data efficiently by capitalizing on these common properties. The most frequently used data structures include stacks, queues, linked lists, and trees. We will emphasize pictorial representations of these data structures and their implementation in Pascal. The algorithms we will study can be programmed in any available programming language; we use Pascal because it has many advantages over other languages.

Before we begin discussing data structures, we should define two important terms. First, a *data type* specifies the kind of information or data a variable may contain. Examples of data types include integer, real, boolean, and char. (See Chapter 1 for definitions of these types.) Second, a *data structure* is a collection of data objects and a set of legal operations to be performed on them.

## 7.2 LINEAR DATA STRUCTURES

### 7.2.1 Arrays

One of the simplest data structures is an **array**, which is a collection of data objects of the same data type. To declare an array in Pascal, we associate a name with it, define its permissible range of subscripts, and specify the type of its elements as shown in Section 1.4.3.3. For example, an array that contains the math SAT

(Scholastic Aptitute Test) scores of 10 students (labeled 1, 2, ..., 10) would be declared as follows:

*mathsatscores :* **array** *[1 .. 10]* **of** *integer;*

We can visualize this structure as in Figure 7.1*a*. Figure 7.1*b* represents the data structure after specific values (integers, in this case) have been stored in the array. One advantage of arrays is that they allow users to store related elements in one handy form and to refer to them with only one name. An array is a particularly simple and convenient data structure for storing "tabular" information.

The two operations associated with arrays are storage and retrieval. A storage operation enters a value into an array at the particular location defined by the subscript. A retrieval operation returns a value stored in a particular location in the array. The storage operation is usually effected by an assignment statement such as:

*mathsatscores[4] := 603*

Both retrieval and storage operations will result in an error if the subscript is outside the prescribed range.

A simple, one-dimensional array can be generalized to an n-dimensional array, each dimension of which has its own set of subscripts. An example of a two-dimensional array declaration in Pascal is:

*satscores :* **array** *[1 .. 10,1 .. 2]* **of** *integer;*

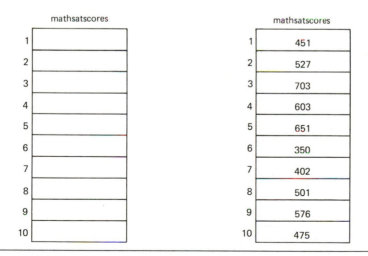

**Figure 7.1a.** Uninitialized one-dimensional array.   **Figure 7.1b.** Initialized array.

We visualize this structure as in Figure 7.2*a*. Figure 7.2*b* represents an initialization of this array. The first column contains the math SAT scores of the 10 students; the second column contains their verbal SAT scores.

We could have used two one-dimensional arrays (one for math scores and the other for verbal scores), but the advantage of the two-dimensional array is that it enables the programmer to keep two related sets of data together. We have labeled the second subscript (column index) with the numbers 1 and 2. These numbers in themselves are meaningless and force programmers to remember that column 1 contains math SAT scores and column 2 contains verbal SAT scores. To enhance the readability of the program, it is preferable to declare the array as follows.

```
type
    scores       = (math,verbal);
var
    satscores    : array [1 .. 10,scores] of integer;
```

This is less taxing on the programmers' memory and will help any future users to understand the program.

Likewise, the first subscript can be replaced with a more meaningful name by adding a constant declaration to denote the maximum number of students; that is:

```
const
    maximumstudents   = 10;
type
    scores            = (math,verbal);
var
    satscores         : array [1 .. maximumstudents,scores] of integer;
```

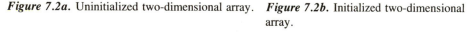

**Figure 7.2a.** Uninitialized two-dimensional array.

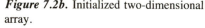

**Figure 7.2b.** Initialized two-dimensional array.

The main advantage of using the constant declaration is that if it becomes necessary to change the number of students, this change can be effected in one step. Instead of going through the program and finding every relevant occurrence of the number 10, programmers need only assign a new value to the constant maximumstudents.

An understanding of how the computer actually stores data in its memory will help programmers to format data more efficiently. The basic problem is how to map any given data structure (e.g., a two-dimensional array) into a computer's memory, which can be visualized as a very large, one-dimensional array, as shown here.

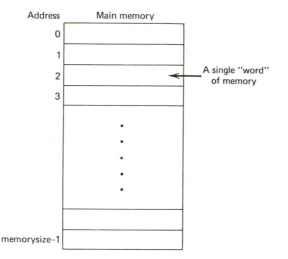

Notice that in this representation the first word of memory has an index (actually an address) of 0. There are "memorysize" words of memory in all and, therefore, the last word in memory is indexed "memorysize-1." The value of "memorysize" in binary computers is usually a power of 2.

We will now consider how one-dimensionl and two-dimensional arrays are mapped into the computer's memory. This mapping should (1) insure that retrieval of elements in the array is efficient, and (2) determine the amount of memory needed to store the array.

For one-dimensional arrays, determining how much memory to allocate is simple. We will assume for now that each element of the array will require one word of memory. The total number of words required to store the array declared as follows:

> **var**
>     *example*    : **array** [*lowerbound . . upperbound*] **of** *integer;*

is:

$$\text{upperbound} - \text{lowerbound} + 1$$

For the previous example of the one-dimensional array of math SAT scores, the number of words required is:

$$10 \text{ (upperbound)} - 1 \text{ (lowerbound)} + 1 = 10$$

(Notice that the number of words required cannot automatically be assigned the same value as the upperbound. These two values are only identical when lowerbound equals 1, which is not always the case. Recall that in Pascal, *any* integer, negative or positive, can be used to represent the lowerbound.)

When an array a is stored in sequential memory locations, the location of the jth element of the array is given by

$$loc(a[j]) = loc(a[\text{lowerbound}]) + j - \text{lowerbound}$$

Once the location of the lowerbound element is established, the location of all other elements is determined in relation to that. This formula applies only when the individual elements of the array each occupy *one* word of memory. However, it can be modified to accommodate array elements requiring more than one word. If each array element requires s words, the total number of words needed to store the array becomes:

$$s * (\text{upperbound} - \text{lowerbound} + 1)$$

The location of the jth element of the array becomes:

$$loc(a[j]) = loc(a[\text{lowerbound}]) + s * (j - \text{lowerbound})$$

(We assume in this case that loc(a[lowerbound]) refers to the first word address in which the array is stored.)

To summarize, for a general one-dimensional array:

a : **array** [*l* . . *u*] **of** *element;*

where each element requires s words, the total number of words occupied by the array is:

$$s * (u - l + 1)$$

and the location of the jth element of the array is:

$$loc(a[j]) = loc(a[l]) + (j - l) * s$$

Consider now how to map the two-dimensional array satscores into the one-dimensional memory of a computer.

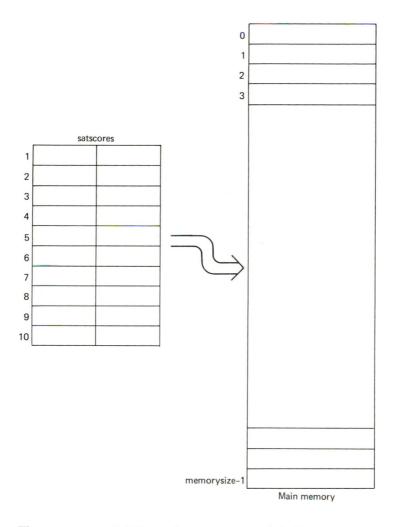

Main memory

There are two straightforward ways to accomplish this mapping. The first method is to store the first column of the array satscores, followed directly by the second column. This is called *column-major ordering*. The second method stores one row right after another and this is called *row-major ordering*. The following picture should help clarify the distinction.

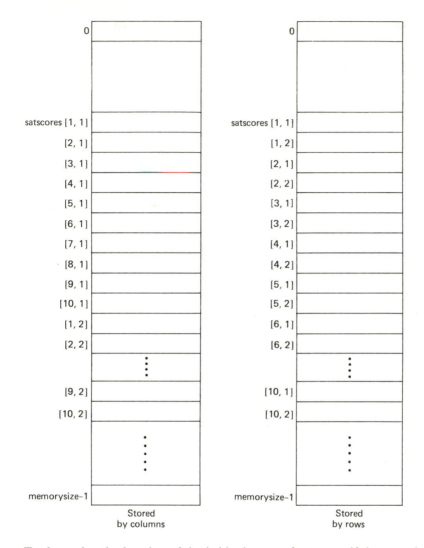

To determine the location of the i, jth element of satscores if the array is stored by columns, we first establish the location of the first element of the array (loc(satscores[1,1])). Each column contains 10 elements. The location of satscores[i,1] would be:

loc(satscores[i,1]) = loc(satscores[1,1]) + (i − 1)

Knowing the location of satscores[i,1], to compute the address of satscores[i,2], we simply add 10 to the loc(satscores[i,1]); that is:

$$loc(satscores[i,2]) = loc(satscores[i,1]) + (i - 1) + 10$$

It is easy to see how to combine the previous two formulas into one that works for $j = 1$ or $j = 2$.

$$loc(satscores[i,j]) = loc(satscores[1,1]) + (i - 1) + 10 * (j - 1)$$

For a two-dimensional array declared:

a    : **array** $[1 .. u_1, 1 .. u_2]$ **of** *integer;*

the preceding formula for computing a[i,j] if the array is stored by columns would be:

$$loc(a[i,j]) = loc(a[1,1]) + (i - 1) + u_2 * (j - 1)$$

We leave the derivation of formulas for arrays declared as:

a    : **array** $[l_1 .. u_1, l_2 .. u_2]$ **of** *integer;*

and higher-dimensional arrays as exercises.

A special kind of array in Pascal is the *string* — a packed array of characters.

**type**
     *string* = **packed array** $[1 .. maxlength]$ **of** *char;*

In some programming languages the string is implemented as a primitive data type of the language. In Pascal, as shown here, the string is implemented using the **array** data type. The primitive operations on strings, such as:

1. *Concatenation*. Merge two strings together into a single string.
2. *Length*. Determine the number of characters in a string.
3. *Pattern Match*. Find the occurrence of a special substring in a larger string.
4. *Substring Change*. Change all occurrences of one substring to another substring in a large string.

are therefore implemented as user-written procedures or functions operating on packed arrays (e.g., see Figure 3.5). We will discuss the processing of strings extensively in The Case Study at the end of this chapter.

### 7.2.2  Records

In section 7.2.1 we used two parallel, one-dimensional arrays (actually, a two-dimensional array) to contain the math and verbal SAT scores of 10 (maxstudents)

students. Each row represented two scores for one student. Pascal also offers another means of representing related data, the **record**. For our example we could put the math and verbal SAT scores into a record and then associate a record with each of the 10 students. We can associate a type name with the record as:

```
const
     maxstudents       = 10;
type
     satscores         =
          record
               math     : 0 . . 800;
               verbal   : 0 . . 800
          end; { of record satscores }
```

Then, to indicate that there are 10 students, each with a math and verbal SAT score, we simply use an array of records.

```
var
     students              : array [1 . . maxstudents] of satscores;
```

We now need a means of accessing the *fields* of the record (i.e., math and verbal). In Pascal we can accomplish this in one of two ways: through the dot notation or through the **with** statement. For example, to assign a math score of 603 and a verbal score of 550 to the first student, we could write:

```
students[1] . math : = 603;
students[1] . verbal : = 550
```

or

```
with students[1] do
begin
     math : = 603;
     verbal : = 550
end
```

Inside the scope of the **with** statement we do not need to include the record variable name, only the field name. For records with many fields the **with** statement provides a very convenient shorthand notation.

The previous example illustrated how to define arrays whose elements are records. You can also define records in which one or more fields are arrays. For example:

```
const
    maxnamelength    = 20;
var
    students            : array [1 . . maxstudents] of
        record
            math        : 0 . . 800;
            verbal      : 0 . . 800;
            name        : array [1 . . maxnamelength] of char
        end; { of record }
```

To select the first character of the third student's name we could write:

```
students[3] . name[1]
```

To write out a list of the names of the students (assuming all this information had already been initialized) we could write:

```
for i : = 1 to maxstudents do
    with students[i] do
    begin
        write(' student ',i:2, ' ');
        for j : = 1 to maxnamelength do write(name[j]);
        writeln
    end
```

It seems logical that you could nest records inside records. And, in fact, you can. For example:

```
var
    students                        : array [1 . . maxstudents] of
        record
            math                    : 0 . . 800;
            verbal                  : 0 . . 800;
            name                    : array [1 . . maxnamelength] of char;
            dateofexam              :
                record
                    month   : 1 . . 12;
                    day     : 1 . . 31;
                    year    : 1900 . . 2000
                end { of record dateofexam }
        end; { of student element record }
```

To initialize the year in which the first student took the SAT, we could write:

>  *students[1].dateofexam.year : = 1980*

or

>  **with** *students[1]* **do**
>  > *dateofexam.year : = 1980*

or

>  **with** *students[1].dateofexam* **do** *year : = 1980*

Arrays and records are different methods of structuring data. The components of an array are all of the same type and are selected by subscripts. The components of a record need not all be the same type and are specified by field names. These two data structures can be combined in any fashion to construct complex data structures.

We will use records often in the programs and program segments in this chapter. To make you more familiar with them, we will discuss a few more examples of how to use records.

Recall that Pascal has two predefined types to express numbers: integer and real. But what if you needed to manipulate complex numbers (i.e., numbers of the form:

$$a + b * i$$

where a and b are real numbers and i is the square root of $-1$). Unlike FORTRAN, Pascal has no predefined way of handling complex numbers. It is easy to remedy this situation with a user-defined type.

>  **type**
>  > *complex*        =
>  > > **record**
>  > > > *a*      *: real; { real part }*
>  > > > *b*      *: real  { imaginary part }*
>  > > **end***; { of record complex }*

(We do not need to represent the i.)

Now we can define variables to be of this new type.

>  **var**
>  > *x*              *: complex;*
>  > *y*              *: complex;*

Arithmetic operations can be performed on complex numbers. For example, the sum of the complex numbers

$$a + bi \text{ and } c + di$$

is

$$(a + c) + (b + d)i$$

We can easily write a procedure to add two complex numbers.

```
procedure complexadd(x : complex; y : complex; var z : complex);
begin
    z.a := x.a + y.a;
    z.b := x.b + y.b
end; { of procedure complexadd }
```

The other arithmetic operations on complex numbers can also be treated easily.

As a second example of using records, consider the following problem. Design a data structure that will help the police department maintain a table of stolen cars by precinct. The following information is required: license plate number, car description (year and make), and whether or not the car was found. If you were to draw a picture of the information required by each precinct it might look like this.

| License plate | Car description | | Found |
|---|---|---|---|
| | Year | Make | |
| | | | |
| | | | |
| | | | |
| | | | |
| | | | |

Now let us try to encode this into a meaningful Pascal data structure step by step. First, since we want the same structure for each precinct, we will declare a constant for the number of precincts:

```
const
    numberofprecincts      = 9;
```

and an array of records (or tables), one for each precinct:

```
type
    bigtable                   = array [1 .. numberofprecincts] of precincttables;
```

Each "precinct table" will be an array that can accommodate up to some maximum number of stolen cars; hence:

```
const
    maxstolencars              = 100;
type
    precincttables             = array [1 .. maxstolencars] of cardescription;
```

Now all we need do is describe what information is required for each stolen car.

```
type
    alfa                       = packed array [1 .. 10] of char;
    cardescription             =
        record
            licenseplate    : alfa;
            description     :
                record
                    year    : integer;
                    make    : alfa
                end; { of record description }
            found               : boolean
        end { of record cardescription };
```

Then, to allocate the space for this data structure in memory, we write:

```
var
    stolencars                 : bigtable;
```

Unfortunately, we cannot just write the description of the data structure down the way we envisioned it because identifiers in Pascal must be declared before they are referenced. (However, see Chapter 8 for an exception to this rule.) Therefore the declarations for the stolen car data structure should read:

```
const
    maxstolencars              = 100;   { max stolen cars per precinct }
    numberofprecincts          = 9;     { number of precincts in city }
```

```
type
    alfa                          = packed array [1 .. 10] of char;
    cardescription                =
        record
            licenseplate     : alfa;
            description      :
                record
                    year     : integer;
                    make     : alfa
                end; { of record description }
            found                : boolean
        end; { of record cardescription }
    precinctables                 = array [1 .. maxstolencars] of cardescription;
    bigtable                      = array [1 .. numberofprecincts] of precincttables;
var
    stolencars                    : bigtable;
```

How do you set the year and make of the third stolen car in precinct five to 1971 and Pontiac, respectively? There are, of course, a number of ways.

```
stolencars[5][3].description.year := 1971;
stolencars[5][3].description.make := 'pontiac'

with stolencars[5,3] do
begin
    description.year := 1971;
    description.make := 'pontiac '
end { of with }

with stolencars[5] [3] do
    with description do
    begin
        year := 1971;
        make := 'pontiac '
    end
```

or

```
with stolencars[5,3].description do
begin
    year := 1971;
    make := 'pontiac '
end { of with }
```

The entire data structure would be laid out in the machine's memory as follows.

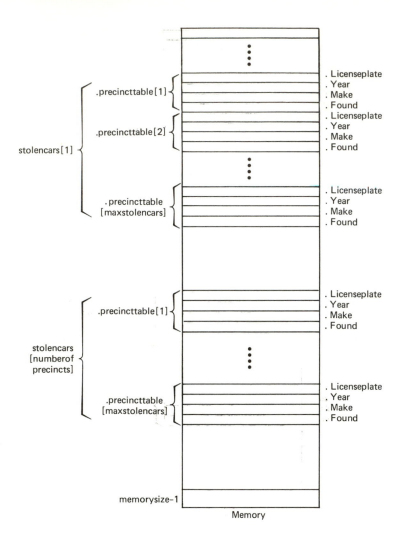

Memory

As a final example of the use of records, we will ask you to design a data structure to be used in playing a type of solitare called "Idiot's Delight." Represent the playing cards as records with one field for the suit and another for the rank.

Idiot's Delight is played by dealing cards face up from a shuffled deck onto a hand (or pile). Whenever the suit of the top card matches that of the fourth card in the pile, the two intervening cards are discarded; when the rank of the top card matches the rank of the fourth card in the pile, all four cards are discarded. The object of the game is to be left with no cards in the hand.

To make sure the rules are understood, we will play an abbreviated game. First, shuffle the deck. Then deal off four cards.

↑
top

Since there is neither a suit nor a rank match, we deal off another card

↑
top

Now, since 5 ♦ and 5 ♣ have the same rank, we discard the top four cards and are left only with the 8 ♥. We deal off three more cards and continue in the same manner.

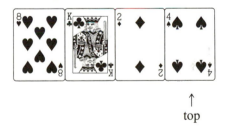

↑
top

At this point, stop and try to write a program that plays, say, 1000 random games of Idiot's Delight. Keep a histogram to show the distribution of the number of undiscarded cards. The choice of a good data structure for the card pile in this program will greatly influence the clarity of the resulting solution. Once you have tried to write the program yourself, read through the following program, which is based on one written by the late Michael Machtey of Purdue University.

```
program idiotsdelight(input,output);
const
    decksize                = 52;          { number of cards in a deck }
    suitsize                = 13;          { number of cards in a suit }
    widthmax                = 78;          { max number of chars across
                                              histogram }

type
    suits                   = (clubs,diamonds,hearts,spades);
    ranks                   = (ace,two,three,four,five,six,seven,eight,nine,ten,
                               jack,queen,king);
    card                    =

        record
            suit            : suits;
            rank            : ranks;
        end; { of record card }
    cards                   = 1 .. decksize;
    deck                    = array [cards] of card;
    histsize                = 0 .. decksize;

var
    cardnumber              : cards:       { one of the fifty-two cards }
    cardsleft               : histsize;    { number of cards left }
    column                  : integer;     { current column of histogram }
    deckofcards             : deck;        { the deck }
    game                    : integer;     { the current game number }
    hand                    : deck;        { deal cards from deck to hand }
    histogram               : array [histsize] of integer;
                                           { keeps track of number of cards left
                                              after each game }
    i                       : integer;     { index variable }
    numberofgames           : integer;     { number of games to play }
    rankofcard              : ranks;       { index variable }
    suitofcard              : suits;       { index variable }
    top                     : histsize;    { points to top card of hand }
    trying                  : boolean;     { true when legal game moves are
                                              possible }

function random : real; extern;
```

```
{ shuffledeck

        shuffle a deck of cards

        entry conditions:
            deckofcards        : array of records representing card deck

        exit conditions:
            the deck is shuffled
}
procedure shuffledeck(var deckofcards : deck);
var
        cardnumber              : cards;        { random card number }
        i                       : cards;        { temporary card index }
        temp                    : card;         { card in deck
begin
        for i := 1 to decksize do
        begin
            { select a random card in deck }
            cardnumber := trunc(decksize * random) + 1;

            { exchange that card with the ith card in deck }
            temp := deckofcards[i];
            deckofcards[i] := deckofcards[cardnumber];
            deckofcards[cardnumber] := temp
        end { of for }
end; { of procedure shuffledeck }

begin { program idiot's delight }
        { initialize the card deck }
        for suitofcard := clubs to spades do
            for rankofcard := ace to king do
            begin
                { convert a suit and rank into a cardnumber }
                cardnumber := ord(rankofcard) + 1 + ord(suitofcard) * suitsize;

                { initialize the suit and rank of this card }
                with deckofcards[cardnumber] do
                begin
                    suit := suitofcard;
                    rank := rankofcard
                end { of with }
            end; { of for }

{ initialize the histogram }
for cardsleft := 0 to decksize do histogram[cardsleft] := 0;
```

```
{ read the number of games to play }
writeln(' input the number of games to play (as an integer)');
read(numberofgames);

writeln(' I will now play ',numberofgames:1, ' games of idiot''s, delight. ');

{ play the games keeping the results in the histogram array }
for game := 1 to numberofgames do
begin
      { shuffle the deck at start of each game }
      shuffledeck(deckofcards);

      { play the game }
      top := 0;
      for cardnumber := 1 to decksize do
      begin
            top := top + 1;
            hand[top] := deckofcards[cardnumber];
            trying := true;

            while trying and (top >= 4) do
            begin
                  if hand[top].suit = hand[top – 3].suit then
                  begin
                        { suits match – discard intervening two cards }
                        hand[top – 2] := hand[top];
                        top := top – 2
                  end
                  else
                  if hand[top].rank = hand[top – 3].rank then
                  begin
                        { ranks match – discard top 4 cards }
                        top := top – 4
                  end
                  else trying := false
            end { of while }
      end; { of for }
      histogram[top] := histogram [top] + 1
end; { of for }
```

```
{ print the histogram }
writeln(' ':11,'out of ',numberofgames:1,' games of idiot's delight played, ');
writeln(' ':11, 'the distribution of the number of undiscarded cards is:');
writeln;
writeln(' cards');
writeln(' left    frequency');
for cardsleft := 0 to decksize div 2 do
begin
    { recall that only an even number of cards can be left }
    write(cardsleft * 2:4,histogram[cardsleft * 2]:6);
    write(' ');
    column := 11; { 4 + 6 + 1 }
    for i := 1 to histogram[cardsleft * 2] do
    begin
        if column >= widthmax then
        begin { start overflow line }
            writeln;
            write(' ':11);
            column := 11
        end; { of line too long }
        write('*');
        column := column + 1
    end; { of for i }
    writeln
end { of for }
end. { of program idiot's delight }
```

### 7.2.3  Stacks

Although arrays are simple and useful and an integral part of every programming language, they cannot solve every data structuring problem. Some data are more amenable to another data structure, the *stack*. A stack of data can be visualized very much like a stack of china plates. Imagine a dishwasher stacking the dishes after washing them, and a busperson taking dishes off the top as they are needed. Although data are not literally stacked in the computer, the effect is the same. As with the china plates, those on top of the stack are accessed first and those on the bottom last (last-in-first-out). At any given point, the dish (or data item) at the top of the stack is the most important because it will be accessed next (see Figure 7.3). Figure 7.3*a* shows a stack with six elements. Figure 7.3*b* shows the same stack with one additional element added. Note that the top pointer always points to the most recently added element.

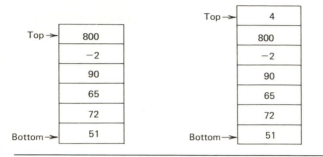

**Figure 7.3a.** A nonempty stack.    **Figure 7.3b.** Same stack with one more element.

It is useful to keep this picture of a stack in mind. There are many ways to implement a stack, but the visual analogy with the china plates applies regardless of which implementation is used.

Stacks occur naturally in the execution of computer programs. Whenever a procedure or function is called, the computer must save the address of the instruction immediately following the procedure call statement (the return address) in order to return to that statement after executing the procedure. When a procedure is called, the return address is placed on an execution time stack. After the procedure is executed, the return address is removed from the stack, and control is transferred to that location.

Stacks are especially useful when several procedures are used in succession (procedure a calls procedure b, which calls procedure c, etc.). In this case, the return addresses appear in the stack in the reverse order (c, b, a). After each procedure is executed, its return address is removed from the stack, and control is returned to this address. This process demonstrates the utility of the last-in-first-out property of stacks.

One way to implement a stack is with a one-dimensional array. In our example, the elements of the array will be of type char. There is a pointer (an index into the array) positioned at the topmost element of the stack. The declarations for the stack are:

```
const
      stacklowerbound   = 1;        { lower bound of stack }
      stackupperbound   = 100;      { upper bound of stack }
      lowerminus1       = 0;        { stacklowerbound − 1 }
type
      stacklimits       = stacklowerbound . . stackupperbound;
      toplimits         = lowerminus1 . . stackupperbound;
      stackarray        = array [stacklimits] of char;
var
      stack             : stackarray;   { the stack }
      top               : toplimits;    { top pointer }
```

Three operations can be performed on a stack: (1) "pushing" an element on top of a stack, (2) "popping" (removing) the topmost element from the stack, and (3) testing whether or not a stack is empty. The procedure to push an element on a stack must first check if the stack is full (i.e., if adding one more element will exceed the capacity of the stack). Likewise, the procedure to pop an element from a stack must first check if there is at least one element left to be removed. Obviously, the "pop" operation cannot be performed on an empty stack.

The push procedure is:

```
procedure push(item : char; var stack : stackarray; var top:
      toplimits; upperbound : stacklimits; var flag : boolean);
begin
      if top < upperbound then          { available space in stack }
      begin
            top := top + 1;             { update top pointer }
            stack[top] := item;         { store item on top of stack }
            flag := true                { push operation successful }
      end
      else flag := false                { indicate overflow }
end; { of procedure push }
```

The stack is said to "overflow" when one attempts to push an item onto a full stack.

The pop operation is implemented as a procedure in which the top pointer is updated and the current top of the stack is returned through the parameter item.

```
procedure pop(var item : char; stack : stackarray; var top:
      toplimits; lowerbound : stacklimits; var flag : boolean);
begin
      if top >= lowerbound then         { stack is not empty }
      begin
            item := stack[top];         { return top of stack }
            top := top - 1;             { update top pointer }
            flag := true                { pop operation successful }
      end
      else flag := false                { indicate underflow }
end; { of procedure pop }
```

With this implementation a stack is empty if $top = lowerbound - 1$; otherwise, it contains at least one element. Attempting to pop off a nonexistent item from the stack is referred to as "underflow."

An alternate way of implementing a stack is to place all the information pertinent to the stack into a Pascal record. This method reduces to two the number of parameters to be passed into the push and pop procedures. The push procedure and its associated declarations become:

```
const
     stacklowerbound       = 1;          { lower bound of stack }
     stackupperbound       = 100;        { upper bound of stack }
     lowerminus1           = 0;          { stacklowerbound − 1 }
type
     stacklimits           = stacklowerbound .. stackupperbound;
     toplimits             = lowerminus1 .. stackupperbound;
     stackarray            = array [stacklimits] of char;
     stackrecord           =
         record
             stacked      : stackarray;  { the stack }
             top          : toplimits;   { top pointer }
             lowerbound   : stacklimits; { stack lower bound }
             upperbound   : stacklimits; { stack upper bound }
             successful   : boolean      { status of operation }
         end; { of record stackrecord }
var
     stack                 : stackrecord;
```

The procedure to push an item onto the stack then becomes:

```
procedure push(item : char; var stack : stackrecord);
begin
     with stack do
     begin
         if top < upperbound then          { available space in stack }
         begin
             top := top + 1;               { update top pointer }
             stacked[top] := item;         { store item on top of stack }
             successful := true            { push operation successful }
         end
         else successful := false          { indicate overflow }
     end { of with stack }
end; { of procedure push }
```

The pop procedure is:

```
procedure pop(var item : char; var stack : stackrecord);
begin
    with stack do
    begin
        if top >= lowerbound then
        begin
            item := stack[top];              { remove item from top of stack }
            top := top - 1;                  { update top pointer }
            successful := true               { pop operation successful }
        end
        else successful := false             { indicate underflow }
    end { of with stack }
end; { of procedure pop }
```

Before procedure push or pop can be called, the stack must be initialized by:

```
with stack do
begin
    top := lowerminus1;
    lowerbound := stacklowerbound;
    upperbound := stackupperbound
end
```

### 7.2.4  Examples of Stacks

A stack can be used to recognize if a particular string is written in a language defined by some grammar. This is what a compiler does when it parses a program to determine if its input string (the program) is syntactically correct. For example, to recognize the set of strings defined by the rule:

some string, w, followed by the character ''c'',
followed by the string w in reverse order

where w is a string of 0 or more characters and w reverse is the string read backward, we can employ a stack. In this language the string 123c321 is legal, but the string 123c312 is not legal because it does not conform to the prescribed pattern.

To determine if a particular string is legal, we read it one character at a time, pushing each character on the stack until we reach the character c. At this point, we continue to read the string and pop off the top element of the stack as each character is read. If the character read is not identical to this top element, the string is illegal. If all the characters match and the stack is empty after the last character has been read, the string is legal.

The following Pascal program reads a string [which is followed by a dollar sign ($) to mark its end] and determines if the string is legal.

```pascal
program wcwreverse(input,output);
const
    stacklowerbound        = 1;          { stack lower bound }
    stackupperbound        = 100;        { stack upper bound }
    lowerminus1            = 0;          { stack lower bound − 1 }
type
    stacklimits           = stacklowerbound .. stackupperbound;
    toplimits             = lowerminus1 .. stackupperbound;
    stackarray            = array [stacklimts] of char;
    stackrecord           =
        record
            stacked       : stackarray;
            top           : toplimits;
            lowerbound    : stacklimits;
            upperbound    : stacklimits;
            successful    : boolean
        end; { of record stackrecord }
var
    ch                    : char;        { character of string }
    match                 : boolean;     { true while string is ok }
    overflow              : boolean;     { true if stack overflow }
    stack                 : stackrecord; { the stack }
    temp                  : char;        { temporary }
    underflow             : boolean;     { true if stack underflow }

procedure push(item : char; var stack : stackrecord);
begin
    with stack do
    begin
        if top < upperbound then
        begin
            top := top + 1;
            stacked[top] := item;
            successful := true
        end
        else successful := false
    end { of with stack }
end; { of procedure push }

procedure pop(var item : char; var stack : stackrecord);
begin
    with stack do
    begin
        if top >= lowerbound then
        begin
            item := stacked[top];
            top := top − 1;
            successful := true
        end
        else successful := false
    end { of with stack }
end; { of procedure pop }
```

```
begin
      { initialization }
      with stack do
      begin
            top := lowerminus1;
            lowerbound := stacklowerbound;
            upperbound := stackupperbound
      end;
      overflow := false;
      underflow := false;

      if not eof then read(ch);
      while (ch <> 'c') and (not overflow) and (not eof) do
      begin
            push(ch,stack);
            overflow := not stack.successful;
            read(ch)
      end; { of while }

      if (not eof) and (not overflow) then
      begin
            read(ch);
            match := true;
            while match and (ch <> '$') and (not eof) and (not underflow) do
            begin
                  pop(temp,stack);
                  underflow := not stack.successful;
                  if not underflow then
                        if temp <> ch then match := false else read(ch)
            end { of while }
      end; { of if }

      if match and (ch = '$') and (stack . top = lowerminus1) then
            writeln(' legal string ');
      if overflow then writeln(' stack overflow; increase stack space');
      if eof then writeln(' end of file encountered; did you leave off the $');
      if underflow or (not match) then writeln(' illegal string ')
end.
```

A more elegant way to solve the same problem is by using a *recursive* function (i.e., a function that calls itself ). Before we write the recursive version of the program, we will briefly review a simple example of a recursive function—one to evaluate n! (called n factorial). n! is defined by:

$$n! = n * (n - 1)!$$
with    $1! = 1$
and    $0! = 1$    by definition.

A Pascal function to evaluate n! is:

```
function nfactorial(n : integer) : integer;
begin
    if (n = 0) or (n = 1) then nfactorial := 1
    else nfactorial := n * nfactorial(n − 1)
end; { of function nfactorial }
```

Execution of nfactorial(3) invokes another instance of nfactorial with an argument of 2, and this second invocation results in a call to nfactorial with the argument 1. The recursive calls stop at this point, since nfactorial(1) evaluates to 1. The second instance of the function (with n = 2) receives a value 1 from the third instance [nfactorial(1)], multiplies it by 2, and returns the product $2 * 1 = 2$ as the value of nfactorial(2) requested by the original instance of nfactorial with the argument 3. Hence the final value of the function will be $3 * 2 = 6$.

In general, every recursive call to a procedure will require us to save the return address and restart execution of the same procedure. When we complete the execution of the procedure, we are not yet through. We must still complete any previously suspended invocations of this procedure, and we must finish them in the *reverse* order in which they were saved.

Thus a stack is the ideal data structure to help us implement recursion in our language. A recursive procedure call [e.g., nfactorial(2) or nfactorial(1) from the preceding] will cause the following operation.

```
push ("the return address on the stack")
```

Whenever we complete execution of the procedure, we would execute the following:

```
if "stack is not empty" then
    pop ("a return address off the stack")
```

and continue execution of the procedure from that point.

It would be a good idea to go back and retrace the execution of nfactorial(3), keeping track of the state of the stack and the top pointer.

Returning to the example of recognizing if a string is of the correct form, study the following program until you can convince yourself that the recursive function match will accept only legal strings. Try to trace through this program using the following strings.

$$aca\$, \ 11c12\$, \ abcdd\$, \ and \ c\$$$

```
program accept(input,output);

{ this is a recursive program to recognize a string of the form: w'c'm$
    where w is some string, 'c' is the character c, and m is the reverse of w. note
    that w may not contain the letter 'c' or a '$'

    jon spear – 8 feb 80
}

function match : boolean;
var
    ch1                    : char;        { first char of a symmetric pair }
    ch2                    : char;        { last char of a symmetric pair }
    t                      : boolean;     { holds value returned by recursive call }
begin
    if not eof then
    begin
        read(ch1);
        if ch1 <> 'c' then
        begin
            t := match;
            if t and not eof then
            begin
                read(ch2);
                match := (ch1 = ch2)
            end
            else match := false
        end
        else match := true { found a c }
    end
    else match := false
end; { of function match }

begin { program accept }
    { note that complete left-to-right conditional expression is assumed since
        match has the side effect of advancing the input pointer. these semantic
        assumptions appear to be consistent with the Pascal standard, but could
        be confusing . . .
    }
    if match then
    begin
        if input ↑ = '$' then writeln(' legal string')
        else writeln(' no dollar sign')
    end
    else writeln(' illegal string')
end. { of program accept }
```

Stacks are also used to convert expressions from infix notation to postfix notation. An infix expression is an algebraic expression in which the operators are located between the operands. In postfix notation the operator immediately follows the operands. Examples of infix expressions are:

$$a * b, f * g - b, d / e * c + 2$$

The corresponding postfix expressions are:

$$ab*, fg*b-, de/c*2+$$

Notice that the operands in a postfix expression occur in the same order as in the corresponding infix expression. Only the position of the operators is changed.

Some operators require two operands (e.g., $+$, $-$, $*$, and $/$). In a postfix expression these operands directly precede their operators. But, more important, the order in which the arithmetic operators are applied is explicit in the postfix expression. As an example, in the infix expression:

$$a + b * c$$

the rules of Pascal dictate that the multiplication $b * c$ is performed before the addition. We say that multiplication takes precedence over additiion. The corresponding postfix expression is:

$$a\ b\ c * +$$

It is now clear that the multiplication operator applies to the operands b and c. The addition operation is then applied to the operand a and the result of $b * c$.

To convert an infix expression to a postfix expression, we use a stack to contain operators that cannot be output until their respective operands are output. In the exercises at the end of this chapter you will be required to design and write an infix to postfix program.

### 7.2.5  Queues

Queues are familiar from everyday life—we can hardly avoid waiting in a queue (or line). Queues develop whenever people, work, computer jobs, and the like, must wait for some service. We used the idea of a queue in the case study at the end of Chapter 6.

The first person to enter a queue is also the first to leave it. Hence queues are often called first-in-first-out (FIFO) lists.

A queue is a linear list for which all insertions are made at one end. All deletions (and usually all accesses) are made at the other end.

This definition implies that two positions in a queue are especially important: the front and the rear. We identify these positions by pointers as in the following queue.

```
a    b  c  d  e  f    g
↑                     ↑
Front                 Rear
```

The next element to be deleted will be a, which would change the picture to:

```
b      c  d  e  f    g
↑                    ↑
Front                Rear
```

One way to implement a queue is with an array. For this implementation we must specify the maximum array length (maximum number of queued elements), the type of the queued elements (in this case, characters), and front and rear pointers (indices into the array). The necessary Pascal declarations follow.

```
const
    queuelowerbound          = 1;          { lower bound of queue }
    queueupperbound          = 100;        { upper bound of queue }
    lowerminus1              = 0;          { queue lower bound − 1 }
type
    limits                   = lowerminus1 . . queueupperbound;
    queuelimits              = queuelowerbound . . queueupperbound;
    queuearray               = array [queuelimits] of char;
    queuerecord              =
        record
            queued           : queuearray;    { the queue }
            front            : limits;        { front pointer }
            rear             : limits;        { rear pointer }
            lowerbound       : queuelimits;   { queue lower bound }
            upperbound       : queuelimits;   { queue upper bound }
            successful       : boolean        { status of operation }
        end; { of record queuerecord }
var
    queue                    : queuerecord;
```

An initially empty queue is represented by front = rear = lowerminus1.

Rear always points to the last element in the queue. Because of our previous assumption, we cannot let the front pointer point to the first element in the queue;

otherwise, a one-element queue would have front and rear pointing to the same element—implying an empty queue. To resolve this, we will assume that front is always *one less* than the actual front of the queue. Our previous example would then have the following pictorial representation.

```
            a  b  c  d  e  f    g
        ↑                       ↑
      Front                   Rear
```

*Enqueueing* is the operation by which an element is added to a queue. Under our assumptions, the procedure is:

```
procedure enqueue(item : char; var queue : queuerecord);
begin
    with queue do
    begin
        if rear < upperbound then
        begin
            rear := rear + 1;              { update pointer to rear of queue }
            queued[rear] := item;          { enter item on queue }
            successful := true             { successful enqueue operation }
        end
        else successful := false           { indicate overflow }
    end { of with queue }
end; { of procedure enqueue }
```

The operation to delete an element from a queue is called *dequeueing* and is implemented as follows.

```
procedure dequeue(var item : char; var queue : queuerecord);
begin
    with queue do
    begin
        if front <> rear then
        begin
            front := front + 1;            { update pointer to front of queue }
            item := queued[front];         { remove item at head of queue }
            successful := true             { successful dequeue operation }
        end
        else successful := false           { indicate underflow }
    end { of with queue }
end; { of procedure dequeue }
```

Before either procedure enqueue or dequeue is called, the queue must be initialized by:

```
with queue do
begin
        front := lowerminus1;
        rear := lowerminus1;
        lowerbound := queuelowerbound;
        upperbound := queueupperbound
end.
```

Unfortunately, this implementation is seriously flawed. Since both front and rear migrate to the end of the array, it is possible for rear = queueupperbound when there may actually be space in the array to contain enqueued elements. For simplicity, let us see what happens to a queue of size 3 (i.e., the constants queuelowerbound and queueupperbound are set to 1 and 3, respectively). If we attempted to enqueue four elements in succession, the queue would overflow, which is exactly what we would want to happen.

But if, instead, we had the following sequence of operations:

```
enqueue('a',queue);
dequeue(item,queue);
enqueue('c',queue);
enqueue('f',queue);
enqueue('g',queue) { This will improperly cause an overflow }
```

we see that in fact there is a free location, pointed to by front, which could be used to contain the 'g.'

We could shift all elements "up" in the array whenever rear = queueupperbound and front<>queueupperbound and front<>queuelowerbound, but this is a time-consuming operation, especially if queueupperbound is large.

It is better to consider an alternate implementation in which we view the queue as a ring (see Figure 7.4). The elements of the queue are still stored in an array whose declarations may be:

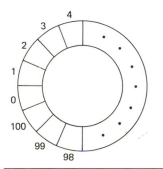

**Figure 7.4.** Queue viewed as a ring.

```
const
    queuelowerbound                    = 0;          { queue lower bound }
    queueupperbound                    = 100;        { queue upper bound }
type
    queuelimits                        = queuelowerbound .. queueupperbound;
    queuerecord                        =
        record
            queued                     : array [queuelimits] of char;
            front                      : queuelimits;
            rear                       : queuelimits;
            lowerbound                 : queuelimits;
            upperbound                 : queuelimits;
            successful                 : boolean
        end; { of record queuerecord }
```

The procedures to add an element to the queue (enqueue) and remove an element from the queue (dequeue) are coded as follows.

```
procedure enqueue(item : char; var queue : queuerecord);
begin
    with queue do
    begin
        if rear = upperbound then rear := lowerbound
        else rear := rear + 1;
        if front <> rear then                       { space available in queue }
        begin
            queued[rear] := item;                   { enter item on queue }
            successful := true                      { successful enqueue }
        end
        else successful := false                    { indicate queue overflow }
    end
end; { of procedure enqueue }

procedure dequeue(var item : char; var queue : queuerecord);
begin
    with queue do
    begin
        if front <> rear then                       { non-empty queue }
        begin
            if front = upperbound then front := lowerbound
            else front := front + 1;
            item := queued[front];                  { remove item from queue }
            successful := true
        end
        else successful := false                    { indicate queue underfow }
    end
end; { of procedure dequeue }
```

Before either procedure enqueue or dequeue is called, the queue must be initialized by:

```
with queue do
begin
      front : = queuelowerbound;
      rear : = queuelowerbound;
      lowerbound : = queuelowerbound;
      upperbound : = queueupperbound
end
```

Notice that in this implementation we have allocated 101 elements for the array queued, but the maximum number of items in a full queue is 100.

### 7.2.6  A Queue-t Example

Consider the following problem. You are given a stack and asked to reverse its contents, as in the following example.

How would you accomplish this task? An elegant solution (although it requires more storage than is actually needed), employs a queue. You simply pop off elements of the stack one at a time and enqueue them. Then dequeue the elements one at a time and push them onto the stack.

```
while stack.top <> lowerminus1 do
begin
      pop(item,stack);
      enqueue(item,queue)
end; { of while }

while queue.front <> queue.rear do
begin
      dequeue(item,queue);
      push(item,stack)
end { of while }
```

## 7.3 CASE STUDY—A LEXICAL SCANNER

In Pascal a program is simply a string of characters. A Pascal compiler reads the string and determines if it is syntactically correct. To accomplish this, the compiler divides the input string into substrings (or *tokens*).

For example, in the program segment:

**var**
         *i : integer;*
**begin**
         *i := 5*
**end**

the compiler would collect the first string of characters (**var**) into a token. The next tokens the compiler finds as it processes the input text are:

*i, :, integer, ;,* **begin**, *i, := , 5,* **end**,

The process by which a program (such as a compiler) processes strings into tokens is usually called *lexical analysis*, or *scanning*.

A lexical scanner has several purposes, including:

1. The conversion from characters representing a number to the internal binary representation of that number; for our purposes the numbers may be integers, reals, or octal numbers.
2. Collecting together the characters of a nonnumeric string.
3. Processing single-character special strings (:, *, + , − , etc.) or double-character special strings (e.g., := ).
4. Recognizing literal strings surrounded by quotation marks.
5. Detecting and recovering from errors in processing illegal strings.

To see how integers are recognized and converted, consider the string:

1675

At this point you may be reading this as the number one thousand six hundred seventy five. But the compiler (or scanner) will read this line *one character* at a time. So it will read the *character* 1, not the *number* 1, the character 6, not the number 6, and so forth. The distinction between the character representation of a number and its numeric value is often blurred, but here it is crucial.

First, we consider this specific example (converting from the characters 1, 6, 7, and 5 to the number 1675). Later we will generalize the process. Assume that we declare:

**var**
        *ch          : char;*
        *number    : integer;*

and issue:

*read(ch)*

where ch = '1'. To convert the character '1' to the number 1, we could use:

*number := ord(ch) − ord('0')*

where ord is a standard Pascal function that yields the ordinal number of the character passed as its argument; on a CDC/Cyber series machine:

ord('0') = 33b   { b denotes an octal, or base 8, number }
ord('1') = 34b
ord('2') = 35b
ord('3') = 36b
        .
        .
        .
ord('9') = 44b

Now you should be able to see why the numeric value of the character '1' can be determined using:

ord('1') − ord('0')      (34b − 33b = 1)

We can read the next character of the number by again executing:

*read(ch)*

Now, if we execute:

*number := ord(ch) − ord('0')*

as before, we lose the value of the previous digit. To retain it and its relative numerical value, we perform:

*number := 10 * number + ord(ch) − ord('0')*

instead. We can continue in the same manner until the entire number has been converted. To recap, we can convert the characters 1, 6, 7, and 5 into the number 1675 by:

```
read(ch); { ch = '1' }
number := ord(ch) − ord('0'); { number = 1 }
read(ch); { ch = '6' }
number := 10 * number + ord(ch) − ord('0'); { number = 16 }
read(ch); { ch = '7' }
number := 10 * number + ord(ch) − ord('0'); { number = 167 }
read(ch); { ch = '5' }
number := 10 * number + ord(ch) − ord('0'); { number = 1675 }
```

It should be obvious that an iterative procedure is called for. The iteration will terminate when we read a character that is not a digit. For now we assume we have a function that performs the following task.

numeric(ch) returns true if ch in '0' . . '9' and false otherwise

Then our iterative process might look like:

```
read(ch);
if numeric(ch) then
begin
    number := 0;
    repeat
        number := 10 * number + ord(ch) − ord('0');
        read(ch)
    until not numeric(ch)
end
```

Do you see anything wrong with this program segment? If file input were empty, the first read(ch) would attempt to read past the end of file—which is a fatal run-time error. Since we do not want always to check the end-of-file status before reading the next character from the input file, we will assume that we have written a procedure getnextchar that does the checking for us; it will return the next character on the input file if there is one, or some type of nonnumeric, nonalphabetic character if the end-of-file or end-of-line has been encountered. A call to getnextchar(ch) would then replace calls to read(ch).

Next, assume that an octal number is followed by the letter b, as in:

17b

This may seem acceptable since, once we fall out of the **repeat** loop, we can check if ch = 'b'. If it does, the value in the variable number will be interpreted as 17 *octal*, not 17 decimal. You might think the simple thing to do would be to write a conversion procedure that would convert the decimal value into an octal value. Unfortunately, this overlooks one major problem: octal numbers (base 8) can only contain the digits 0 to 7. Hence a number such as 1825b is illegal. Our scanner must be designed to recognize such illegal values. It would be nice if we could read the input file backward to see if the number had an 8 or a 9, but this is not possible.

A simple alternative is, once a digit is recognized, to place it and any succeeding digits into an array that will be rescanned after we know whether or not the string of digits is followed by a 'b'; that is:

```
getnextchar(ch);
if numeric(ch) then
begin
     i := 0;
     repeat
          i := i + 1;
          { note : we are converting characters representing a digit to the digit
             itself here }
          digit[i] : = ord(ch) − ord('0');
          getnextchar(ch)
     until not numeric(ch);

     number : = 0;
     if ch = 'b' then                        { convert as an octal number }
     begin
          k := 1;
          repeat
               if digit[k] in [8,9] then
               begin
                    error : = true;
                    k := i
               end
               else number : = 8 * number + digit[k];
               k := k + 1
          until k > i
     end
     else
     begin
          k := 1;
          repeat
               number : = 10 * number + digit[k];
               k := k + 1
          until k > i
     end
end
```

Study this program segment, noting the difference between converting octal numbers and decimal numbers. After carefully reviewing the code, can you spot any other potential problems? What happens if the numeric value of the number being scanned is greater than maxint? A simple solution is to check that the value in the variable number does not exceed maxint. Consult the full scanner program at the end of this chapter for one way of doing this.

We have been describing a method for recognizing (and converting) integers and octal numbers. The process can be diagrammed as:

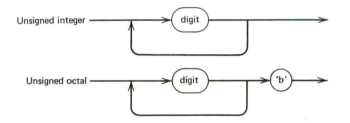

To be complete, our scanner must be able to recognize real numbers as well. Recall (see Section 1.2.1.2) that the following are all legal real numbers.

1.0     1.1E+5     6.7E–6     15E5

The diagram for real numbers looks like this.

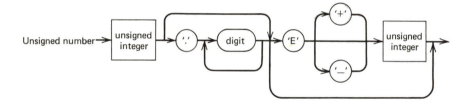

Because the techniques we use to recognize reals are very similar to the process just described for scanning integers and octal numbers, you should study the scanner program at the end of the chapter.

The scanner should also recognize tokens in the form of identifiers (i.e., alphabetic characters followed by 0 or more alphabetic or numeric characters).

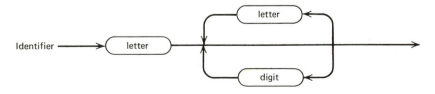

The scanner will return these characters in an array that is dimensioned to hold tokenlenmax characters.

```
const
    tokenlenmax    = 80;
var
    tokenbuffer    : packed array [0 .. tokenlenmax] of char;
    j              : 0 .. tokenlenmax;
```

If we assume the function alphabetic has been defined as:

alphabetic(ch) returns true if ch in 'a' . . 'z', false otherwise

the portion of the scanner that recognizes identifiers is simple.

```
getnextchar(ch);
if alphabetic(ch) then
begin
    j := 0;
    repeat
        if j < tokenlenmax then
        begin
            j := j + 1;
            tokenbuffer[j] := ch
        end;
        getnextchar(ch)
    until not (alphabetic(ch) or numeric(ch))
end
```

Notice that if an indentifier is longer than tokenlenmax characters, those characters will be read and ignored.

The scanner we have implemented incorporates an error flagging mechanism similar to that of the Pascal compiler. For example, any line containing errors will be printed followed by a line with arrows pointing to the symbols in error.

```
58b    17. 695   abc
 ↑      ↑
```

To accomplish this, we read each input line into a buffer (array) in the procedure getnextline. The procedure getnextchar takes characters from the line buffer and updates a pointer (or, in this case, an index into the array) for each character processed.

As before, instead of having many parameters to the scanner (the procedure called getnextsymbol), we use a record to collect all the information that the scanner needs.

The following program incorporates many of the preceding ideas. The scanner we have written actually is much more powerful and versatile than what we have been describing. The code is modular and, we feel, very easy to read and understand.

```
program scanner(input,output);
const
    debug                = false;              { debug flag }
    maxcharsperline      = 140;                { max characters per line }
    maxexponent          = 200;                { allowable exponent for
                                                 real numbers }
    quote                = ' ' ' ';            { for literal strings }
    tokenlenmax          = 80;                 { token buffer size }
    version = ' scanner 0.4 – a basic lexical scanner. jon l spear, 03 mar 80';

type
    tokenclass           =
        (delimiter, identifier, integerconstant, literal, realconstant, tendoffile,
         tendofline);
    tokenrec             =
        record
            blankptr        : 0 .. tokenlenmax;    { used to blank fill buffer }
            tbptr           : 0 .. tokenlenmax;    { index of last char added }
            tokenbuffer     : packed array [1 .. tokenlenmax] of char;
            case class      : tokenclass of
                integerconstant   : (integervalue : integer);
                realconstant      : (realvalue : real)
        end; { of case and record tokenrec }
    errorclass           =
        (errnone, erroct, errnodigit, errbigint, errexposize, errexpochar,
         errmissingquote, errlongliteral, errlast);
    lineindex            = 0 .. maxcharsperline;
    linebufrec           =
        record
            ch          : char;           { the line buffer char }
            charptr     : lineindex;      { next char to be processed }
            echo        : boolean;        { true → echo each line to output }
            endoffile   : boolean;        { true → at end of file }
            endofline   : boolean;        { true → at end of line }
            errorline   : array [lineindex] of errorclass;
            errorset    : set of errorclass; { for the whole file }
            fileerror   : boolean;        { true if the file had an error }
            length      : lineindex;      { length of line }
            line        : array [lineindex] of char; { the one line buffer }
            linecount   : integer;        { counts input lines }
            lineerror   : boolean;        { set true if an error is found on the
                                            line }
            pfrac       : lineindex;      { ptr to first digit of frac part }
            pint        : lineindex;      { ptr to first nonzero char of number }
            pnum        : lineindex       { ptr to the first char of a number }
        end; { of record linebufrec }
```

```
var
    linebuffer              : linebufrec;     { a one line buffer }
    token                   : tokenrec;       { holds a lexical token }
```

{ initialize

    initialize line buffer
}

```
procedure initialize(var linebuffer : linebufrec);
var
    i    : lineindex;              { loop index }
begin
    if debug then writeln(' initializing line buffer');
    with linebuffer do
    begin
        echo := true;                              { we will echo the input lines }
        lineerror := false;
        linecount := 0;
        for i := 0 to maxcharsperline do
        begin
            line[i] := ' ';
            errorline[i] := errnone
        end; { of for }

        errorset := [ ];
        fileerror := false;
        endoffile := false;
        endofline := true;
        pnum := 0
    end { of with }
end; { of procedure initialize }
```

{ getnextline

    read a new line into the linebuffer
}

```
procedure getnextline(var linebuffer : linebufrec);

    { printline

        write a line and it's line number to output

    }
```

```
procedure printline(var linebuffer : linebufrec);
var
    i   : lineindex;   { loop index }
begin
    with linebuffer do
    begin
        write(linecount:6, ' ');
        for i := 1 to length do write(line[i]);
        writeln
    end
end; { of procedure printline }

{ printerrorline

    print pointers to errors, add to errorset, clear lineerror
}

procedure printerrorline(var linebuffer : linebufrec);
var
    column              : integer;          { output column number }
    i                   : integer;          { loop index }
    j                   : integer;          { loop index }
    num                 : integer;          { ord(errclass) }
begin
    column := 0;
    with linebuffer do
    begin
        printline(linebuffer); { this could be removed later }
        write(' *****':6,' '); { space over line number }
        for i := 1 to length + 1 do
            if errorline[i] <> errnone then
            begin
                errorset := errorset + [errorline[i]];
                num := ord(errorline[i]); { errornumber }
                if i > column then
                begin
                    for j := column + 2 to i do write(' '); { tab }
                    write(' ↑ ');
                    column := i
                end
                else
                begin
                    write(',');
                    column := column + 1
                end;
```

```
                    write(num:1); { use a 1 or 2 char field }
                    column := column + 1;
                    if num > 9 then column := column + 1;
                    errorline[i] := errnone
                end; {of if and for }
            writeln;
            lineerror := false;
            fileerror := true
        end { of with }
    end; { of procedure printerrorline }


begin { of procedure getnextline }
    if debug then writeln(' getting new line');
    with linebuffer do
    begin
        if lineerror then printerrorline(linebuffer); { last line had errors }
        if not eof(input) then { read the line }
        begin
            length := 0;
            while not eoln(input) do { line overflow assumed impossible }
            begin
                length := length + 1;
                read(input,line[length])
            end;
            readln(input); { get next line so eof can be checked }

            { delete any trailing blanks }
            line[0] := '*';
            while line[length] = ' ' do length := length - 1;
            line[length+1] := ' '; { ensure endofline returns blank }

            linecount := linecount + 1;
            if echo then printline(linebuffer);

            charptr := 1;
            ch := line[charptr];
            endofline := (charptr > length)
        end { not eof }
        else endoffile := true
    end { with }
end; { of procedure getnextline }

{ alphabetic

    function to determine if a character is a letter
}
function alphabetic(ch : char) : boolean;
begin
    alphabetic := ch in ['a' .. 'z']
end; { of function alphabetic }
```

```
{ numeric

    function to determine if a character is a digit
}
function numeric(ch : char) : boolean;
begin
    numeric : = ch in ['0' . . '9']
end; { of function numeric }

{ getnextsymbol

    find next token in linebuffer
}
procedure getnextsymbol(var linebuffer : linebufrec; var token : tokenrec);

    { puterror

        place error message at the current buffer pointer
    }

    procedure puterror(error : errorclass; var linebuffer : linebufrec);
    begin
        with linebuffer do
        begin
            lineerror : = true;
            errorline[charptr] : = error
        end
    end; { of procedure puterror }

{ blankfill

    ensure that the token buffer is blank filled
}
procedure blankfill(var token : tokenrec);
begin
    with token do
    begin
        while blankptr > tbptr do
        begin
            tokenbuffer[blankptr] = ' ';
            blankptr : = blankptr − 1
        end;
        blankptr : = tbptr
    end { of with }
end; { of procedure blankfill }

{ getnextcharacter

    read next character from line buffer and advance pointer
}
```

```
procedure getnextchar(var linebuffer : linebufrec);
begin
    with linebuffer do
    begin
        if endofline then
            if eof(input) then endoffile := true
            else getnextline(linebuffer)
        else
        begin
            charptr := charptr + 1;
            if charptr > length then endofline := true
        end;
        ch := line[charptr]
    end { of with }
end; { of procedure getnextchar }

{ scanidentifier

    scan alphanumeric characters (copying them to tokenbuffer)
}

procedure scanidentifier(var linebuffer : linebufrec; var token : tokenrec);
begin
    if debug then writeln(' scanning identifier');
    with linebuffer , token do
    begin
        class := identifier;
        tbptr := 0;
            repeat                          { first char is known to be alphabetic }
                if tbptr < tokenlenmax then
                begin
                    tbptr := tbptr + 1;
                    tokenbuffer[tbptr] := ch
                end;
                getnextchar(linebuffer)
            until not (alphabetic(ch) or numeric(ch))
    end { with }
end; { of procedure scanidentifier }
```

```
{ scannumber

   convert a decimal or octal number, or a real to internal form
}

procedure scannumber(var linebuffer : linebufrec; var token : tokenrec);
var
    i     : integer;

    { convinteger

       convert part of linebuffer to an integer (with no overflow)
    }

    procedure convinteger(var linebuffer : linebufrec; base : integer;
          maxint : integer; first, last : lineindex; var n : integer);
    var
        digit   : 0 .. 9;    { holds a single digit's worth }
        i       : integer;   { loop index }
        x       : real;      { used to check for overflow }
    begin
        n := 0;
        x := 0.0;
        i := first;
        while i < last do
        begin
            digit := ord(linebuffer.line[i]) − ord('0');
            if digit >= base then
            begin
                puterror(erroct,linebuffer);
                i := last { terminate loop }
            end;

            x := x * base + digit;
            if x <= maxint then n := n * base + digit
            else
            begin
                puterror(errbigint,linebuffer);
                i := last
            end;
            i := i + 1
        end { of while }
    end; { of procedure convinteger }
```

```
{ scaninteger

    scan a decimal integer
}

procedure scaninteger(var linebuffer : linebufrec; var token : tokenrec);
begin
    with linebuffer , token do
    begin
        class : = integerconstant;
        convinteger(linebuffer, 10, maxint, pint, charptr, integervalue)
    end
end; { of procedure scaninteger }

{ scanoctal

    scan an octal number
}

procedure scanoctal(var linebuffer : linebufrec; var token : tokenrec);
begin
    with linebuffer , token do
    begin
        class : = integerconstant;
        convinteger(linebuffer, 8, maxint, pint, charptr, integervalue);
        getnextchar(linebuffer) { skip 'b' }
    end
end; { of procedure scanoctal }

{ scanreal

    scan a real number with/without exponent
}

procedure scanreal(var linebuffer : linebufrec; var token : tokenrec);
var
    i           : integer;
    expo        : integer;
    nexpo       : integer;         { normalized exponent
    scale       : integer;
    x           : real;            { accumulator }
    negexp      : boolean;         { true if exponent is < 0 }
    r           : real;            { used to compute power of 10 }
    fac         : real;            { used to computer power of 10 }
```

```
begin
     if debug then writeln(' scanning real number ');
     with linebuffer , token do
     begin
          class := realconstant;

          { do integer part. overflow assumed impossible }
          x := 0.0;
          expo := 0;
          for i := pint to charptr − 1 do
               x = x ∗ 10.0 + ord(line[i]) − ord('0');

          nexpo := charptr − pint;
          scale := 0;
          if ch = '.' then
          begin
               getnextchar(linebuffer); { skip '.' }
               pfrac := charptr;
               if numeric(ch) then
                    repeat
                         scale := scale − 1;
                         x := x ∗ 10.0 + ord(ch) − ord('0');
                         getnextchar(linebuffer)
                    until not numeric(ch)
               else puterror(errnodigit,linebuffer);

               { check if we must find first nonzero digit }
               if nexpo = 0 then { integer part was zero }
               begin
                    i := pfrac;
                    while line[i] = '0' do i := i + 1;
                    nexpo := pfrac − i { = trunc(log10(x)) }
               end
          end; { fractional part }

          { do we have an exponent? }
          if ch = 'e' then
          begin
               negexp := false;
               getnextchar(linebuffer); { skip 'e' }
               if ch = '−' then
               begin
                    negexp := true;
                    getnextchar(linebuffer) { skip '−' }
               end
               else if ch = '+' then getnextchar(linebuffer);
```

```
        { build exponent }
        if numeric(ch) then
        begin
            repeat
                expo := expo * 10 + ord(ch) − ord('0');
                getnextchar(linebuffer)
            until not numeric(ch);

            { adjust scale and nexpo }
            if negexp then
            begin
                scale := scale − expo;
                nexpo := scale − expo
            end
            else
            begin
                scale := scale + expo;
                nexpo := nexpo + expo
            end
        end
        else puterror(errexpochar,linebuffer)
    end; { exponent }

    { compute 10 * scale using right to left binary method }
    if abs(nexpo) <= maxexponent then
        if scale <> 0 then { must adjust exponent }
        begin
            r := 1.0;
            negexp := scale < 0;
            scale := abs(scale);
            fac := 10.0;
            repeat
                if odd(scale) then r := r * fac;
                fac := sqr(fac);
                scale := scale div 2
            until scale = 0;
            if negexp then realvalue := x / r
            else realvalue := x * r
        end { apply exponent }
        else realvalue := x
    else puterror(errexposize,linebuffer)
end { of with }
end; { of procedure scanreal }
```

```
begin { of procedure scannumber }
    if debug then writeln(' scanning number');
    with linebuffer , token do
    begin
        tbptr := 0; { reset token buffer pointer }
        pnum := charptr; { insures no read past eoln }

        { skip leading zeros }
        while ch = '0' do getnextchar(linebuffer);
        pint := charptr; { first non-zero char }

        { scan integer part }
        while numeric(ch) do getnextchar(linebuffer);

        if ch <> 'b' then
        begin
            if not ((ch = '.') or (ch = 'e')) then scaninteger(linebuffer,token)
            else scanreal(linebuffer,token)
        end
        else scanoctal(linebuffer,token);

        { copy number into token buffer }
        i := pnum;
        tbptr := 0;
        while (i < charptr) and (tbptr < tokenlenmax) do
        begin
            tbptr := tbptr + 1;
            tokenbuffer[tbptr] := line[i];
            i := i + 1
        end; { of copy }

        pnum : = 0      { enable getnextline }
    end { with }
end; { of procedure scannumber }

{ scanliteral

    read in a literal string
}

procedure scanliteral(var linebuffer : linebufrec; var token : tokenrec);
    var
        working   : boolean;       { true if the closing quote has not been
                                     found }
begin
    if debug then writeln(' scanning literal');
    with linebuffer , token do
    begin
        class := literal;
        tbptr := 0;
        getnextchar(linebuffer); { skip first quote }
        working := true ;
```

```
        while working and not endofline do
        begin
            if ch = quote then { is it two in a row? }
            begin
                getnextchar(linebuffer);
                { if ch is a quote, continue since it is an imbedded one }
                working := ch = quote
            end;
            if working then
            begin
                if tbptr < tokenlenmax then
                begin
                    tbptr := tbptr + 1;
                    tokenbuffer[tbptr] := ch;
                    getnextchar(linebuffer)
                end
                else { string too long }
                begin
                    puterror(errlongliteral,linebuffer);
                    while (ch <> quote) and not endofline do
                        getnextchar(linebuffer); { skip over string }
                    if ch = quote then getnextchar(linebuffer);
                    working := false
                end { overflow }
            end { of if working }
        end; { of while }
        if working then puterror(errmissingquote,linebuffer)
    end { with }
end; { of procedure scanliteral }

{ scandelimiter

    put ch into token buffer and advance
}

procedure scandelimiter(var linebuffer : linebufrec; var token : tokenrec);
begin
    if debug then writeln(' scanning delimiter');
    token.class := delimiter;
    token.tbptr := 1;
    token.tokenbuffer[token.tbptr] := linebuffer.ch;
    getnextchar(linebuffer)
end; { of procedure scandelimiter }
```

```
{ scanendofline

    return end of line status

}

procedure scanendofline(var linebuffer : linebufrec; var token : tokenrec);
begin
    if debug then writeln(' scanning end of line');
    token.class := tendofline;
    token.tbptr := 0
end; { of procedure scan endofline }

{ scanfileend

    return end of file status

}

procedure scanfileend(var linebuffer : linebufrec; var token : tokenrec);
begin
    if debug then writeln(' scanning end of file');
    token.class := tendoffile;
    token.tbptr := 0
end; { of procedure scanfileend }

begin { of procedure getnextsymbol }
    if debug then writeln(' getting next symbol. (ch=',linebuffer.ch,')');
    if (token.class = tendofline) or (token.class = tendoffile) then
            getnextline(linebuffer);
    with linebuffer do
    begin
        { scan leading blanks }
        while (ch = ' ') and not endofline do getnextchar(linebuffer);

        { classify token based on its first char }
        if alphabetic(ch) then scanidentifier(linebuffer,token)
        else
        if numeric(ch) then scannumber(linebuffer,token)
        else
        if ch = quote then scanliteral(linebuffer,token)
        else
        if not endofline then scandelimiter(linebuffer,token)
        else
        if not endoffile then scanendofline(linebuffer,token)
        else
        if endoffile then scanfileend(linebuffer,token)
        else
                halt              ·
    end; { with }
    blankfill(token)     { follow token with blanks }
end; { of procedure getnextsymbol }
```

```
{ reporterrors

     write a list of errors that have been found in the file
}

procedure reporterrors(var linebuffer : linebufrec);
var
     err    : errorclass;       { loop index }
begin
     writeln(' ***** errors in file:');
     writeln;
     for err := succ(errnone) to pred(errlast) do
          if err in linebuffer.errorset then
          begin
               write(ord(err):8, ' : ' );
               case err of
                    erroct             : write('digit 8 or 9 in octal',
                                               'constant');
                    errbigint          : write('integer constant >',
                                               'maxint(=',maxint:1,')');
                    errexposize        : write('abs(real exponent) >'
                                               'maxexponent(=',
                                               maxexponent:1,')');
                    errexpochar        : write('digit expected in'
                                               'exponent');
                    errnodigit         : write('digit expected after ".." ');
                    errmissingquote    : write('no closing quote in literal');
                    errlongliteral     : write('literal too long (max is',
                                               tokenlenmax:1, ' chars)')
               end; { of case }
               writeln
          end; { of error and for loop }
     writeln;
     writeln(' end of error list')
end; { of procedure reporterrors }
```

```
begin { of program scanner }
    page(output);
    writeln(version);
    initialize(linebuffer);
    getnextline(linebuffer); { read the first line }
    if not linebuffer.endoffile then token.class : = delimiter
    else token.class : = tendoffile;
    token.blankptr : = tokenlenmax;

    while token.class <> tendoffile do
    begin
        getnextsymbol(linebuffer,token);
        write(' ',token.tokenbuffer : 20, ' → ');
        with token do
            case class of
                identifier              : write('ident');
                integerconstant         : write('integer=',integervalue);
                realconstant            : write('real = ',realvalue);
                delimiter               : write('delimiter');
                literal                 : write('literal');
                tendofline              : write('end of line');
                tendoffile              : write('end of file')
            end; { of case and with }
        writeln;
    end; { of while }
    writeln;
    if linebuffer.fileerror then reporterrors(linebuffer);
    writeln(' execution of scanner complete');
end. { program scanner }
```

## EXERCISES FOR CHAPTER 7

1. Derive an expression for the location of the i, jth element of an array with the following declaration.

   **var**
       $a$ : **array** $[l_1 .. u_1, l_2 .. u_2]$ **of** integer;

   Assume the array is stored by columns; $l_1$, $l_2$, $u_1$, and $u_2$ are defined as constants with the following restrictions.

$$l_1 <= u_1 \quad \text{and} \quad l_2 <= u_2$$

2. Repeat Exercise 1 assuming that the array a is stored by rows.

3. Generalize the results in Exercises 1 and 2 to handle n-dimensional arrays. That is:

**var**

   $a$ : **array** $[l_1 .. u_1, l_2 .. u_2, l_3 .. u_3, ..., l_n .. u_n]$ **of** *integer;*

4. Given the declaration:

**var**

| *students* | : **array** $[1 .. 100]$ **of** |
|---|---|
| **record** | |
| *math* | : *integer;* |
| *verbal* | : *integer;* |
| *name* | : **array** $[1 .. 10]$ **of** *char* |
| **end**; { *of record* } | |

derive an expression for:

*loc(students[i] . name[j])*

where $1 <= i <= 100$ and $1 <= j <= 10$.

5. Write a complete Pascal declaration (using records) to describe the name, height, eye color, age, sex, and weight of a person.

6. Write a Pascal declaration that contains information regarding a student's academic history. Include name, student number, courses taken, grades, and current address.

7. Modify the Idiot's Delight program to output the rank and suit of the card at the top of the hand. Use an array to store the names of the ranks (ace, two, three, . . . kings), and suits (clubs, spades, hearts, diamonds).

8. Convert the following infix expression to postfix.

   $(a*b) * (c+d) \uparrow e * (f-g) \uparrow h - i - j * k/l$

   ($\uparrow$ is used to denote exponentiation.)

9. Ackerman's function $A(m,n)$ is defined as follows.

$$A(m,n) = \begin{cases} n + 1, \text{ if } m = 0 \\ A(m-1,1), \text{ if } n = 0 \\ A(m-1,A(m,n-1)), \text{ otherwise} \end{cases}$$

Write a recursive Pascal function to compute this function. What is the value of $A(3,2)$?

10. Binomial coefficients are common in mathematics. They are defined as:

$$\binom{n}{m} = \frac{n!}{m!\,(n-m)!}$$

and can can be computed recursively using the following relation.

$$\binom{n}{m} = \binom{n-1}{m} + \binom{n-1}{m-1}$$

Write a recursive Pascal function to evaluate binomial coefficients.

11. The transpose of a given matrix is defined so that the i, jth element of the original matrix is placed in the j, ith element of the transpose matrix. For example:

$$A = \begin{bmatrix} 1 & 6 & 19 \\ 5 & 1 & 30 \\ 7 & 5 & 16 \end{bmatrix} \qquad A^T = \begin{bmatrix} 1 & 5 & 7 \\ 6 & 1 & 5 \\ 19 & 30 & 16 \end{bmatrix}$$

Notice that we are simply interchanging rows and columns. Write a Pascal procedure to compute the transpose of a matrix in place (i.e., use only the storage reserved for the original array).

12. Modify the program scanner of Section 7.3 so that:

(a) It accepts and correctly converts binary, or base 2, values. The syntax of these values is:

dddd . . . dy

where "d" can be only the digits 0 or 1.

(b) It recognizes a special class of identifier called "reserved words." The list of reserved words is in a special table and, if one is found, the program prints out:

reserved word =

and the identifier.

# Chapter 8

# LINKED LISTS AND TREES

## 8.1 INTRODUCTION

In Chapter 7 we discussed arrays and some of their uses (implementing stacks or queues, and manipulating strings). Although arrays are useful in many contexts, they are not appropriate for all situations. For example, arrays are not necessarily the best way to store an ordered list, especially if frequent insertions or deletions are necessary. Consider the editor we began outlining in Chapter 5. Assume that the entire file that we are to edit is to be read into an array. Our first decision is whether to store one character per word in the array or to pack as many characters as possible into a word. To economize on space, we will choose the latter. A possible array declaration would look like this.

```
const
    alfaleng    = 10;        { number of characters per word }
    maxwords    = 1000;      { maximum number of words to allocate for text
                               buffer }
type
    alfa        = packed array [1. .alfaleng] of char;
var
    textbuffer  : array [1. .maxwords] of alfa;
```

There are at least three problems with using an array for this application. First, the fixed size of the array (maxwords) imposes a limit on the size of the file we can edit ($<=$ alfaleng * maxwords characters). Second, we must decide how to treat the end of each line of our text file. A simple but inelegant solution is to place at the end of each line two characters that would be very unlikely to occur in the file (e.g., ##). (If we had used only one character to denote the end-of-line, we would have been unable to use that character anywhere in the text file.) Third, the most serious problem occurs when one attempts to insert or delete a line of text in the array. If the file to be edited contains the lines:

6 + 6 =
14 for large
values of six.
pascal was a mathematician

using our data representation, the contents of the array would be:

textbuffer[1]     = '6 + 6 = ## '
textbuffer[2]     = '14 for lar    '
textbuffer[3]     = 'ge##         '
textbuffer[4]     = 'values of     '
textbuffer[5]     = 'six.##        '
textbuffer[6]     = 'pascal was    '
textbuffer[7]     = 'a mathemat    '
textbuffer[8]     = 'ician##       '

.

.

textbuffer[1000]   = ?

If we were now to insert the line:

there are no pianos in japan.

after the third line in the file ("values of six."), we would have to: (1) find the line "values of six." in the array, (2) determine the storage requirement of the line to be inserted (three words), and (3) shift all lines after "values of six." down three words before inserting the new line.

To delete a line of text may require moving blocks of words "up" in the array. The worst case occurs when we need to delete the first line, in which case every line after that will have to be shifted up.

For larger files, the insertion and deletion operations can be time consuming. The data structure just given would only be practical for very short files, if then.

We obviously need a better way to represent the data. The following is a pictorial representation of another data structure for our edit file. (We will describe the necessary implementation details shortly.)

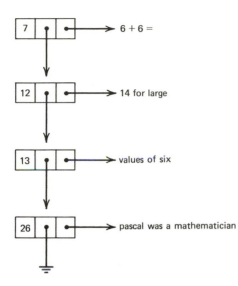

In this representation, every line of the edit file is preceded by a *node* containing three *fields*. A node is a collection of data items, each of which may be a data field or a link field. The link fields are used to store *pointers* (arrows).

The first field of each of the nodes contains a count of the number of characters in the line that is "pointed to" by the third field in the node. The middle field of each node (except the last one) contains a pointer (or *link*) to the next node in the data structure. The middle link field of the last node has a special pointer (the ground pointer) that signifies that there are no nodes following it (hence there are no more lines in the edit file).

Now let us reconsider the problem of inserting:

there are no pianos in japan.

after the line "values of six." It should be clear that all we have to do is update the picture as follows. We must (1) create a "new" node, (2) set the first field of the node to 29, (3) set the middle field to point to the node that the line "values of six." used to point to, (4) reset that node to point to the "new" node, and (5) set the third field to point to the string to be inserted.

We will gradually build up to an implementation, but what we want to stress here is that you should *always* draw a picture of a data structure before attempting to implement it.

To summarize the problem with using an array to represent the text of the edit file: the data movement is caused by storing *logically* consecutive elements of an ordered list into *physically* consecutive storage locations. Conceptually, there is no reason for the elements of an ordered list to reside in consecutive locations. Assigning elements to arbitrary locations does present a problem: how do we preserve the *logical ordering?* The solution, as we have seen, is to include explicit pointers to the next element in the logical sequence. Then any element may be accessed by following the links, provided we know the location of the *first* element. Storing an ordered list in this way eliminates data movement during insertion and deletion operations.

## 8.2   LINKED LISTS

The data structure introduced in the previous section is a *linked list*. In essence, a linked list is an ordered list of nodes. The *first* node is accessed by a special pointer (often called the *head* pointer). The *last* node contains a *null* link field that indicates the end of the list. A general diagram of a linked list is shown in Figure 8.1 on the next page ($d_i$ refers to elements in the data field of a node).

### 8.2.1   Pointers in Pascal

Storage locations for a linked list are usually created *dynamically* (while the program is executing). Usually, however, variables are declared and referenced by an identifier name. For example:

```
var
    a               : integer;
    b               : real;
    c               : char;
```

declares three *static* global variables. Each global variable occupies one fixed (known) memory location.

Suppose we want to set up a linked list with nodes that contain two fields: a social security number and a link field to point to the next node in the list. Pictorially:

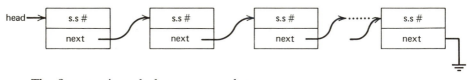

The first step is to declare a type such as:

```
type
    person          =  ↑ personrecord;
```

which defines person as a *pointer type*. Each value of type person will be a pointer to an undetermined storage area that contains information of type personrecord. (No-

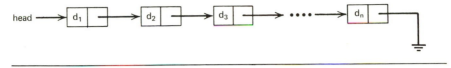

**Figure 8.1** Linked list diagram.

tice that personrecord has not been declared. This is an exception to the rule that all identifiers must be declared before they are used.) The item of type personrecord is a two-field node declared using the following **record** type.

**type**
    *personrecord =*
        **record**
            *socialsecurity*       *: integer;*
            *next*                *: person*
        **end***; { of record personrecord }*

The field socialsecurity will contain a person's social security number stored as an integer, and the next field will contain a pointer to the next node in the linked list. This declaration has not allocated *any* space.

    Each entry, or node, in the list (except the first) is referenced by a pointer from the previous node. To reference the first node, we must declare a variable (head):

**var**
    *head*                 *: person;*

in which we can store a pointer to the first node.

    This declaration creates a storage location for head *only;* its value is undefined. It does not point to any storage location.

    The predefined Pascal procedure new is used to allocate dynamically storage locations to instances of data types. For example:

    *new(head)*

creates a new node that is *referenced by* head:

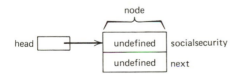

We can store a value into the socialsecurity field of the newly created node by:

    *head ↑ .socialsecurity : =  123456789*

Head ↑ is the first node, which is a record that has two fields. The field identifer .socialsecurity is used to select the socialsecurity field of that record.

In Pascal the reserved word **nil** is used to set the final pointer in the last node of the list to point to nothing. We can assign the constant value **nil** to any *pointer* field, regardless of type.

We also use **nil** to denote empty lists with no entries or nodes. The following statement creates an empty list of personrecord.

head : = **nil**

or

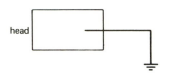

Thus the name head refers to the actual value of the pointer variable, while head ↑ refers to the object being pointed at by head. The difference made by the ↑ cannot be overemphasized.

These ideas are now combined into a simple program that creates a linked list of social security numbers read from file input until an end of file is encountered.

```
program linkedlist(input,output);
type
     person                          = ↑ personrecord;
     personrecord                    =
          record
               socialsecurity        : integer;
               next                  : person
          end; { of record personrecord }
var
     head                            : person;      { pointer to head of list }
     node                            : person;      { temporary }
     ssnumber                        : integer;     { social security number }
begin
     { initialize list to be empty }
     head : = nil;
```

```
{ process the social security numbers until an end of file }
read(ssnumber);
while not eof do
begin
        new(node);                              { allocate a new node }
        node ↑ .socialsecurity := ssnumber;     { initialize social security
                                                  number }

        node ↑ .next := head;                   { link new node to
                                                  previous head of list }

        head := node;                           { make new node the first
                                                  in the list }

        read(ssnumber)
end; { of while }
        .
        .
        .
end.
```

It is instructive to trace through this program statement by statement to see how it works. Assume file input contains the following data.

*111111111*
*555555555*
*777777777*

The first executable statement creates an empty list with the following picture.

Next 111111111 is read into ssnumber. Since we have not reached the end of file, we create a new node.

The next two assignment statements change the picture as shown.

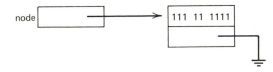

Then the assignment head := node makes head point to the first (and only) element of the list.

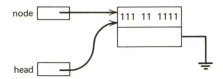

The next value is read into ssnumber and the **while** loop is executed again. Now the picture is updated to:

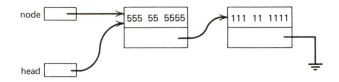

Finally, the last value is read and the picture will look like this.

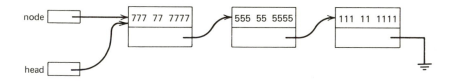

Before you read on, be sure you can trace through the execution of this program and understand how this list was formed. If you understand this example, what follows should be easy.

A typical operation on linked lists is to search for nodes with a particular attribute. As an example, there are two ways we can find the social security number 555555555. First, since we know what the linked list looks like, the node we are searching for is simply:

head ↑ .next ↑

More generally (but still with the assumption that 555555555 appears in the list), we could access this node through another pointer variable.

```
var
    pointer                                    : person;
    .
    .
    .
    pointer : = head;
    while pointer ↑ .socialsecurity <> 555555555 do
        pointer : = pointer ↑ .next
```

When this **while** loop terminates, pointer will point to the node whose social security field contains 555555555.

The search method we have just described works only if the item is in the list. If we do not know if the item is in the list, the search procedure must be changed so as not to run off the end of the list. For example, to find a social security number (which is in the variable ssnumber) in the list, we could proceed in this way.

```
    pointer : = head;
    while (pointer <> nil) and
        (pointer ↑ .socialsecurity <> ssnumber) do
            pointer : = pointer ↑ .next
```

There is a bug in this algorithm; try to locate it before continuing. The problem is that Pascal will evaluate both alternatives of the conjunction. Hence, even when pointer = **nil** becomes true, making the first alternative false, Pascal will evaluate pointer ↑ .socialsecurity. But **nil** ↑ .socialsecurity does not exist. This will result in an execution time error that could be avoided by using a boolean variable.

```
    pointer : = head;
    found : = false;
    while (pointer <> nil) and (not found) do
        if pointer ↑ .socialsecurity = ssnumber then found : = true
        else pointer : = pointer ↑ .next
```

When the **while** loop terminates, we can determine if ssnumber is in the list by checking the value of the boolean variable found.

### 8.2.2  Stacks as Linked Lists

In Chapter 7 we saw that a stack can be implemented using an array. Sometimes it is better to implement a stack using a linked list, as depicted here.

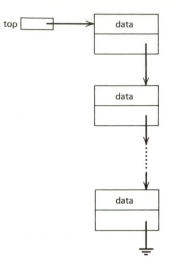

To do so, we first declare the type:

> **type**
>> *stackpointer*                    = ↑ *stackelement;*

which defines stackpointer to be a pointer type. Each element of the stack will have two fields: a data field (for our purposes it will be used to hold items of type char) and a pointer field used to contain a pointer to the next node in the stack.

> **type**
>> *stackelement*                =
>>> **record**
>>>> *data*                : *char;*
>>>> *next*                : *stackpointer*
>>> **end***; { of record stackelement }*

The first entry in the stack is pointed to by the stack pointer top.

> **var**
>> *top*                          : *stackpointer;*

Initially:

> *top* : = **nil**

is used to denote an empty stack. To push an item onto the top of the stack, we call the procedure push, which is modified to read:

```
procedure push(item : char; var top : stackpointer);
var
    node                    : stackpointer;
begin
    new(node);              { acquire new element for stack }
    node ↑.data := item;    { insert item into node }
    node ↑.next := top;     { make this node point to old top }
    top := node             { establish new top of stack }
end; { of procedure push }
```

Unfortunately, this approach does not handle stack overflow. We cannot ascertain whether the procedure new will be able to "find enough space" for the new node. There is no standard method in Pascal to determine if all the available space for new node creation has been exhausted. If we do run out of space, the procedure new will "blow up" instead of returning an error flag. That is, a postmortem dump will be produced, and the program will halt.

How do we pop an element off the stack? Try to write the procedure pop before you read further.

```
procedure pop(var item : char; var top : stackpointer;
    var successful : boolean);
begin
    if top <> nil then              { at least one element on stack }
    begin
        item := top ↑.data;         { set item to element at top of stack }
        top := top ↑.next;          { update top pointer }
        successful := true          { pop operation successful }
    end
    else successful := false        { indicate stack underflow }
end; { of procedure pop }
```

This procedure works, with one exception. Consider what happens when procedure pop is called on this list.

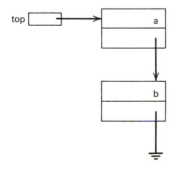

Since top<>**nil,** we execute the compound statement after the **then.** This will change the data structure (picture) to look like this.

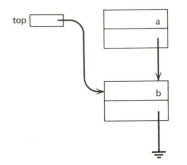

No problem? Look again. Now there is no way to reference the node whose data field contains the character a. Pascal permits us to return nodes like this to the list of allocatable nodes by using the predefined procedure dispose. To take advantage of this facility in the procedure pop, we declare the local variable:

```
var
      temp                    : stackpointer;
```

and change the compound statement to read:

```
begin
      item := top ↑ .data;          { set item to element at top of stack }
      temp := top;                  { save pointer to top of stack }
      top := top ↑ .next;           { update top pointer }
      dispose(temp);                { return node to available space }
      successful := true            { pop operation successful }
end
```

The temporary is used to hold another pointer to the node to be disposed. (*Note.* Some Pascal compilers do not implement the procedure dispose.)

### 8.2.3   Queues as Linked Lists

We can also view a queue as a linked list. This is exactly what we did in the simulation case study of Chapter 6.

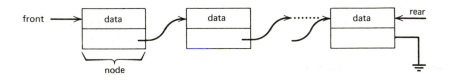

It is easier to implement a queue with the linked list representation than with the array representation. Try it. The node structure is the same as for the stack, although we change the names of the (field) identifiers for clarity.

```
type
    queuepointer              = ↑ queueelement;
    queueelement              =
        record    ,
            data              : char;
            next              : queuepointer
        end; { of record queueelement }
var
    front                     : queuepointer;
    rear                      : queuepointer;
```

An empty queue is initialized using:

```
front := nil
```

The procedure to add a new element to the rear of the queue is:

```
procedure enqueue(item : char; var front : queuepointer;
        var rear : queuepointer);
var
    node                      : queuepointer;
begin
    { create new queue element and initialize its fields }
    new(node);
    with node ↑ do
    begin
        data := item;
        next := nil
    end;


    { add new node to rear of queue }
    if front = nil then              { queue is empty }
    begin
        front := node;               { front and rear will point }
        rear := node                 { to the only element in the queue }
    end
    else                             { queue not empty }
    begin
        rear ↑ .next := node;        { add new node to rear of queue }
        rear := node                 { establish a new rear of queue }
    end
end; { of procedure enqueue }
```

The corresponding procedure to delete an element from the front of the queue is:

```
procedure dequeue(var item : char; var front : queuepointer;
        var successful : boolean);
var
        node                              : queuepointer;
begin
        if front <> nil then              { non-empty queue }
        begin
            item := front ↑ .data;        { return item at head of queue }
            node := front;                { save pointer to head of queue }
            front := front ↑ .next;       { update front pointer }
            dispose(node);                { return node to available space }
            successful := true            { dequeue operation successful }
        end
        else successful := false          { indicate queue underflow }
end; { of procedure dequeue }
```

### 8.2.4  Simple Operations on Linked Lists

The preceding sections showed two examples of using linked lists to implement stacks and queues; later sections will describe several other applications of linked lists. First, we will present some examples to help you gain more experience manipulating linked lists. For these examples we will assume that the node structure looks like:

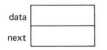

where the data field is of type char and the next field contains a pointer to the next node in the list. We assume the following declarations.

```
type
    nodepointer                   = ↑ node;
    node                          =
        record
            data                  : char;
            next                  : nodepointer
        end; { of record node }
```

Suppose you are given two lists, as shown:

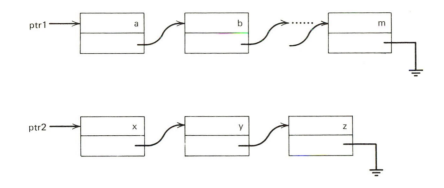

and are asked to concatenate them to produce one list that looks like:

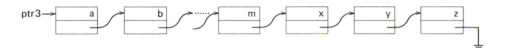

One possible solution is the following procedure of three parameters.

```
procedure concatenate(list1 : nodepointer; list2 : nodepointer;
    var list3 : nodepointer);
var
    ptr1                        : nodepointer;    { pointer to nodes of list 1 }
begin
    { find last node of list 1 }
    ptr1 := list1;
    while ptr1 ↑ .next <> nil do ptr1 := ptr1 ↑ .next;

    { note : ptr1 should now point to last node in list 1; link list 1 to list 2 }
    ptr1 ↑ .next := list2;

    { return pointer to first list }
    list3 := list1
end; { of procedure concatenate }
```

Does this procedure meet our guidelines for robustness (Chapters 3 and 4)? Hardly. It will work if presented with the preceding two lists, but what if list1 were the following empty list?

Then the condition in the **while** loop would generate a run-time error. It is vital *always to check the boundary conditions* of an algorithm. When working with lists, always check that an algorithm will work when presented with an empty list. Also, make sure that list algorithms work properly for the last node in a list.

Try to correct the concatenation procedure before looking at our revised version.

```
procedure concatenate(list1 : nodepointer; list2 : nodepointer;
    var list3 : nodepointer);
var
    ptr1                              : nodepointer;    { pointer to nodes of list 1 }
begin
    if list1 <> nil then                     { first list not empty }
    begin
        list3 := list1;
        if list2 <> nil then                 { second list not empty }
        begin
            { search for end of list 1 }
            ptr1 := list1;
            while ptr1 ↑ .next <> nil do ptr1 := ptr1 ↑ .next;
            ptr1 ↑ .next := list2
        end
    end
    else list3 := list2                      { return pointer to list 2 }
end; { of procedure concatenate }
```

In this procedure the **while** loop is used to "walk" down a list until we come to the last node. This can be a very time-consuming operation when the list is long. To speed up such a search, it is often a good idea to keep a pointer to the last element in the list.

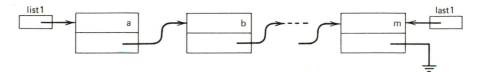

(We now have a structure identical to a queue, although we will be performing operations other than enqueue and dequeue to the structure.) It is not really necessary to have botn pointers. Consider the following representation of the list.

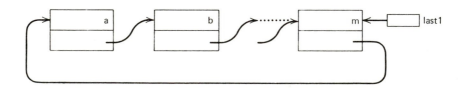

This defines a *circular list*. We no longer have an explicit pointer to the head of the list (although it is easy to locate, provided last1<>**nil**; then the first node in the list is last1 ↑ .next). Make the necessary modifications to the concatenation procedure to handle *circular lists* passed as parameters.

Now consider the problem of determining how many nodes there are in a circular list. If we are not careful, this procedure can easily run around in circles. (This is known as a dynamic halt.)

```
function numberofnodes(list : nodepointer) : integer;
var
      count                    : integer;        { counts the number of nodes }
      lastnode                 : nodepointer;    { holds pointer to last node }
      pointer                  : nodepointer;    { used to walk through list }
begin
      { initialize }
      count := 0;

      if list <> nil then                        { at least one node in list }
      begin
          { save pointer to last node in circular list }
          lastnode := list;

          { walk through list counting number of nodes }
          pointer := list;
          repeat
              count := count + 1;
              pointer := pointer ↑ .next
          until pointer = lastnode
      end;

      { return length }
      numberofnodes := count
end; { of function numberofnodes }
```

As a final example, suppose you are given the list:

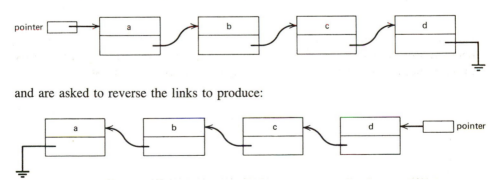

and are asked to reverse the links to produce:

This is more challenging than it looks. Try it. We will develop the solution in the exercises at the end of the chapter.

## 8.3   SENTINEL NODES

When working with linked lists, we are often confronted with the problem of locating a record with a specific key in a linked list. If the key is not present, then we must add a new node to the list with this key value. As we have seen before, this requires searching the list until we encounter either a node with the correct key value or the end of the list. This task can be simplified if we know that the key is in the list. That is the idea behind using a *sentinel node*. Before searching the list for the key, we put the key value into a node (designated the sentinel node) that is the last node of the list. The data structure for an empty list would look like:

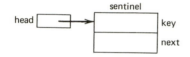

If we had to search for a node containing a key value of 10, the first step would be to put 10 into the sentinel node:

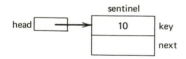

and then search the list starting at the node pointed to by head. If, after searching the list, we find the key in the sentinel node, we will add a new node to the list before the sentinel node.

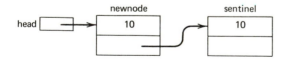

If later we tried to locate the key 20 in the list, we would first put 20 into the sentinel node:

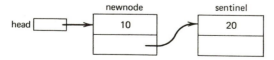

then search the list starting from its head. Since the search will find 20 in the sentinel node, we add a new node, as shown.

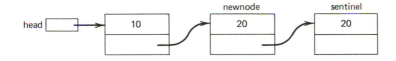

Try to set up the necessary declarations and write a search procedure that conforms to these specifications. Here is ours.

```
program find(input, output);
type
    nodepointer                    = ↑ node;
    node                           =
        record
            key                    : integer;
            next                   : nodepointer
        end; { of record node }
var
    head                           : nodepointer;    { pointer to head of list }
    keyvalue                       : integer;        { key to search for }
    sentinel                       : nodepointer;    { pointer to sentinel node }

procedure search(keyvalue : integer; var head : nodepointer;
    sentinel : nodepointer);
var
    currentnode                    : nodepointer;    { for walking through list }
    node                           : nodepointer;    { used to allocate new
                                                       node }
    previousnode                   : nodepointer;    { for walking through list }
begin
    { put searched for key into the sentinel node }
    sentinel ↑ .key : = keyvalue;

    { we walk through the list using two pointers: one to the previous node and
        the other to the current node we are inspecting }
    previousnode : = head;
    currentnode : = head ↑ .next;
    while currentnode ↑ .key <> keyvalue do
    begin
        previousnode : = currentnode;
        currentnode : = currentnode ↑ .next
    end; { of while }
```

```
        { currentnode is the node with a key of keyvalue; if it is the sentinel node,
          set up a new node and place it before the sentinel node but after
          previousnode }
        if currentnode = sentinel then
        begin
            new(node);                          { allocate new node }
            node ↑ .key := keyvalue;            { update key field }
            node ↑ .next := sentinel;           { make it point to sentinel node }
            if previousnode := head
            then head := node
            else previousnode ↑ .next := node   { previous node points to node }
        end { of if }
end; { of procedure search }

begin
    { initialize }
    new(sentinel);
    new(head);
    head ↑ .next := sentinel;

    { read in a key value; search for it in the list; stop reading when a negative
      value is encountered }
    read(keyvalue);
    while keyvalue >= 0 do
    begin
        search(keyvalue,head,sentinel);
        read(keyvalue)
    end { of while }
end.
```

The use of sentinel nodes can reduce the number of special cases that must be checked to find an element in a list. We will be using sentinel nodes in many of our examples, especially in the text editor case study in Chapter 9.

## 8.4  DOUBLY LINKED LISTS

The list structures we have discussed so far contain only one pointer field and generally look like this.

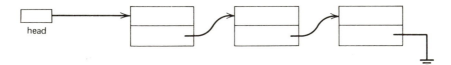

This type of structure allows us to move through the list only in the "forward" direction. If we are somewhere in the middle of a long list and perhaps need to look at a node three to the "left" of the current node, we would have to retraverse the list from the beginning to locate the node. A better solution is to design a data structure that includes both forward *and* backward links, such as:

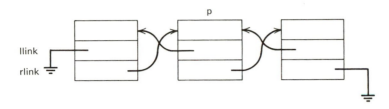

We have named the two pointer fields llink (for pointers to nodes to the left of the current node) and rlink (which corresponds to our usual usage of a pointer field). Notice also that we now have two **nil** pointers in the list to indicate the two "ends" of the list.

   In our previous diagram we set up a pointer p to the middle node in the list. For this doubly linked structure the following condition holds.

$$p = p \uparrow .rlink \uparrow .llink = p \uparrow .llink \uparrow .rlink$$

   A doubly linked list can be converted into a circular doubly linked list in this way.

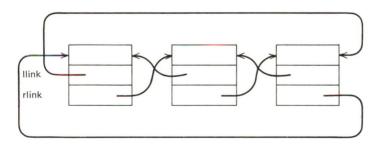

This condition now applies to the circular list, no matter which node p is pointing to. Deleting a node from a doubly linked list is easy. Before a node deletion takes place, our list looks like:

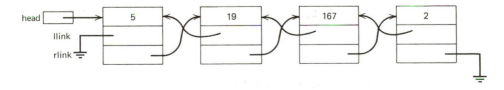

If we delete the node with data field 19, the resulting picture will be:

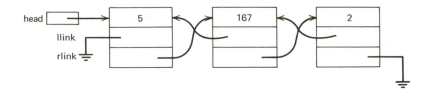

Given a pointer to the list head, write a procedure to delete an arbitrary node; then look at our solution. We assume the following declarations.

```
type
    nodepointer                          = ↑ node;
    node                                 =
        record
            data                    : integer;
            llink                   : nodepointer;
            rlink                   : nodepointer
        end; { of record node }
procedure delete(var ptr : nodepointer; var head : nodepointer);
begin
    if ptr <> head then { not deleting list head }
    begin
        { there must be node to left of current node }
        ptr ↑ .llink ↑ .rlink := ptr ↑ .rlink;

        { last node in list has rlink = nil }
        if ptr ↑ .rlink <> nil then
            ptr ↑ .rlink ↑ .llink := ptr ↑ .llink;
        dispose(ptr)
    end
    else
    begin
        if head ↑ .rlink <> nil then { make that node new head }
        begin
            head ↑ .rlink ↑ .llink := nil;
            head := head ↑ .rlink;
            dispose(ptr)
        end
    end
end; { of procedure delete }
```

As you can see, the procedure is complicated by many different special cases. We can simplify the procedure if we change the representation of our list. Let an empty list be denoted by a header node with the following structure:

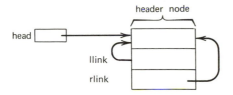

A nonempty list would be represented by:

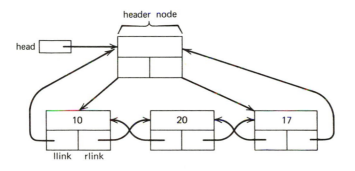

If p is a pointer to any node in this type of doubly linked list, then:

$$p = p \uparrow .rlink \uparrow .llink = p \uparrow .llink \uparrow .rlink$$

Compare the previous deletion procedure with this one.

```
procedure delete(ptr : nodepointer; var head : nodepointer);
begin
    if ptr <> head then
    begin
        ptr ↑ .llink ↑ .rlink := ptr ↑ .rlink;
        ptr ↑ .rlink ↑ .llink := ptr ↑ .llink;
        dispose(ptr)
    end
end; { of procedure delete }
```

The simplicity of this routine is a result of including just one extra node (the head node) in an empty list.

## 8.5   APPLICATIONS OF LINKED LISTS

Linked lists can be used in many different applications. This section will describe some of them. In the exercises at the end of the chapter you will develop and implement algorithms for some of these applications.

Matrices have many applications in all branches of science. They occur naturally in the solution of simultaneous linear equations. For the following system of equations:

$$x + y = 1$$
$$2x + y = 0$$

the coefficients of the variables x and y are grouped together in a matrix.

$$\begin{bmatrix} 1 & 1 \\ 2 & 1 \end{bmatrix}$$

The matrix can be represented by a two-dimensional array.

Problems from physics might require solving systems with thousands of variables. A system of 1000 equations with 1000 unknowns could require a coefficient matrix with 1 million entries! This exceeds the main memory capacity of many large-scale computers.

Large systems often contain many zeroes. The corresponding matrices would also therefore contain many zeroes. Such matrices are called *sparse*. Instead of representing a sparse matrix with a two-dimensional array, we can design a data structure that would represent only the nonzero elements of the matrix. Our data structure will have one node per nonzero coefficient; we include in this node the row and column of the coefficient.

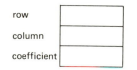

Obviously this node should be linked to something. One method uses header nodes—one node for each column and row—and links the nonzero coefficient entries to other elements in a row or column list or to a header node. For example, the matrix:

$$\begin{bmatrix} 3 & 2 & 0 \\ 0 & 1 & 6 \\ 1 & 0 & 4 \end{bmatrix}$$

might be encoded:

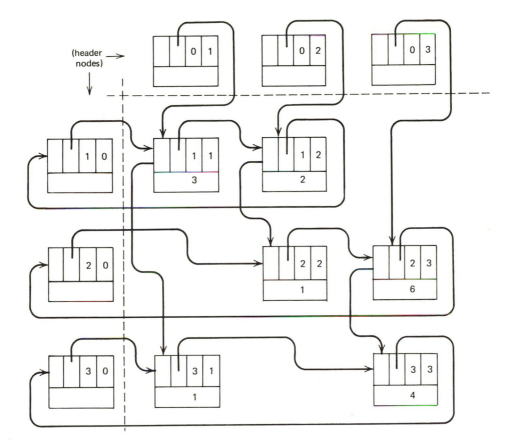

In the exercises you will be asked to write procedures that add and subtract sparse matrices and print the elements of the matrix in a readable form.

Another classic example of the use of linked lists is the symbolic manipulation of polynomials. A polynomial of the form:

$$a_n x^n + a_{n-1} x^{n-1} + \ldots + a_1 x^1 + a_0$$

where the $a_i$ are real coefficients, can be represented (term by term) by a linked list using the following node structure.

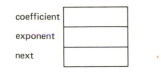

The polynomial:

$5x^3 + 2x - 5$

would be represented by this linked list.

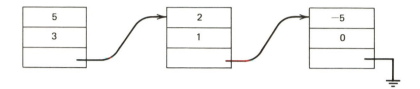

In the exercises you will be asked to implement algorithms to add, subtract, and multiply two polynomials using this linked list representation.

## 8.6  TREES

Trees are familiar from everyday life. Everybody has a family tree.

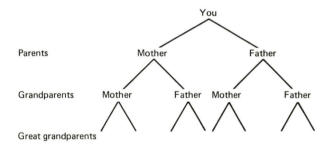

Sporting competitions are often displayed in a treelike manner.

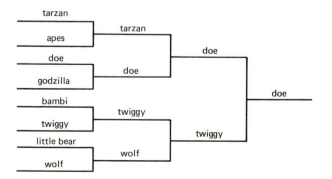

We are also familiar with parse trees.

The man bit the dog.

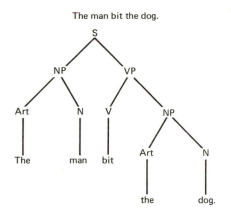

Trees also have many computer-related applications. For example, we can draw a parse tree for the expression a ∗ b + c.

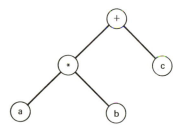

When a compiler parses a program to determine if it is syntactically correct, it can generate a parse tree. A simple expression grammar might look like:

| | | |
|---|---|---|
| expr | : : = term | { (" + " | " − ") term } |
| term | : : = factor | { "∗" factor } |
| factor | : : = letter | "(" expr ")" |
| letter | : : = "a" | "b" | "c" |

where : : = means "is defined as," { } means 0 or more repetitions of the information enclosed inside the braces, and | means alternation, i.e., selection of one of the alternatives. Symbols enclosed in double quotes (") are called terminal symbols and these symbols are the only ones allowed in expressions. These expressions would produce the following parse tree.

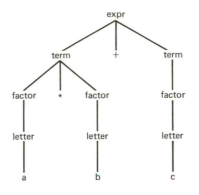

### 8.6.1   Definition

A tree has a precise mathematical definition that you are urged to study. As we describe some properties of trees, keep the following definition in mind.

*Definition.* A tree T is a finite set of one or more nodes such that there is a specially designated node t ε T (called the *root* of T) and T − {t} is partitioned into disjoint subsets $T_1$, $T_2$, . . . , $T_n$, each of which is itself a tree (called a *subtree* of the root t).

This definition implies that every node of a tree is a root of some subtree contained in the whole tree. Let us look at this definition more carefully using the following tree.

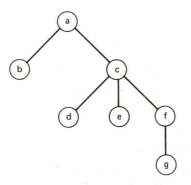

This tree is composed of seven nodes.

T = { a, b, c, d, e, f, g }

Its root is a. It has two subtrees.

$$T_1 = \{\, b \,\} \qquad T_2 = \{\, c, d, e, f, g \,\}$$

The number of subtrees of a node is called the node's *degree*. Hence node a is of degree 2. Nodes with no subtrees (b, d, e, g) are called *leaf* or *terminal nodes*. All other are *nonterminal nodes*. The nonterminal nodes are a, c, and f; they are often called *internal* or *branch nodes*.

In discussing trees we often use familial terms. For example, node a is the *parent* of nodes b and c, and nodes b and c are the *children* of node a.

Before we get mired in more terminology, we should consider how to implement a tree structure. For our example each node will have a char field to represent the node name (a − g) and from 0 to 3 pointer fields.

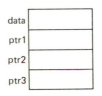

This node structure is adequate for the present example, but not for:

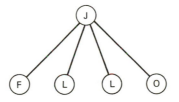

In order to avoid such representational problems we will restrict our attention to a special type of structure called a *binary tree*.

### 8.6.2 Binary Trees

Binary trees are distinct from generalized trees in two important ways. Nodes in a binary tree have degrees less than or equal to two (i.e., have two or less children), and binary trees can be empty (contain no nodes at all). Consider how binary trees are defined.

*Definition.* A *binary tree* is a finite set of nodes that is either empty or consists of a root and two disjoint binary trees called the *left* and *right subtrees*.

Here are some examples of binary trees.

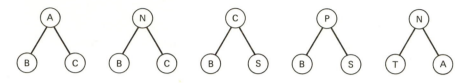

All these binary trees are identical except for their data fields. The following trees are *not* identical.

The binary tree on the left has an empty *right* subtree whereas the one on the right has a empty *left* subtree.

    The definitions of a tree and a binary tree are *recursive*; that is, they define an object in terms of other objects of its class. As we will see, the recursive definition gives rise to many recursive algorithms dealing with these structures.

    We can use the following node structure to represent binary trees.

```
type
    nodepointer              = ↑ node;
    node                     =
        record
            data             : char;
            leftsubtree      : nodepointer;
            rightsubtree     : nodepointer
        end; { of record node }
var
    root                     : nodepointer;
```

Then the tree:

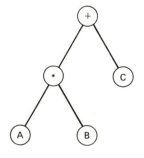

is represented by the data structure:

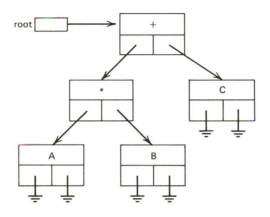

    To determine the number of nodes in a tree, we will have to "walk through" the tree structure, counting the nodes as we go. This is known as *traversing* a tree, or *visiting* each node of the tree exactly once.

    There are a number of ways to do this. One method starts at the root and visits it; in our case we add one to a counter, keeping track of how many nodes we have visited so far. Next we have a choice: we can visit either the nodes in the left subtree (*, a, b) or those in the right subtree (c). If we choose to visit the left subtree first, we traverse it in the same way as the original tree. We visit the root (of the subtree) and then visit all the nodes of the left subtree followed by all the nodes of the right subtree.

    Let us see exactly what this means for our previous example.

1: visit the root : (' + ')
2: visit the left subtree: ('*' 'a' 'b')
3: visit the right subtree: ('c')

Steps 2 and 3 each involve visiting a subtree. We can visit each node of a subtree by reapplying all three steps:

    1:   visit the root ' + '

    2:   visit the left subtree
2.1:   visit the root '*'
2.2:   visit the left subtree
2.3:   visit the right subtree

    3:   visit the right subtree
3.1:   visit the root 'c'

3.2:    visit the left subtree
3.3:    visit the right subtree

When a subtree is empty (Steps 3.2 and 3.3), we go on to the next step.

1:    visit the root ' + '

2:    visit the left subtree
2.1:    visit the root '*'
2.2:    visit the left subtree
2.2.1:    visit the root 'a'
2.2.2:    visit the left subtree (empty)
2.2.3:    visit the right subtree (empty)
2.3:    visit the right subtree
2.3.1:    visit the root 'b'
2.3.2:    visit the right subtree (empty)
2.3.3:    visit the left subtree (empty)

3:    visit the right subtree
3.1:    visit the root 'c'
3.2:    visit the left subtree (empty)
3.3:    visit the right subtree (empty)

You may think that all of this will be difficult to program, but it is actually a trivial application of a recursive procedure.

```
procedure preorder(root : nodepointer; var count : integer);
begin
    if root <> nil then
    begin
        { visit the root }
        count := count + 1;

        { visit the left subtree }
        preorder(root↑.leftsubtree);

        { visit the right subtree }
        preorder(root↑.rightsubtree)
    end
end; { of procedure preorder }
```

Take a few minutes to step through the actions of this procedure on the example tree. If you understand how the recursion is working, you will have no problem understanding how any other algorithm in this chapter works.

There are a number of other ways to traverse a tree. As you might have guessed from the procedure name, the traversal scheme we have been describing is called *preorder traversal*. It walks through the tree by:

1.  Visiting the root.
2.  Visiting the left subtree.
3.  Visiting the right subtree.

Other traversal methods can be arrived at by permuting the order of the preceding operations. Two permutations of the most interest result in *postorder traversal*:

1.  Visit the left subtree.
2.  Visit the right subtree.
3.  Visit the root.

and *inorder traversal*:

1.  Visit the left subtree.
2.  Visit the root.
3.  Visit the right subtree.

There is a natural correspondence between the preorder, postorder, and inorder traversals of a tree and the prefix, postfix, and infix forms of an expression. Using our previous tree and assuming by "visit the root" that we mean output the contents of the data field:

| | | |
|---|---|---|
| *prefix* expression for a*b+c: | +*abc | (operators precede operands) |
| *preorder* traversal of tree: | +*abc | |
| *postfix* expression for a*b+c: | ab*c+ | (operators follow operands) |
| *postorder* traversal of tree: | ab*c+ | |
| *infix* expression for a*b+c: | a*b+c | |
| *inorder* traversal of tree: | a*b+c | |

For completeness we will write the postorder and inorder traversal procedures; we code the step "visit the root" as a call on procedure visit with a pointer to the root.

```
procedure postorder(root : nodepointer);
begin
    if root <> nil then
    begin
        { visit the left subtree }
        postorder(root↑.leftsubtree);

        { visit the right subtree }
        postorder(root↑.rightsubtree);

        { visit the root }
        visit(root)
    end
end; { of procedure postorder }

procedure inorder(root : nodepointer);
begin
    if root <> nil then
    begin
        { visit the left subtree }
        inorder(root↑.leftsubtree);

        { visit the root }
        visit(root);

        { visit the right subtree }
        inorder(root↑.rightsubtree)
    end
end; { of procedure inorder }
```

There is a convenient way to remember the order in which nodes are visited in each of the traversal methods. We start by ''tracing out'' the form of the tree as indicated.

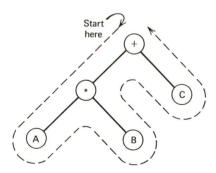

For a preorder traversal we visit a node as we pass it on the left, tracing out the tree. Hence the nodes will be visited: +*abc.

For a postorder traversal we visit a node as we pass it on the right going up. Hence the nodes will be visited: ab*c+.

Finally, for an inorder traversal we visit a node as we pass underneath it. Hence the nodes will be visited: a*b+c.

Another scheme for remembering the order in which nodes are visited in the three traversal methods is:

> while tracing out the tree
> > in *pre*order you visit a node *before* visiting its children;
> > in *post*order you visit a node *after* visiting its children; and
> > in *in*order you visit a node in the *middle* of visiting its children.

Now consider the following problem. You are given pointers to two binary trees. Determine if they are "equal," that is, if they have the same structure, and if the contents of the data field of corresponding nodes in the trees are the same. By this definition, the two trees:

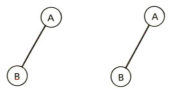

are equal but these two trees are not.

Try to write a recursive algorithm to solve this problem. Remember the boundary conditions. If you need help solving this problem, look at the exercises at the end of this chapter.

As another instructive exercise, consider the problem of making an exact copy of a binary tree. Given a pointer to the tree to be copied:

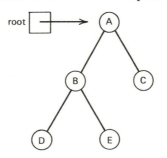

make an exact duplicate of it, letting root1 point to the copy.

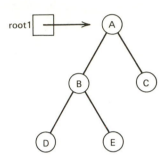

We will use a modified postorder traversal of the original tree; it works like this.

1. Copy the left subtree.
2. Copy the right subtree.
3. Allocate a new node.

```
procedure copy(root : nodepointer; var root1 : nodepointer);
var
      lefttree                              : nodepointer;
      righttree                             : nodepointer;
begin
      root1 := nil;
      if root <> nil then
      begin
          { copy left subtree-lefttree will point to left subtree of node being
             copied }
          copy(root ↑ .leftsubtree,lefttree);

          { copy right subtree-righttree will point to right subtree of node being
             copied }
          copy(root ↑ .rightsubtree,righttree);

          { allocate a new node and set its fields }
          new(root1);
          root1 ↑ .leftsubtree := lefttree;
          root1 ↑ .rightsubtree := righttree;
          root1 ↑ .data := root ↑ .data
      end
end; { of procedure copy }
```

We strongly urge you to trace through this procedure with the following tree:

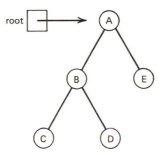

to see exactly how the copy operation accomplishes its task.

### 8.6.3 Threaded Binary Trees

With the representation we have chosen for binary trees there are many **nil** pointers. If there are n nodes in a tree, there will be 2n link fields. n + 1 of these will contain **nil** pointers, which can be used to point to other nodes in the tree. These new pointers are called *threads*. In particular we will let the leftsubtree field point to the predecessor of the current node in an inorder traversal and the rightsubtree field point to the successor of the current node in an inorder traversal. Using this scheme, the tree:

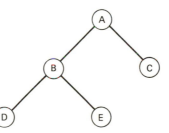

would have the representation:

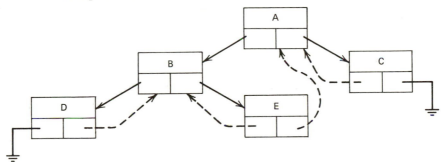

where we have represented the threads by dashed links. Unfortunately, this suffers from one simple flaw: the program cannot determine when a link field contains a pointer or a thread. To distinguish these two cases, we will add two ''bit fields'' (i.e., boolean) to resolve the problem. The node structure will look like:

| data | |
|---|---|
| leftsubtree | |
| rightsubtree | |
| leftthread | |
| rightthread | |

where:

leftthread = true if leftsubtree field contains a thread, false otherwise

and

rightthread = true if rightsubtree field contains a thread, false otherwise

To save space and insure that the leftthread and rightthread fields do not occupy one word apiece, we will use a **packed record** to implement the node structure.

```
type
    node                              =
        packed record
            data                : char;
            leftsubtree         : nodepointer;
            rightsubtree        : nodepointer;
            leftthread          : boolean;
            rightthread         : boolean
        end; { of packed record node }
```

If you look back at the internal representation of the tree, you will see that there are still two nodes containing **nil** pointers. As before, we will add one more node (the head node) to the tree structure. An empty tree will contain this node and be represented as:

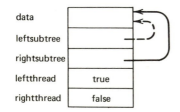

| data | |
|---|---|
| leftsubtree | |
| rightsubtree | |
| leftthread | true |
| rightthread | false |

The leftsubtree field of the head node will point to the root of a nonempty tree, as this picture suggests.

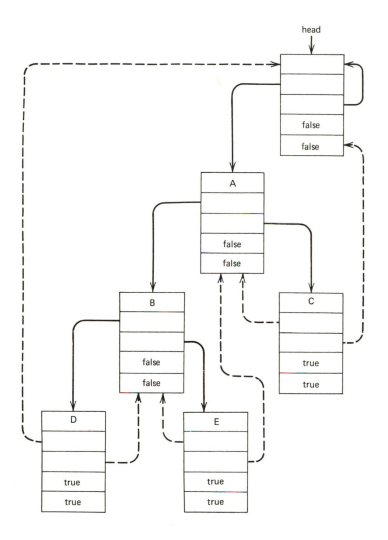

The exercises will explore a nonrecursive method for traversing a threaded binary tree in inorder.

### 8.6.4  Binary Tree Representation of Trees

We covered binary trees in such depth because every tree can be represented as a binary tree. As an example, consider this tree.

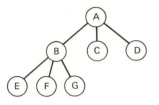

Recall that a node in a binary tree has only two link fields: one to point to the leftsubtree of that node and the other to point to the rightsubtree of that node. In this tree consider removing all but the leftmost pointer of a node, as shown. Now we use the "free" rightsubtree pointers to link nodes on the same "line."

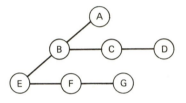

This may not look like a binary tree, but it is. (Try tilting the figure 45 degrees.)

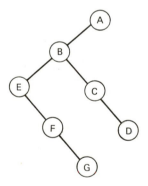

This simple transformation scheme is significant because it circumvents the problem of not knowing how many link fields to allocate in a node used in a tree. Using this technique, we can always convert any arbitrary tree to a binary tree.


## 8.7  CONCLUSION

In this chapter we have described two widely used data structures: linked lists and trees. The exercises reinforce and expand on the material presented. It is not essential to work out all the details of every exercise, but it is important to at least try to sketch out the data structure you would use to solve each problem.

An important point that we have made in the last two chapters bears repeating here. It is extremely important to have a good pictorial representation of a data structure. Draw the picture of the data structure before you try to use it. Study the picture to see if the operations to be performed on the data structure can be simplified if, for example, you were to use an extra pointer field or header node. Then use the picture to guide you in the implementation. The final picture of any data structure used in a program should always be included in the documentation (or user's manual) for the program.

## EXERCISES FOR CHAPTER 8

1. The following singly linked list represents a queue with front and rear pointers.

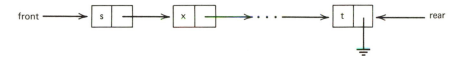

Show that only one of the pointers front and rear is necessary if the queue is represented as a singly linked *circular* list. Write a complete Pascal procedure to dequeue an element from the "circular" queue. Draw a picture first.

Write a Pascal function that counts the number of nodes in a list. The parameter to the function should be a pointer to some element in the list. Your function should work properly whether the list is circular or one-way with a **nil** pointer in the last node.

2. Write a Pascal procedure Merge (p1,p2) that merges two ordered lists, pointed to by p1 and p2, respectively. The program should merge the nodes of p2 into p1 so p1 remains an ordered list, as illustrated. Duplicate nodes in list p2 should not be merged into p1 but should remain in list p2. Thus, when the merge is complete, list p1 will be the union of lists p1 and p2, and list p2 will be the intersection.

Before merge:

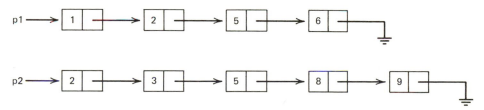

After merge:

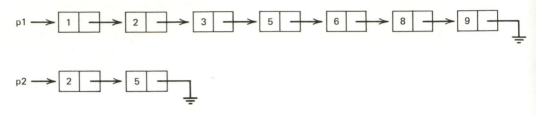

3. Write a Pascal procedure to reverse the order of nodes in a list with ptr as a pointer to the first node.

Before operation:

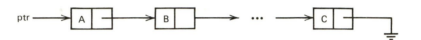

After operation:

(*Hint.* Use three pointers walking through the list.)

4. Write a Pascal procedure to sort a list into increasing order.

Before operation:

After operation:

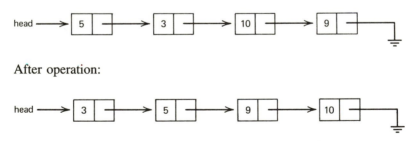

We have emphasized throughout this chapter that you must always check boundary conditions on list processing algorithms. Such checking can be reduced at the expense of a little storage by including appropriate head nodes and making lists circular. The following exercises ask you to rework some of the examples in the text by including head nodes.

5. Section 8.2.3 describes a procedure to enqueue an item in a queue represented as a linked list. Insert a head node in the data structure used to implement a queue. This implies that an empty queue will contain one node (the head node). Rewrite procedures enqueue and dequeue with this assumption.

6. In the implementation of procedure dequeue we returned the contents of the data field of the node at the head of the list. Is it generally more desirable to return a pointer to the record removed from the queue?

7. Add a head node to the lists used in the concatenation procedure. Will this simplify the concatenation procedure? Does your head node contain two pointer fields?

8. Section 8.3 described the use of sentinel nodes. An alternative to this is using a circularly linked list with a head node. Initially, it looks like:

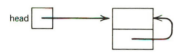

```
{ create an empty list }
new(head);
head ↑ .next : = head
```

Searching for a node is simple.

```
head ↑ .data : = item;      { head node used as sentinel }
ptr : = head ↑ .next;
while ptr ↑ .data <> item do ptr : = ptr ↑ .next
```

Ptr now points to a node containing the searched for item (it may be pointing to the head node.) Note that there are no special cases in this search. Try to insert a new node in such a list. There should be no special cases.

9. Set up the data structure described in the text to represent an n-x-n sparse matrix (where n = 50).

10. Write a procedure that reads three numbers (two indices and a value) and constructs a new node with these three values. Write another procedure to link this node into the data structure at the appropriate place.

11. Write a procedure that will output the values in the sparse matrix. The output should be designed to be as readable as possible and take as little paper as possible.

12. Next write a procedure to add two sparse matrices A and B and generate a third, C.

13. Repeat Exercise 12 performing a subtraction instead of an addition.

14. Using the node structure described in the text, build the data structure for the polynomial:

$$x^3 + 2x^2 - x + 10$$

Write a procedure that inputs a coefficient, exponent pair and builds the data structure dynamically.

15. Write a procedure that outputs the polynomial encoded in the data structure described in the text. Use ↑ to denote exponentiation.

16. Write a procedure to add two polynomials.

17. Write a procedure to subtract two polynomials.

18. Write a procedure to eliminate any unnecessary terms in the polynomials above. For example:

        eliminate        0x,      0x², . . etc.
    and   change            1x↑0  to 1

You should have only one constant term at the end of the polynomial.

19. Write a procedure to multiply two polynomials. Reduce your answer.

20. What are the terminal and nonterminal nodes in this binary tree?

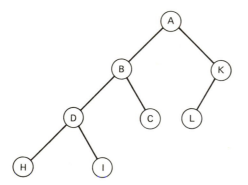

21. Design and implement a method of inputting and constructing a binary tree.

22. In the tree in Exercise 20, node A is at *level* 1; nodes B and K are at level 2; nodes D, C, and L are at level 3; and nodes H and I are at level 4. Write a Pascal procedure to output the nodes of a binary tree by increasing level number. Try to write it recursively.

23. Write a procedure that counts the number of leaf nodes in a binary tree.

24. Two binary trees are ''equal'' if they have the same structure and if the contents of the data field of corresponding nodes in the tree are the same. Write a function of two arguments that traverses each binary tree in preorder. The function returns true if the trees are equal and false otherwise.

25. Given a threaded binary tree, write a procedure to find the inorder successor of any node in the tree.

26. Using your procedure from Exercise 25, write a nonrecursive algorithm to traverse a threaded binary tree in inorder.

# Chapter 9

# FILES AND A CASE STUDY

## 9.1 INTRODUCTION

In this chapter we discuss sequential files and a large program ( a text editor) that uses them. Any program that uses input or output statements also uses files. The read statement reads data from a file named input, and the write statement writes data to a file named output. Pascal allows you to create and manipulate files other than input and output, as we will see next.

## 9.2 FILES

Like an array, a file in Pascal contains components of the same type. But there are two important differences between arrays and files.

1. An array has a fixed size. A file may initially contain zero components and expand as more components are added. (The number of components in a file is called its length.)

2. Elements of an array can be accessed in any order. For a file, however, components must be accessed sequentially.

We visualize a file like this.

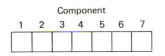

The two operations that can be performed on files are reading and writing. A *file pointer* is associated with each file. If the file pointer is positioned as shown:

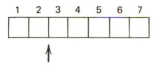

the third component will be read next by a read operation on the file. You cannot read past the last component in a file. After reading the seventh component, the picture looks like:

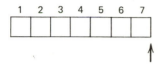

Attempting another read operation on this file causes a run-time error (attempt to read past the end-of-file). The boolean function eof determines if the file pointer is at the end-of-file.

The situation is different for the write operation. Here the file pointer points to the current end-of-file (unless it has been reset).

The next write operation adds a fifth component and also moves the file pointer.

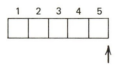

Every file in Pascal except input and output must be defined in a **var** declaration. The file is given a name and a type in this declaration (like any other variable). Its type is:

**file of** *componenttype*

where componenttype specifies the type of the components of the file. Here are some example file declarations.

```
var
    temp                        : file of char;
    x                           : file of integer;
    y                           : file of real;
```

Files input and output are predefined **file of** char and should not appear in **var** declarations. We can associate a type name with the file type as follows:

```
type
    text                   = file of char;
var
    temp                   : text;
```

In Pascal the name text is a predefined type identifier with the preceding meaning. Hence files input and output are of type text.

The components of a file need not be restricted to simple types, as the following example illustrates.

```
type
    employeerecord         =
        record
            name            : alfa;
            age             : 0..150;
            sex             : (male,female)
        end; { of record employeerecord }
var
    workers                : file of employeerecord;
```

Files other than input and output must be prepared for writing or reading by *rewrite* and *reset* operations. These are predefined procedures.

Before writing the first component to a file called f, we must execute:

*rewrite(f)*

This prepares file f for subsequent write operations. After the rewrite operation, file f will have zero components; that is, it will be an empty file. To write components to the file, we use a modified form of the write statement.

*write(f,x)*

Notice that the first parameter to the procedure is the file name (if the file name is omitted from the argument list of procedure write, "output" is assumed). This statement appends a component (whose value is that of the expression x) to the end of file

f. The type of the expression x must be compatible with the declared component type of file f.

The effect of the write operation is:

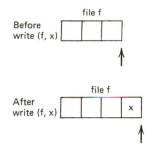

More than one component can be added to a file using the write statement.

*write(f,a,b,c,d)*

The updated file looks like:

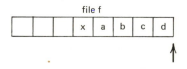

A file created by a program can be read by the same program (or other programs). Before issuing the first read statement, we must execute:

*reset(f)*

This prepares file f for subsequent read operations. After the reset operation, the file pointer will precede the first component of the file.

To read the first component from the file, we use a modified form of the read statement.

*read(f,x)*

The first parameter to the procedure is the file name. If the file name is omitted from the argument list of procedure read, the file name "input" is assumed. The read statement copies the next component of file f into the variable x and moves the file pointer past this component. The type of the variable x must be compatible with the declared component type of file f.

The effect of the read operation is illustrated by the following diagram.

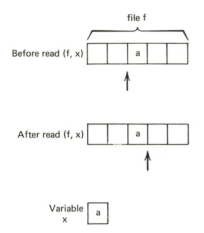

Files that are listed in the program header of a Pascal program either already exist (before the program is run) or are created by the program and are to be retained after it terminates (in some system-dependent manner). Any other files that are used as temporary (or scratch) files need not be declared in the **program** statement. All files other than input and output must be declared in the **var** declaration.

### 9.2.1  The File Buffer Variable

Every file has a special variable called the *buffer variable*. For a file named f the buffer variable is called f ↑ . The buffer variable can be used just like any other variable. When a file is read, the buffer variable contains the component immediately to the right of the file pointer. For example, if file f is reset by:

*reset(f)*

a picture of file f might look like:

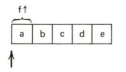

Then the value of f ↑ is a. The file pointer can be moved one position to the right by using a get operation on the file.

   *get(f)*

Its effect is:

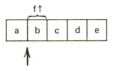

Now the value of f ↑ is b. The read operation is defined in terms of the get operation and the buffer variable.

$$read(f,x) \qquad = \qquad \begin{aligned} &\textbf{begin} \\ &\quad x := f\!\uparrow; \\ &\quad get(f) \\ &\textbf{end} \end{aligned}$$

This definition for read can sometimes cause problems. When data are read from a file, the file pointer is positioned at the component following the one just read. If there is no component there, the file pointer will be pointing to the end-of-file. As a result, if a program is structured in such a way that an eof test is performed after a read statement, a valid data item from the file may not get processed.

We can append a component to a file by using the buffer variable and a put operation. For example, assume the buffer variable has been assigned a value through the assignment statement:

   *f↑ := a*

Then this value may be appended to the file f by using the put(f) command, which writes the contents of the buffer variable to the end of the file.

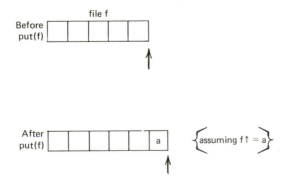

After the put operation, the value of the buffer variable (f ↑ ) is undefined.
It should be clear that the write statement is equivalent to:

$$write(f,x) \quad \doteq \quad \begin{array}{l} \textbf{begin} \\ \quad f\uparrow := x; \\ \quad put(f) \\ \textbf{end} \end{array}$$

### 9.2.2   Text Files

Any file to be printed on a line printer must be of type text. Recall that:

**type**
       *text*       = **file of** *char;*

Text files have one distinctive property. They are composed of *lines* terminated
by an *end-of-line* marker. The end-of-line marker is not treated as a character and
therefore cannot be manipulated as such. The end-of-line marker can be tested for by
the boolean function eoln(f), where f is the name of some text file. (If no file name
is specified, ''input'' is assumed.) The function returns true if the file pointer is
positioned immediately before the end-of-line marker. In the following example, we
let eol stand for the end-of-line marker. A text file can be pictured as:

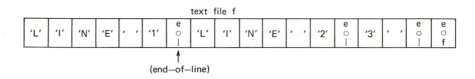

text  file  f

(end—of—line)

To read the first character from the file f, we perform:

    *reset(f);*       { *which positions the file pointer to*
                    *the left of the first character and fills*
                    *the buffer variable* }

    *read(f,ch)*       { *which sets ch = 'L' and moves the*
                    *file pointer to the left of the second*
                    *character* }

If the file pointer is positioned immediately to the left of an eol marker and the
following read operation is performed:

    *read(f,ch)*

a blank is placed into the variable ch (ch = ' ') and the file pointer is shifted past the eol marker.

The effect of the readln(f) operation is to move the file pointer past the *next* eol marker in the file. Hence f ↑ will contain the first character of the next line (if there is one).

Eol markers can be inserted in text files that are being written by issuing a

*writeln(f)*

This appends the eol marker to the text file and updates the file pointer in the usual manner.

The following program shows how to make a copy of a text file.

```
program copy(file1,file2,output);
var
    ch                  : char;      { temporary }
    file1               : text;      { file to be copied }
    file2               : text;      { will be a copy of file1 }
begin
    { prepare to read from file1 }
    reset(file1);

    { prepare to write to file2 }
    rewrite(file2);

    { loop through file1 reading (and writing) one line at a time }
    while not eof(file1) do
    begin
        { copy next line }
        while not eoln(file1) do
        begin
            read(file1,ch);
            write(file2,ch)
        end; { of while }

        { read past eol marker }
        readln(file1);

        { mark eol on file2 }
        writeln(file2)
    end { of while }
end.
```

Text files will be used in the text editor described in the following sections.

## 9.3  CASE STUDY: A TEXT EDITOR

If you have been using an interactive system, you have probably been using a text editor to prepare programs. The text editor allows you to maintain and update a file easily. The editor is "driven" by commands from the user. Here is a typical set of commands for a simple editor.

1. *Print*. Print a line (or lines) of the file being edited.
2. *Delete*. Delete a line (or lines) from the file being edited.
3. *Insert*. Insert a line (or lines) into the file being edited.
4. *Change*. Change a string on a line to a new string.
5. *Stop*. Leave the text editor and update the file being edited.

The editor we will describe and implement is more complicated than this; the syntax of the editor's commands and the commands themselves are based on an editor developed by Dan Dorrough of Purdue University.

The editor is a string- (or context-) oriented editor. Line numbers are not an intrinsic part of the file to be edited so, to print the nth line in the edit file, you try to locate the line by using a command to search for a string that occurs on that line.

Most editor commands are performed on the "current line." When the editor is entered, the first line of the file becomes the current line. The editor prompts users for a command by printing a #. If users issue the print command (abbreviated by the letter P), the current line is printed. If users type P2, the editor will print the current line and then move the current line pointer to the next line and print it as well. Notice that this resets the current line pointer. Any time the editor is in command mode, you can determine the current line either by issuing the P command or by typing a carriage return. This implies that the editor uses a default command when users do not supply one; that command is P.

We mentioned that the editor is string oriented. Strings are enclosed in delimiters [such as a slash (/)], as in:

/string/

Strings are used with many of the editor's commands. For example, we can instruct the editor to locate the first occurrence of a string by using the "find" command (abbreviated F).

F/string/

The editor starts its search for the string on the line *following* the current one. If it finds the string in the file, it will print the line containing that string *and* reset the current line pointer to that location. If the editor does not find the string, the message ''string not found'' will be printed, and the editor will *not* reset the current line pointer.

The string delimiters can be any nonalphabetic, nonnumeric character that does not appear in the string. Some examples of strings enclosed in delimiters are:

/THIS IS A STRING/
;SO IS THIS;
?THIS TOO?

If two delimiters are used, they must be identical. On many of the commands, the trailing delimiter is not necessary; for example, the find command can look like:

F/string

(if / is not part of the string). Since this editor was designed with lazy people in mind, it is sufficient to type:

/string

to find the string in the edit file. Remember that the editor's default command is Print. Typing:

/string

instructs the editor to print the first line after the current line that contains the string.

The find command locates strings ''forward'' in the file (i.e., between the current line and the end-of-file). If you need to find a string *before* the current line, precede the string by a minus ($-$).

F$-$/string ,

If the string is found, that line will be printed and the current line pointer will be updated accordingly.

The insertion command allows you to insert any number of lines *after* the current line. Users type an I (for insert) when the editor is in command mode (i.e., when the editor has typed the #). This places the editor in insertion mode, which it distinguishes from command mode by typing an equal sign ($=$). Users can return to command mode by typing a # as the first character directly after the $=$.

To clarify the commands described so far, we will present a simple editing session. The edit file contains these lines.

THIS IS A TEST.
FOR THE NEXT 60 SECONDS
THIS STATION
WILL CONDUCT A TEST

If the editor is invoked with this as the edit file, the current line pointer will be set to the first line of the file. The editor will print its command mode prompt.

#

To print the current line, the user could type a P following the # sign.

#P

followed by a carriage return to send the command to the editor. The editor responds by printing the current line:

THIS IS A TEST.

followed by the command mode prompt on the next line.

#

If we want to add some text after the last line of the file, we first search for a string on the last line.

#F/WILL

The editor prints:

WILL CONDUCT A TEST

and makes this the current line. It then signifies that it is ready for another command by typing:

#

Now we type I followed by a carriage return to enter insertion mode (to insert *after* the current line).

#I

The editor responds with equal signs (=) until we type a # after the =.

    = OF THE EMERGENCY
    = BROADCAST SYSTEM.
    = THIS IS ONLY A
    = TEST.
    = #
    #

When we return to command mode, the "current line" is the last line entered in insertion mode.

    TEST.

The editor attaches special significance to the two symbols:

    ↑ (or ∧ on some terminals)

and

    ↓ (or ! on some terminals)

↑ denotes the first line of the file, and ↓ denotes the last line of the file. For example, the command:

    ↑ P

will print the first line of the edit file, and:

    ↑ P ↓

will print the entire contents of the edit file (leaving the current line pointer at the last line).

The next section contains a complete description of all the commands in the editor. We will then present an extended example of how to use the editor. After that we will discuss how to implement the editor. A complete implementation is presented in the last section of this chapter.

## 9.3.1  Editor Commands

All editor commands are recognized by their first letter. Users have the choice of using the first letter or typing the complete command, as in:

F or FIND

In the command descriptions we will enclose any optional characters in square brackets [    ].

### 9.3.1.1    Commands to Change Current Line

Command:     B[OTTOM]
Meaning:     Reset the current line pointer to point to the last line in the edit file. The last line is printed. The message "empty file" is printed if there are no lines in the edit file.
Examples:    #B
             #BOTTOM

Command:     F[IND][—]/string[/]
Meaning:     Find the first occurrence of the string in the edit file, starting the search at the line following the current line. If the string is found, print the line containing this string. If the string is not found, print the message "string not found" and do not reset the current line pointer. If the string is preceded by a — , the search is performed "backward" in the file.
Examples:    #F/THIS STRING/
             #F/A LONG STRING WITH TRAILING BLANKS  /
             #F;JIMMY CRACK CORN;
             #F — ?I AM NOT A CROOK?
             #FIND/.

Command:     N[EXT] [<NUMBER>]
Meaning:     In the simplest form (just N), the line following the current one is made the new current line and is printed. If there is a positive or negative integer in the command the current line is moved forward or backward that many lines. If there is no "next" line, the message "end-of-file" or "top-of-file" is printed and the current line pointer is not changed.
Examples:    #N
             #N3    (If it exists, the third line following the current line becomes the new current line.)
             #N–2   (If it exists, the second line before the current line becomes the new current line.)

Command:     T[OP]
Meaning:     Make the first line of the edit file the current line and print it.

The message ''empty file'' is printed if there are no lines in the edit file.

Examples:   #T
          #TOP

## 9.3.1.2   Edit Directives

Command:   A[PPEND]/STRING/[<COLUMN NUMBER>]
Meaning:   Append the string to the end of the current line (if no column number is specified) and print the line. If a column number is specified, append the string to the current line starting in that column. Any characters in that column and beyond from the original line are discarded.
Examples:   Assume the current line is:
          MARY HAD A LITTLE
          #A/LA/
          MARY HAD A LITTLE LA
          #A/MB/
          MARY HAD A LITTLE LAMB
          #APPEND/, ITS FLEECE/
          MARY HAD A LITTLE LAMB, ITS FLEECE
          #A/./5
          MARY.

Command:   C[HANGE]/OLD STRING/NEW STRING/
Meaning:   Change the first occurrence of the old string on the current line to the new string. If the old string is not present on the current line, the message ''no such string'' is printed. If the new string is not specified, the old string is replaced by the null string (i.e., it is removed from the current line). We assume that a null string precedes each line of the edit file. Hence we can add characters to the beginning of a line by changing the null string (specified by / /) to the characters to be added.
Examples:   Assume the current line is:
          JACK BE MIMBLE, JACK BE QUACK
          #C/MI/NI/
          JACK BE NIMBLE, JACK BE QUACK
          #C/UA/UI/
          JACK BE NIMBLE, JACK BE QUICK
          #C/,/ /
          JACK BE NIMBLE JACK BE QUICK
          #C/ /BLACK /

BLACKJACK BE NIMBLE JACK BE QUICK
#CHANGE/BLACKJACK / /
BE NIMBLE JACK BE QUICK


Command:    D[ELETE] [<STRING> OR <NUMBER>]

Meaning:    Delete the current line (if optional string or number is not specified) and reset the current line pointer to the line following the deleted one. If the last line (bottom line) of the edit file is deleted, the message "bottom-of-file" is printed and the new bottom line becomes the current line. A specified number of lines can be deleted (starting from the current line) by using D<NUMBER>. Starting from the current line, lines can be deleted forward or backward in the edit file until a specified string is found (the line containing that string is deleted as well) by using D<STRING> or D-<STRING>.

Examples:    #D (delete the current line only)
#D2 (delete current line and the one following it)
#D-2 (delete current line and the one preceding it)
#D/IT/ (delete current line and all lines until the occurrence of the string 'IT')
#D-/THIS/
#D ↓ (delete all lines from current line to end-of-edit file)
# ↑ D ↓ (delete all lines in the edit file; this command is to be used at your own risk)


Command:    I[NSERT]

Meaning:    Enter the insertion mode; lines are inserted in the file after the current line. The editor prompts users by printing an equal sign (=). To leave insertion mode, users type a # after the = sign prompt. The last line typed becomes the current line.

Examples:    #I
=THIS WILL BE THE CURRENT LINE
= #
#INSERT
=THE LAST LINE
=TYPED
=IS THE CURRENT
=LINE
= #
#P
LINE

Users can also type a command after the # sign used to leave insertion mode.

#I
= FOR EXAMPLE,
= TO FIND THE FIRST
= LINE THAT THE
= USER INSERTED:
= #F—/FOR
FOR EXAMPLE,
#

| | |
|---|---|
| Command: | P[RINT] [<STRING> OR <NUMBER>] |
| Meaning: | In the simplest form (P), just the current line is printed (if it exists). A specified number of lines can be printed (starting with the current line) by using P<NUMBER>. P<STRING> prints lines, starting with the current one, until a line containing the string is found and printed. P-<STRING> prints lines going backward through the edit file until the line containing the string is found and printed. |
| Examples: | #P |

#P2
#P/THIS
#P-/THAT
#P ↓ (prints current line and all lines after it)
# ↑ P2 (prints first two lines of edit file)
# ↑ P ↓ (prints entire edit file)

The current line becomes the last line printed by the command.

| | |
|---|---|
| Command: | R[EPLACE] |
| Meaning: | The current line is deleted and the editor enters insertion mode. All lines entered by users in insertion mode replace the current line deleted. The last line entered by users in insertion mode becomes the current line. |
| Examples: | #R |

= FIRST LINE
= SECOND LINE
= LAST LINE
= #
#REPLACE
= ONE
= #P
ONE

Command:      = <text>
Meaning:      The = command replaces the current line with the text
              following the equal sign. The first character of the replacement
              text is the one directly following the = . The current line is not
              moved; it points to the replacement line.
Example:      = THIS LINE REPLACES THE CURRENT LINE.

Command:      (null)
Meaning:      The null command (which results from typing a carriage return
              with no other information on the line) causes the current line to
              be printed.

### 9.3.1.3   System Directives

Command:      ABORT
Meaning:      Terminate the editing session. Do *not* update the edit file; that is,
              the edit file is not changed from its contents upon entry to the
              editor.
Example:      ABORT

Command:      S[TOP]
Meaning:      Terminate the editing session. Write the internal buffer out to the
              edit file.
Examples:     S
              STOP

### 9.3.1.4   Environment Directives

Command:      H[EADER]<COLUMNS>
Meaning:      See Exercise 13 at the end of this chapter.

Command:      V[ERIFY]
Meaning:      The verify flag is toggled "on" or "off." The verify flag is
              used by directives such as "CHANGE" and "NEXT" to display
              their results. The verify flag is "on" when the editor is initially
              entered.
Examples:     V { Turn verification off if it is currently on. }
              VERIFY { Turn it back on. }

Any edit directive (see Section 9.3.1.2) can be preceded by plus or minus some
number. This will add (or subtract) that number to (or from) the current line pointer
before applying the edit directive. For example:

# + 2P   prints the second line past the current line

# − 1I   insertion mode is entered; the insertion takes place before the current
line

Although the editor does not store any internal line numbers for lines in the edit
file, you can still refer to any line by a line number. For example:

#2P              will print the second line in the edit file

#10C/IT/IS       changes IT to IS on line 10.

## 9.3.2  Example of Editor Usage

This section contains a sample text editing session that uses most of the commands
of the editor. We will be working with the following edit file.

AVOID CLICHES LIKE THE PLAQUE.
THIS IS A SENTENCE
NOONE PERSON IS EXACTLY ALIKE.
THINGS DON'T SEEM TO BE WHAT THEY APPEAR.
THREE ATTEMPTS AT HUMER, NO COMPLETIONS.
6 + 6 = 14

There are some obvious errors in this file that we will correct. When we enter the
text editor, the first line becomes the current line. The editor also responds with its
command mode prompt.

#

If we type a carriage return, the editor will respond:

AVOID CLICHES LIKE THE PLAQUE.
#

We fix the misspelling on the current line using the change command:

#C/Q/G
AVOID CLICHES LIKE THE PLAGUE.
#

The next line is missing a period. Each of the following command sequences could
be used to add the period (we are assuming that each time a command sequence
begins, the current line is the first line in the edit file).

#N
THIS IS A SENTENCE
#C/CE/CE.
THIS IS A SENTENCE.

#N
THIS IS A SENTENCE
#A/.
THIS IS A SENTENCE.

# + 1P
THIS IS A SENTENCE
#A/.
THIS IS A SENTENCE.

# + 1A/.
THIS IS A SENTENCE.

#/ENC
THIS IS A SENTENCE
#A/.
THIS IS A SENTENCE.

As you can see, there are many ways of accomplishing the same task. Let us go to the last line in the edit file and fix it.

#B
6 + 6 = 14
#A/ FOR LARGE VALUES OF SIX.
6 + 6 = 14 FOR LARGE VALUES OF SIX.

Now we will correct the spelling error on the preceding line.

# − 1C/ME/MO
THREE ATTEMPTS AT HUMOR, NO COMPLETIONS.

We now correct the line starting with NOONE.

#F − /NOONE
NOONE PERSON IS EXACTLY ALIKE.
#C/O/O /
NO ONE PERSON IS EXACTLY ALIKE.

We can insert a line after this one by:

    #I
    = NO TWO PEOPLE ARE EXACTLY DIFFERENT.
    = #
    #

To delete the line just inserted, we say:

    #D

Finally, to print the entire edit file, we use:

    # ↑ P ↓
    AVOID CLICHES LIKE THE PLAGUE.
    THIS IS A SENTENCE.
    NO ONE PERSON IS EXACTLY ALIKE.
    THINGS DON'T SEEM TO BE WHAT THEY APPEAR.
    THREE ATTEMPTS AT HUMOR, NO COMPLETIONS.
    6 + 6 = 14 FOR LARGE VALUES OF SIX.

To leave the editor, we use the STOP command (abbreviated S).

    #S

This command will rewrite the edit file with its modified contents.

Section 9.3.3 will describe our implementation of the text editor.

### 9.3.3   Data Structures for the Text Editor

We have implemented the text editor described in Sections 9.3.1 to 9.3.2. The implementation follows the guidelines of top-down program design described in earlier chapters. Instead of detailing all of the design decisions and refinements we went through, as we did for the simulator in Chapter 6, we will concentrate on describing the major data structures employed in our implementation.

One of the major design decisions we made very early was that the editor would be an "in-core" editor. That is, the entire contents of the edit file would initially be read into memory. All commands would then act on the data in memory and not modify the contents of the file. After leaving the editor through the STOP command, the edit file would be updated to reflect any changes.

This decision limits the editor to processing small or medium-sized files (with less than 2500 lines). This decision considerably simplified the design of the editor.

It was made for another reason as well. Some of the editor commands work going either "forward" or "backward" from the current line. From our discussion of files in Section 9.2, we know that we can read (or write) a file only in the forward direction. It could happen that the user instructs the editor to print the last line of the file (by using the BOTTOM command). The user then asks for the line preceding the last line (by using the N − 1 command). There would be no choice but to reset the edit file and read through it until that line was encountered. Therefore, for efficiency, we sacrificed our ability to process large files. This was an acceptable compromise for our purposes (which is to demonstrate the inner workings of a text editor) but would probably not be acceptable for a production editor for a large community of users.

As we discussed previously, the array is an inadequate data structure to store the text of the edit file. We need a data structure for which insertions, deletions, and forward and backward searching are easy and efficient.

The data structure best suited to these requirements is a doubly linked list. Each line of the text file will be preceded by a header node that looks like:

| length | |
|---|---|
| nextline | |
| previousline | |
| firstnode | |

where the fields of the record have the following interpretations.

| Field | Meaning |
|---|---|
| length | The number of characters in the line pointed to by this header node |
| nextline | Contains a pointer to the header node of the next line in the edit file |
| previousline | Contains a pointer to the header node of the previous line in the edit file |
| firstnode | Contains a pointer to the first node used to represent the characters on this line |

To save space, the actual line of text may be represented by one or more nodes. Consider the problem. To be flexible, the editor should be able to handle lines with as many characters as there are on a data card (80). It would be better to accommodate lines of length 140, the number of characters that can be printed by a line printer. We could design a node structure that could contain up to 140 characters, but this is extremely wasteful, since we anticipate that the average line length is about 34 characters. For this reason we have chosen to represent the characters of a line as a linked list of nodes whose declarations are:

```
const
    charspernode                = 34;  { number of characters per
                                          node }
type
    linecharptr                 = ↑ linecharnode;
    linecharnode                =
        record
            nextnode                : linecharptr;
            chars                   :packed array [1..charspernode] of
                                        char
        end; { of record linecharnode }
```

We allocate only as many nodes as are necessary to accommodate the characters on the line. For example, the line:

*FOUR SCORE AND SEVEN YEARS AGO OUR FOREFATHERS*

would be represented as (including the header node):

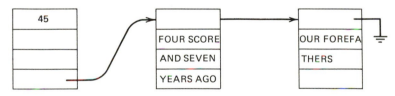

(We are assuming 10 characters can be packed into one word.)

Our data structure will also use a sentinel node (see Chapter 8). This has several advantages. First, it allows us to check quickly if the edit file is empty. The sentinel node is a header node whose nextline field points to the first header node for the first line of the edit file and whose previousline field points to the last line of the edit file. An empty edit file is represented by:

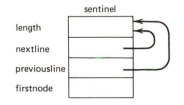

Before executing any command, we check if the edit file is empty.

**if** *sentinel ↑ .nextline = sentinel* **then** { *empty edit file* }

Second, the sentinel node allows us to access quickly the last line of the edit file when the BOTTOM command is executed. If the edit file is not empty, the header node for the bottom line is referenced by

*sentinel ↑ .previousline*

Third, the sentinel node allows us quick access to the first line of the edit file.

*sentinel ↑ .nextline*

Each line of the edit file is represented by one or more nodes. A header node is also associated with each line; it contains the current number of characters on the line, pointers to the previous and next lines in the edit file, and a pointer to the first node, which holds the characters of the currentline. The type declaration for these header nodes is:

```
type
    lineptr                    = ↑ lineptrnode;
    lineptrnode                =
        record
            length             : 0..maxchars;
            nextline           : lineptr;
            previousline       : lineptr;
            firstnode          : linecharptr
        end;
```

An edit file containing:

*6+6=*
*14 FOR LARGE*
*VALUES OF SIX.*
*PASCAL WAS A MATHEMATICIAN.*

would be represented internally by the following data structure.

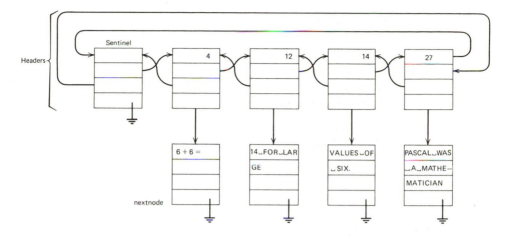

The data structure we chose makes many of the editor commands efficient with respect to memory space and execution time. Packing the characters of a line into one or more nodes in a linked list makes the FIND command somewhat cumbersome. The easiest solution to this problem is to unpack the characters of a line into an array with one character per word. Then checking for a string becomes as easy as the pattern match routine we developed in Figure 3.5.

We also employ a data structure (a **record**) that simplifies the processing of user commands. We read the characters of the command into a (packed) array. The number of characters in the command is stored in the same record. Since we will be scanning the command line (see Section 7.3), we also maintain a pointer to the next character of the command line to be processed. This data structure is called commanddef and its **type** declaration is shown in the listing in the next section.

Commands are looked up in one of two tables. One table contains the single-letter abbreviations for the commands, and the other contains the full command names. If a user command does not match one from these tables, the editor prints the message "no such command" and prompts users for another command. These tables are contained in the record structure called commandtable. The **type** declaration is given in the listing in the next section.

### 9.3.4 Editor Listing

This section contains the full listing of our implementation of the editor. Study this program for three reasons.

1. The editor was developed using the structural and stylistic guidelines we presented in Chapters 2 to 6. No module in the program is more than 100 lines long. Each module has a simple purpose and is implemented as a single-entry, single-exit piece of code.

2.   The editor uses and manipulates a variety of the data structures we described in Chapters 7 and 8. It illustrates the use of linked lists, sentinel nodes, arrays, and scanning, all in one program.

3.   The editor is not a trivial program. Much can be learned by studying its inner workings. Many of the techniques employed in this program will be useful in developing other programs.

In the lists that follow the number in parentheses after a procedure name is the page number on which that procedure is described.

*Subsidiary Routines*

min (381)
readline (381)
insertline (381–382)
packline (382)
readfile (383)
errormessage (384)
checkempty (384)
removetrailingblanks (384–385)
readcommand (385)
skipblanks (386)
movelinepointer (386–387)
numeric (387)
getnumber(387–388)
processprefix (389–390)
alphabetic (390)
getcommand (390–392)
commandordinal (392–393)
endparse (393)
getstring(393–394)
unpackline (394–395)
stringin(395–396)
locate (397)
getparameter (397–398)
printline (398–399)
printpackedline (399)
freetext (399)
deleteline (400)
readataline (400)

*Commands*

abort (401)
append (401–402)
bottom (403)
change (403–405)
delete (405)
equal (406)
find (406–407)
header (407–408)
insert (408–409)
next (409)
print (409–410)
replace (410–411)
stop (411–412)
top (412)
verifyflag (412–413)
main program (413–415)

```
program editor(edfile,input,output);
{ text editing program

written and copyright by:
          steven collins and
          steven c. bruell
          university of minnesota
}
const
    charspernode            = 34;     { number of characters per node }
    debug                   = true;   { debug flag }
    maxcharp1               = 201;    { maximum length of edit file line plus
                                        one }
    maxchars                = 200;    { maximum length of edit file line }
    maxcommandlength        = 7;      { maximum length of command
                                        word }
    numberlongcommands      = 17;     { number of long commands }
    numbershortcommands     = 18;     { number of one letter commands }
    off                     = false;  { verify flag definition }
    on                      = true;   { verify flag definition }

type
    { header node description; one header node precedes each line of text }
    linecharptr             = ↑ linecharnode;
    lineptr                 = ↑ lineptrnode;
    lineptrnode             =
        record
            length              : 0. .maxchars;
            nextline            : lineptr;
            previousline        : lineptr;
            firstnode           : linecharptr
        end; { of record lineptrnode }

    { node structure for characters in line of text; more than one node is
      necessary for lines longer than charspernode }
    linecharnode
        record
            nextnode            : linecharptr;
            chars               : packed array [1. .charspernode] of char
        end; { of record linecharnode }

    { unpacked representation of a line of text }
    linelengthdef           = 0. .maxchars;
    linedef                 =
        record
            length              : linelengthdef;
            position            : 0. .maxcharp1;
            chars               : array [1. .maxcharp1] of char
        end; { of record linedef }
```

```
{ used for error message }
messagetype                = packed array [1. .30] of char;

{ node structure used to encode user commands }
commanddef                 =
    record
        length                 : linelengthdef;
        position               : 0. .maxcharp1;
        chars                  : packed array [1. .maxcommandlength] of char
    end; { of record commanddef }

{ node structure to represent strings that are part of source editor
  commands }
stringdef                  =
    record
        first                  : 0. .maxcharp1;
        last                   : linelengthdef;
        length                 : linelengthdef
    end; { of record stringdef }

{ describes a table for editor commands }
commandtable               =
    record
        shortcommands          : array [1. .numbershortcommands] of char;
        longcommands           : array [1. .numberlongcommands] of packed
                                     array [1. .maxcommandlength] of char
    end; { of record commandtable }

var
    command                : commanddef;    { command currently being
                                               processed }
    commandline            : linedef;       { last command line entered by
                                               user }
    currentline            : lineptr;       { line of file at which the editor
                                               is currently positioned }
    edfile                 : text;          { the file to be edited }
    isitacommand           : boolean;       { true if command escape (#)
                                               is entered as data input line }
    legalcommand           : boolean;       { true if legal command read }
    noerror                : boolean;       { false if any errors are
                                               detected while processing a
                                               command }
    ordinal                : integer;       { number of command }
    running                : boolean;       { set false on stop or abort
                                               command }
    sentinel               : lineptr;       { sentinel node from which
                                               data structure is built }
    tablecommands          : commandtable;  { table of valid commands }
    verify                 : boolean;       { true when in verify mode }
```

{ *min*

  *entry conditions:*
  
        *x,y*                        *: integers to be tested*

    *exit conditions*
  
        *min*                       *: minimum of x and y*
}

**function** *min(x : integer; y : integer) : integer;*
**begin**
    **if** *x* < *y* **then** *min* := *x* **else** *min* := *y*
**end**; { *of function min* }

{ *readline*

  *read one line from data file*

  *entry conditions:*
  
        *f*                          *: the edit file*

  *exit conditions:*
  
        *line*                    *: next line read from edit file*
}

**procedure** *readline(**var** f : text; **var** line : linedef);*
**begin**
    **with** *line* **do**
    **begin**
        *length* := *0;*
        **while not** *eoln(f)* **do**
        **begin**
            *length* := *length* + *1;*
            *read(f,chars[length])*
        **end** { *of while not eoln* }
    **end**; { *of with* }
    *readln(f)*
**end**; { *of procedure readfile* }

{ *insertline*

  *insert new line into data structure*

  *entry conditions:*
  
          *currentline*           *: pointer to line after which to insert newline*
          *newline*               *: pointer to line to be inserted*

  *exit conditions:*
  
        *pointers set to insert newline after currentline*
}

```
procedure insertline(currentline : lineptr; newline : lineptr);
begin
    newline ↑ .nextline : = currentline ↑ .nextline;
    newline ↑ .previousline : = currentline;
    newline ↑ .nextline ↑ .previousline : = newline;
    currentline ↑ .nextline : = newline
end; { of procedure insertline }
```

```
{ packline

  pack text line into internal data structure

  entry conditions:
        line                        : a text line in unpacked form

  exit conditions:
        line packed into packedline
}
```

```
procedure packline(line : linedef; packedline : lineptr);
var
    charnum         : 1. .charspernode;    { loop index }
    charspacked     : integer;             { number of characters packed so
                                             far }
    node            : linecharptr;         { node being packed into }
    oldnode         : linecharptr;         { last node packed into }
begin
    packedline ↑ .length : = line.length;
    if line.length <> 0 then
    begin
        { pack initial node }
        new(node);
        packedline ↑ .firstnode : = node;
        for charnum : = 1 to min(line.length, charspernode) do
            node ↑ .chars[charnum] : = line.chars[charnum];
        charspacked : = charspernode;

        { pack the remainder of the line }
        while charspacked < line.length do
        begin
            oldnode : = node;
            new(node);
            oldnode ↑ .nextnode : = node;
            for charnum : = 1 to min(line.length − charspacked, charspernode) do
                node ↑ .chars[charnum] : = line.chars[charspacked + charnum];
            charspacked : = charspacked + charspernode
        end; { of while }
        node ↑ .nextnode : = nil
    end
    else packedline ↑ .firstnode : = nil
end; { of procedure packline }
```

```
{ readfile

  read edit file and create internal data structure

  entry conditions:
        f                          : the edit file

  exit conditions:
        currentline                : pointer to the current line
        sentinel                   : pointer to the sentinel node of the data
                                     structure
}


procedure readfile(var f : text; var currentline : lineptr;
      var sentinel : lineptr);
var
      line                  : linedef;        { scratch line buffer }
      newline               : lineptr;        { new line being created in
                                                data structure }
begin
      reset (f);
      new(currentline);
      sentinel : = currentline;

      { initialize the sentinel node }
      with sentinel ↑ do
      begin
          length : = 0;
          previousline : = currentline;
          nextline : = currentline;
          firstnode : = nil
      end; { of with }

      { read the file and set up the data structure }
      while not eof(f) do
      begin
          readline(f,line);
          new(newline);
          insertline(currentline,newline);
          currentline : = newline;
          packline(line,currentline)
      end { of while }
end; { of procedure readfile }
```

{ *errormessage*

  *print error message and set noerror flag to false*

  *entry conditions:*
     *message*                    *: error message to be printed*

  *exit conditions:*
     *noerror*                     *: set to false*
     *error message printed*
}

**procedure** *errormessage(***var** *noerror : boolean; message : messagetype);*
**begin**
     *writeln(' ',message);*
     *noerror :* = *false*
**end**; { *of procedure errormessage* }

{ *checkempty*

  *check if edit file is empty*

  *entry conditions:*
        *sentinel*                 *: pointer to sentinel node*
        *data structure for edit file has been built at this point*

  *exit conditions:*
        *noerror*                    *: set false if file is empty*
        *error message printed if file is empty*
}

**procedure** *checkempty(sentinel : lineptr;* **var** *noerror : boolean);*
**begin**
     **if** *noerror* **and** *(sentinel ↑ .nextline = sentinel)* **then**
        *errormessage(noerror,'empty edit file.              ')*
**end**; { *of procedure checkempty* }

{ *removetrailingblanks*

  *remove trailing blanks from a text line*

  *entry conditions:*
        *line*                     *: line from which blanks are to be removed*

  *exit conditions:*
        *line.length reset to ignore trailing blanks*
}

```
procedure removetrailingblanks(var line : linedef);
var
    index                        : integer;        { character index }
begin
    with line do
    begin
        for index := 1 to length do
            if chars[index] <> ' ' then length := index;
        position := 1;
        chars[length + 1] := ' ';
        if length = 0 then length := 1
    end { of with }
end; { of procedure removetrailingblanks }

{ readcommand

  read next command line
        this is a system dependent routine

  entry conditions:
        called from main loop of editor

  exit conditions:
        commandline             : next user command line read from input
}

procedure readcommand(prompt : char; var line : linedef);
begin
    with line do
    begin
        writeln(prompt,col,col,'a');
        { eos is a boolean function which returns true if we are at the end of a
          segment }
        if eos then getseg(input);
        readln;
        length := 0;
        while not eoln(input) do
        begin
            length := length + 1;
            read(chars[length])
        end; { of while }
        removetrailingblanks(line);
    end { of with }
end; { of procedure readcommand }
```

```
{ skipblanks

  skip blanks on line

  entry conditions:
        line                        : line in which blanks are to be skipped

  exit conditions:
        position set to next nonblank character
}
procedure skipblanks(var line : linedef);
begin
    with line do
    begin
        while (position <= length) and (chars[position] = ' ') do
            position := position + 1
    end { of with }
end; { of procedure skipblanks }

{ movelinepointer

  move current line pointer

  entry conditions:
        currentline              : pointer to current line in file
        linestomove              : number of lines to move (forward or
                                     backward)
        sentinel                 : pointer to sentinel node

  exit conditions:
        currentline              : pointer to line after movement
        noerror                  : updated appropriately
}

procedure movelinepointer(var currentline : lineptr; linestomove : integer;
    sentinel : lineptr; var noerror : boolean);
var
    bottomoffile              : lineptr;         { pointer to last line in edit
                                                   file }
    topoffile                 : lineptr;         { pointer to first line in edit
                                                   file }
begin
    checkempty(sentinel,noerror);
    if noerror then
    begin
        topoffile := sentinel↑.nextline;
        bottomoffile := sentinel↑.previousline;
```

```
        while ((currentline <> topoffile) and (linestomove < 0)) or
               ((currentline <> bottomoffile) and (linestomove > 0)) do
        begin
             if linestomove < 0 then                    { move backward }
             begin
                  linestomove := linestomove + 1;
                  currentline := currentline ↑ .previousline
             end
             else                                       { move forward }
             begin
                  linestomove := linestomove − 1;
                  currentline := currentline ↑ .nextline
             end { of if }
        end; { of while }

        if linestomove <> 0 then
             if linestomove > 0 then
                  errormessage(noerror, 'end of file. ')
             else
                  errormessage(noerror, 'top of file. ')
     end { of if noerror }
end; { of procedure movelinepointer }

{ numeric

  test if a character is numeric

  entry conditions:
        ch                           : character to be tested

  exit conditions:
        numeric                      : true if ch is numeric, false otherwise
}

function numeric(ch : char) : boolean;
begin
     numeric := (ch >= '0') and (ch <= '9')
end; { of function numeric }

{ getnumber

  get numeric parameter from command line

  entry conditions:
        line                         : line from which to obtain number

  exit conditions:
        number                       : value of number found
        legalnumber                  : true if valid number found, false otherwise
}
```

```
procedure getnumber(var line : linedef; var number : integer;
    var legalnumber : boolean);
var
    sign                              : integer;          { set to 1 for a plus sign and
                                                            − 1 for a minus sign }
begin
    { initialize }
    number := 0;
    legalnumber := false;
    skipblanks(line);
    with line do
    begin
        if position <= length then
        begin
            if chars[position] = '!' then                { process exclamation point }
            begin                                        { as infinity }
                position := position + 1;
                number := maxint;
                legalnumber := true
            end
            else                                         { process numeric parameter }
            begin
                { check for sign }
                sign := 1;
                if chars[position] = '−' then
                begin
                    sign := − 1;
                    position := position + 1
                end
                else
                if chars[position] = '+' then
                begin
                    sign := 1;
                    position := position + 1
                end; { of if }

                { build numeric value from characters representing number }
                while (position <= length) and numeric(chars[position]) do
                begin
                    number := 10 * number + ord(chars[position]) − ord('0');
                    position := position + 1;
                    legalnumber := true
                end; { of while }

                number := sign * number
            end { of if chars }
        end { of if }
    end { of with }
end; { of procedure getnumber }
```

{ *processprefix*

*process any command prefix characters*

*entry conditions:*
>        *commandline*          *: last command user entered*
>        *currentline*          *: pointer to currentline*
>        *sentinel*             *: pointer to sentinel node*

*exit conditions:*
>        *currentline*          *: pointer to currentline updated*
>        *noerror*              *: updated appropriately*
}

**procedure** *processprefix(***var** *commandline : linedef;* **var** *currentline : lineptr;*
        *sentinel : lineptr;* **var** *noerror : boolean);*
**var**
>        *bottomoffile*            *: lineptr;*        { *pointer to last line of file* }
>        *legalnumber*             *: boolean;*        { *true if getnumber returns a*
>                                                          *valid number* }
>        *number*                  *: integer;*        { *line count for plus and minus*
>                                                          *prefix* }
>        *stillprefix*             *: boolean;*        { *false if at end of command*
>                                                          *prefix characters* }
>        *topoffile*               *: lineptr;*        { *pointer to top of file* }
**begin**
>    { *initialize* }
>    *bottomoffile := sentinel↑.previousline;*
>    *topoffile := sentinel↑.nextline;*
>    *skipblanks(commandline);*
>
>    **with** *commandline* **do**
>    **begin**
>        **if** *(position <= length)* **and** *(chars[position] <> '=')* **then**
>        **begin**
>            *stillprefix := true;*
>
>            { *process all prefix characters* }
>            **while** *(position <= length)* **and** *stillprefix* **and** *noerror* **do**
>            **begin**
>                **if** *chars[position] = '!'* **then**
>                **begin**
>                    *currentline := bottomoffile;*
>                    *checkempty(sentinel,noerror)*
>                **end**

```
            else
            if (chars[position] = '+') or (chars[position] = '−') then
            begin
                getnumber(commandline,number,legalnumber);
                if legalnumber then
                    movelinepointer(currentline,number,sentinel,noerror)
                else
                    errormessage(noerror,'illegal character.              ');
                stillprefix := false
            end
            else
            if chars[position] = '↑' then
            begin
                checkempty(sentinel,noerror);
                currentline := topoffile
            end
            else
            if chars[position] <> ' ' then stillprefix := false;

            if stillprefix then position := position + 1
        end { of while }
    end { of if }
  end { of with }
end; { of procedure processprefix }

{ alphabetic

  test if character is alphabetic

  entry conditions:
        ch                      : character to be tested

  exit conditions:
        alphabetic              : true if character is alphabetic, false otherwise
}
function alphabetic(ch : char) : boolean;
begin
    alphabetic := (ch >= 'a') and (ch <= 'z')
end; { of function alphabetic }

{ getcommand

  get command from command line

  entry conditions:
        commandline             : line from which to obtain command

  exit conditions:
        command                 : command from command line
        legalcommand            : true if legal command scanned
        noerror                 : updated appropriately
}
```

```
procedure getcommand(var commandline : linedef; var command : commanddef;
    var legalcommand : boolean; var noerror : boolean);
var
    commandchar                  : integer;        { command character position
                                                     index }
begin
    { initialize }
    command.length : = 0;
    skipblanks(commandline);
    legalcommand : = true;

    { clear command }
    for commandchar : = 1 to maxcommandlength do
        command.chars[commandchar] : = ' ';

    with commandline do
    begin
        if position > length then
        begin
            { if empty command assume print command }
            legalcommand : = true;
            command.chars[1] : = 'p';
            command.length : = 1
        end
        else
        if not (alphabetic(chars[position]) or numeric(chars[position])) then
        begin
            { if delimiter assume find command }
            legalcommand : = true;
            command.chars[1] : = 'f';
            command.length : = 1
        end
        else
        if chars[position] = '=' then
        begin
            { process equal command }
            legalcommand : = true;
            command.chars[1] : = '=';
            command.length : = 1;
            position : = position + 1
        end
        else
```

```
    begin
        { build normal command }
        while alphabetic(chars[position]) and (position <= length) and
            noerror do
        begin
            if command.length < maxcommandlength then
            begin
                { add next character to the command }
                command.length := command.length + 1;
                command.chars[command.length] := chars[position];
                position := position + 1;
                legalcommand := true
            end { of if }
            else errormessage(noerror, 'no such command.              ')
        end { of while }
    end { of if }
    end { of with }
end; { of procedure getcommand }

{ commandordinal

  return the ordinal of a command

  entry conditions:
        command              : command to return ordinal for
        tablecommands        : table of the legal commands

  exit conditions:
        noerror              : updated appropriately
        ordinal              : ordinal of command
}

procedure commandordinal(command : commanddef; var ordinal : integer;
    var tablecommands : commandtable; var noerror : boolean);
var
    index                        : integer;        { loop index
begin
    index := 1;
    if command.length = 1 then
    begin
        tablecommands.shortcommands[numbershortcommands] :=
            command.chars[1];
        while command.chars[1] <> tablecommands.shortcommands[index] do
            index := index + 1;
        if index = numbershortcommands then
            errormessage(noerror, 'no such command.              ')
    end
```

```
      else
      begin
          tablecommands.longcommands[numberlongcommands] : =
              command.chars;
          while command.chars <> tablecommands.longcommands[index] do
              index : = index + 1;
          if index = numberlongcommands then
              errormessage(noerror, 'no such command.                    ')
      end; { of if }
      ordinal : = index
end; { of procedure commandordinal }
```

{ *endparse*

  *check for superfluous characters on command*

  *entry conditions:*
          *commandline*              *: current command line*

  *exit conditions:*
          *noerror*                  *: set false if any nonblank characters are after*
                                     *the current position of line*

}

```
procedure endparse(commandline : linedef; var noerror : boolean);
begin
      if noerror then
      begin
          skipblanks(commandline);
          if commandline.position <= commandline.length then
              errormessage(noerror, 'invalid parameter.                  ')
      end { of if }
end; { of procedure endparse }
```

{ *getstring*

  *get string parameters from current line*

  *entry conditions:*
          *commandline*              *: obtain string parameters from this line*

  *exit conditions:*
          *legalstring*             *: true if valid delimited string found, false*
                                    *otherwise*
          *string*                  *: pointers to string in line*

}

```
procedure getstring(var commandline : linedef; var string : stringdef;
    var legalstring : boolean);
var
    delimiter                    : char;            { string delimiter character }
begin
    { initialize }
    skipblanks(commandline);
    legalstring := false;
    string.length := 0;

    with commandline do
    begin
        if position <= length then
        begin
            if (not alphabetic(chars[position])) and
                (not numeric(chars[position])) and (chars[position] <> '−') and
                (chars[position] <> '+') and (chars[position] <> '!') then
            begin
                delimiter := chars[position];
                legalstring := true;
                position := position + 1;
                string.first := position;

                { find second delimiter }
                while (chars[position] <> delimiter) and (position <= length)
                do
                    position := position + 1;
                string.last := position − 1;
                string.length := string.last − string.first + 1
            end { of if }
        end { of if position }
    end; { of with }

    if string.length = 0 then
    begin
        string.first := 1;
        string.last := 0
    end { of if }
end; { of procedure getstring }

{ unpackline

  unpack a packed line

  entry conditions:
        pline                    : pointer to line to be unpacked

  exit conditions:
        line                     : pline is unpacked into line
}
```

```
procedure unpackline(var line : linedef; pline : lineptr);
var
      charnum           : 1..charspernode;        { character position index }
      node              : linecharptr;            { pointer to node being unpacked }
      unpackcount       : integer;                { unpacked character count }
begin
      with line do
      begin
          length := pline ↑ .length;
          if length <> 0 then
          begin
              node := pline ↑ .firstnode;
              unpackcount := 0;

              { unpack each node }
              repeat
                  for charnum := 1 to min(charspernode,length-unpackcount) do
                      chars[unpackcount + charnum] := node ↑ .chars[charnum];
                  unpackcount := unpackcount + charspernode;
                  node := node ↑ .nextnode
              until node = nil
          end { of if }
      end { of with }
end; { of procedure unpackline }

{ stringin

  test if string is contained in line

  entry conditions:
          commandline          : string appears in command line
          line                 : line to search for string
          string               : pointers to string to search for

  exit conditions:
          found                    : true if string contained in line, false otherwise
          line is positioned at first character in string
}
```

```
procedure stringin(var line : linedef; string : stringdef;
    var commandline : linedef; var found : boolean);
var
    done                    : boolean;        { true if string cannot be found }
    index                   : integer;        { character position index }
    stringthere             : boolean;        { true if string has been found }
begin
    line.position : = 0;

    { we assume the null string is at beginning of each line }
    if string.length = 0 then stringthere : = true
    else
    begin
        { search for string in the line }
        with line do
        begin
            stringthere : = false;
            done : = false;
            chars[length + 1] : = commandline.chars[string.first];

            repeat
                position : = position + 1;
                if (position + string.length − 1) > length then
                begin
                    { we went past end of line; no match }
                    done : = true
                end
                else                            { check if entire string matches }
                begin
                    stringthere : = true;
                    for index : = string.first to string.last do
                    begin
                        if commandline.chars[index] < >
                            line.chars[line.position + index − string.first]
                        then stringthere : = false
                    end { of for }
                end { of if }
            until done or stringthere
        end { of with }
    end; { of if }
    found : = stringthere
end; { of procedure stringin }
```

{ *locate*

  *find next occurrence of a string in the file*

  *entry conditions:*

|  |  |
|---|---|
| *commandline* | : *current command line* |
| *increment* | : *direction to move in file* |
| *sentinel* | : *pointer to sentinel node* |
| *string* | : *pointers to string to locate* |

  *exit conditions:*

|  |  |
|---|---|
| *count* | : *line count to first occurrence of string* |
| *noerror* | : *updated appropriately* |
| *pline* | : *pointer to line containing string* |

}

**procedure** *locate(string : stringdef;* **var** *pline : lineptr;*
    **var** *count : integer; increment : integer; sentinel : lineptr;*
    **var** *commandline : linedef;* **var** *noerror : boolean);*
**var**

|  |  |  |
|---|---|---|
| *found* | : *boolean;* | { *true if string found* } |
| *scratchline* | : *linedef;* | { *scratch line buffer* } |

**begin**
    { *initialize* }
    *found := false;*
    *count := increment;*

    { *scan entire file until string is found* }
    **repeat**
        *movelinepointer(pline,increment,sentinel,noerror);*
        *count := count + increment;*
        **if** *noerror* **then**
        **begin**
            *unpackline(scratchline,pline);*
            *stringin(scratchline,string,commandline,found)*
        **end**
    **until** *found* **or** (**not** *noerror*)
**end**; { *of procedure locate* }

{ *getparameter*

  *evaluate a string/numeric command parameter*

  *entry conditions:*

|  |  |
|---|---|
| *commandline* | : *obtain parameter from this line* |
| *sentinel* | : *pointer to sentinel node* |

  *exit conditions:*

|  |  |
|---|---|
| *count* | : *value of numeric parameter or number of lines to next occurrence of string parameter* |
| *noerror* | : *updated appropriately* |

}

```
procedure getparameter(var commandline : linedef; sentinel : lineptr;
    var count : integer; var noerror : boolean);
var
    legalnumber        : boolean;     { true if legal number scanned }
    legalstring        : boolean;     { true if legal string scanned }
    pline              : lineptr;     { pointer to line being scanned }
    sign               : integer;     { sign of count parameter }
    string             : stringdef;   { string parameter to search for }
begin
    with commandline do
    begin
        if position <= length then
        begin
            { process count parameter on command }
            if chars[position] = '-' then
            begin
                sign := -1;
                position := position + 1
            end
            else sign := 1;
            getstring(commandline,string,legalstring);
            if legalstring then
            begin
                { process deliminted string parameter }
                position := position + 1;
                pline := currentline;
                locate(string,pline,count,sign,sentinel,commandline,noerror)
            end
            else
            begin
                { process numeric parameter }
                getnumber(commandline,count,legalnumber);
                if legalnumber then count := count * sign else count := sign
            end { of if }
        end { of if }
        else count := 1
    end { of with }
end; { of procedure getparameter }

{ printline

  print unpacked data line

  entry conditions:
        line                    : line to be printed

  exit conditions:
        line printed on output
}
```

```
procedure printline(line : linedef);
var
    charnum                         : linelengthdef; { character position index }
begin
    for charnum := 1 to line.length do write(line.chars[charnum]);
    writeln
end; { of procedure printline }

{ printpackedline

  print packed data line

  entry conditions:
        pline                       : pointer to line to be printed

  exit conditions:
        line printed on output
}

procedure printpackedline(pline : lineptr);
var
        index                       : linelengthdef;   { character position index }
        scratchline                 : linedef;         { scratch line buffer }
begin
    unpackline(scratchline,pline);
    printline(scratchline)
end; { of procedure printpackedline }

{ freetext

  free text nodes of a line

  entry conditions:
        pline                       : pointer to line whose nodes we want to return
                                      to the available space list

  exit conditions:
        all memory reserved for text nodes pointed to by pline returned to
        available space list
}

procedure freetext(pline : lineptr);
var
        node                        : linecharptr;     { next node to dispose of }
        nodegone                    : linecharptr;     { node to dispose of }
begin
    node := pline ↑ .firstnode;
    pline ↑ .length := 0;
    pline ↑ .firstnode := nil;

    while node <> nil do
    begin
        nodegone := node;
        node := nodegone ↑ .nextnode;
        dispose(nodegone)
    end { of while }
end; { of procedure freetext }
```

{ *deleteline*

   *delete line from internal data file structure*

   *entry conditions:*
               *pline*                              *: pointer to line to be deleted*

   *exit conditions:*
               *line deleted; pointers updated; and memory returned to available space*
               *list*
}

**procedure** *deleteline(pline : lineptr);*
**begin**
        *pline ↑ .previousline ↑ .nextline : = pline ↑ .nextline;*
        *pline ↑ .nextline ↑ .previousline : = pline ↑ .previousline;*
        *freetext(pline);*
        *dispose(pline)*
**end**; { *of deleteline* }

{ *readdataline*

   *read next line from terminal*

   *entry conditions:*
            *called by procedure insert*

   *exit conditions:*
            *commandline*              *: command is returned if first character on line*
                                          *is #*
            *isitacommand*             *: true if data line contains a command*
            *line*                     *: next data line read from input*
}

**procedure** *readdataline(***var** *line : linedef;* **var** *commandline : linedef;*
        **var** *isitacommand : boolean);*
**begin**
        *readcommand('= ',line);*

        { *check for command mode prefix* }
        **if** *(line.length > 0)* **and** *(line.chars[1] = '#')* **then**
        **begin**
                *isitacommand : = true;*
                *line.position : = 2;*
                *commandline : = line*
        **end** { *of if* }
**end**; { *of procedure readdataline* }

```
{ abort

  process the abort command

  entry conditions:
        commandline              : line containing abort command

  exit conditions:
        noerror                  : updated appropriately
        running                  : set to false to terminate editor
        message printed and editor is left without updating edit file
}
```

**procedure** *abort(commandline : linedef;* **var** *noerror : boolean;*
    **var** *running : boolean);*
**begin**
    *endparse(commandline,noerror);*
    **if** *noerror* **then**
    **begin**
      *running := false;*
      *errormessage(noerror, 'editor aborted.*              ')*
    **end** *{ of if }*
**end***; { of procedure abort }*

```
{ append

  process the append command

  entry conditions:
            commandline          : line containing the append command
            currentline          : pointer to current line
            sentinel             : pointer to sentinel node
            verify               : verify flag

  exit conditions:
        noerror                  : updated appropriately
        string appended in appropriate column of currentline
}
```

**procedure** *append(***var** *commandline : linedef; currentline : lineptr;*
    *sentinel : lineptr; verify : boolean;* **var** *noerror : boolean);*
**var**
    *charnum*      : *linelengthdef;*   *{ character position index }*
    *column*       : *integer;*        *{ column in which to begin appending }*
    *legalnumber*  : *boolean;*      *{ true if legal number scanned }*
    *legalstring*   : *boolean;*      *{ true if legal string scanned }*
    *scratchline*   : *linedef;*       *{ scratch line buffer }*
    *string*       : *stringdef;*      *{ string in append command }*
    *truncated*    : *integer;*       *{ number of characters truncated }*

```
begin
    unpackline(scratchline,currentline);
    getstring(commandline,string,legalstring);
    if legalstring then
    begin
        { obtain column in which to begin appending the string }
        with commandline do
            if position <= length then position := position + 1;
            getnumber(commandline,column,legalnumber);
            if not legalnumber then column := scratchline.length + 1
    end
    else errormessage(noerror, 'string field missing.                    ');

    endparse(commandline,noerror);
    checkempty(sentinel,noerror);
    if noerror and ((column <= 0) or (column > maxchars)) then
        errormessage(noerror, 'column out of range.                    ');

    if noerror and (string.length > 0) then
    begin
        if (column + string.length − 1) > maxchars then
        begin
            { process truncation (non-fatal error) }
            truncated := column + string.length − 1 − maxchars;
            string.last := string.last − truncated;
            string.length := string.length − truncated;
            errormessage(noerror, 'truncation.                    ');
            noerror := true
        end; { of if }

        if string.length > 0 then
        begin
            { blank fill to column }
            for charnum := scratchline.length + 1 to column do
                scratchline.chars[charnum] := ' ';

            { append the string to the line }
            for charnum := 0 to scratchline.length − 1 do
                scratchline.chars[column + charnum] :=
                    commandline.chars[string.first + charnum];

            scratchline.length := column + string.length − 1;
            packline(scratchline,currentline)
        end; { of if }

        if verify then printline(scratchline)
    end { of if }
end; { of procedure append }
```

{ *bottom*

   *process bottom command*

   *entry conditions:*

| | |
|---|---|
| *commandline* | *: line containing bottom command* |
| *currentline* | *: pointer to current line* |
| *sentinel* | *: pointer to sentinel node* |
| *verify* | *: verify flag* |

   *exit conditions:*

| | |
|---|---|
| *currentline* | *: set to point to last line in file* |
| *noerror* | *: updated appropriately* |

}

**procedure** *bottom(***var** *commandline : linedef;* **var** *currentline : lineptr;*
    *sentinel : lineptr; verify : boolean;* **var** *noerror : boolean);*
**begin**
    *endparse(commandline,noerror);*
    *checkempty(sentinel,noerror);*
    **if** *noerror* **then**
    **begin**
       *currentline := sentinel ↑ .previousline;*
       **if** *verify* **then** *printpackedline(currentline)*
    **end** { *of if* }
**end**; { *of procedure bottom* }

{ *change*

   *process the change command*

   *entry conditions:*

| | |
|---|---|
| *commandline* | *: line containing change command* |
| *currentline* | *: pointer to current line* |
| *sentinel* | *: pointer to sentinel node* |
| *verify* | *: verify flag* |

   *exit conditions:*

| | |
|---|---|
| *noerror* | *: updated appropriately* |

}

**procedure** *change(***var** *commandline : linedef; currentline : lineptr;*
    *sentinel : lineptr; verify : boolean;* **var** *noerror : boolean);*
**var**

| | | |
|---|---|---|
| *index* | *: integer;* | { *character position index* } |
| *legal1string* | *: boolean;* | { *true if old string is legal* } |
| *legal2string* | *: boolean;* | { *true if new string is legal* } |
| *scratch1line* | *: linedef;* | { *scratch line buffer* } |
| *scratch2line* | *: linedef;* | { *scratch line buffer* } |
| *stringthere* | *: boolean;* | { *true if string in file* } |
| *string1* | *: stringdef;* | { *old string* } |
| *string2* | *: stringdef;* | { *new string* } |

```
begin
    { get old string and new string }
    getstring(commandline,string1,legal1string);
    getstring(commandline,string2,legal2string);
    if legal1string and legal2string then
    begin
        commandline.position := commandline.position + 1;
        endparse(commandline,noerror);
        checkempty(sentinel,noerror);
        if noerror then
        begin
            unpackline(scratch1line,currentline);
            stringin(scratch1line,string1,commandline,stringthere);
            if stringthere then
            begin
                { change the string }
                if scratch1line.position > 1 then
                begin
                    { transfer all characters before the old string }
                    for index := 1 to scratch1line.position − 1 do
                        scratch2line.chars[index] := scratch1line.chars[index];
                    scratch2line.position := scratch1line.position − 1
                end
                else scratch2line.position := 0;
                if string2.length > 0 then
                begin
                    { transfer the new string }
                    for index := string2.first to string2.last do
                    begin
                        scratch2line.position := scratch2line.position + 1;
                        scratch2line.chars[scratch2line.position] :=
                            commandline.chars[index]
                    end { of for }
                end; { of if }
                scratch1line.position := scratch1line.position + string1.length;
                if scratch1line.position = 0 then scratch1line.position := 1;

                { transfer all characters after the old string }
                while scratch1line.position <= scratch1line.length do
                begin
                    scratch2line.position := scratch2line.position + 1;
                    scratch2line.chars[scratch2line.position] :=
                        scratch1line.chars[scratch1line.position];
                    scratch1line.position := scratch1line.position + 1
                end; { of while }

                { clean up }
                scratch2line.length := scratch2line.position;
                packline(scratch2line,currentline);
                if verify then printline(scratch2line)
            end
```

                    **else** *errormessage(noerror, 'string not found.*                              *')*
                **end** { *of if* }
            **end**
            **else** *errormessage(noerror, 'invalid parameter.*                              *')*
        **end**; { *of procedure change* }

{ *delete*

  *process the delete command*

  *entry conditions:*
            *commandline*              : *line containing the delete command*
            *currentline*              : *pointer to current line*
            *sentinel*                 : *pointer to sentinel node*
            *verify*                   : *verify flag*

  *exit conditions:*
            *currentline*              : *set to point to line after line(s) deleted*
            *noerror*                  : *updated appropriately*
}

**procedure** *delete(***var** *commandline : linedef;* **var** *currentline : lineptr;*
        *sentinel : lineptr; verify : boolean;* **var** *noerror : boolean);*
    **var**
        *count*                    : *integer;*          { *number of lines to delete* }
        *increment*                : *integer;*          { *direction to move* }
        *pline*                    : *lineptr;*          { *pointer to line being deleted* }
    **begin**
        *getparameter(commandline, sentinel, count, noerror);*
        *endparse(commandline, noerror);*
        *checkempty(sentinel, noerror);*
        **if** *noerror* **then**
        **begin**
            **if** *count* > *0* **then** *increment* : = *1* **else** *increment* : = − *1;*

            **while** *(count* <> *0)* **and** *noerror* **do**
            **begin**
                *pline* : = *currentline;*
                **if** *verify* **then** *printpackedline(pline);*
                *movelinepointer(currentline, increment, sentinel, noerror);*
                *count* : = *count* − *increment;*
                *deleteline(pline)*
            **end**; { *of while* }

            **if not** *noerror* **then**
                **if** *increment* > *0* **then** *currentline* : = *sentinel* ↑ *.previousline*
                **else** *currentline* : = *sentinel* ↑ *.nextline*
        **end** { *of if* }
    **end**; { *of procedure delete* }

```
{ equal

  process the equal command .

  entry conditions:
        commandline              : contains line to replace currentline

  exit conditions:
        commandline, excluding = sign, replaces currentline
}
procedure equal(var commandline : linedef; var currentline : lineptr);
var
      index                      : linelengthdef;   { character position index }
      newposition                : linelengthdef;   { character position index in
                                                       data line }
      pline                      : lineptr;         { pointer to packed data line }
begin
      with commandline do
      begin
          if position <= length then
          begin
              newposition := 0;
              for index := position to length do
              begin
                  newposition := newposition + 1;
                  chars[newposition] := chars[index]
              end;
              length := newposition
          end
          else length := 0;

          { insert data line into the file }
          new(pline);
          packline(commandline,pline);
          insertline(currentline,pline);
          currentline := pline
      end { of with }
end; { of procedure equal }

{ find

  process the find command

  entry conditions:
          commandline              : line containing the find command
          currentline              : pointer to current line
          sentinel                 : pointer to sentinel node
          verify                   : verify flag

  exit conditions:
          currentline              : set to point to line containing string
          noerror                  : updated appropriately
}
```

```
procedure find(var commandline : linedef; var currentline : lineptr;
    sentinel : lineptr; verify : boolean; var noerror : boolean);
var
    count          : integer;      { number of lines to move }
    increment      : integer;      { direction to move }
    legalstring    : boolean;      { true if legal string on find command }
    pline          : lineptr;      { pointer to line being searched }
    string         : stringdef;    { temporary string buffer }
begin
    with commandline do
    begin
        { check for minus sign }
        if (chars[position] = '-') and (position <= length) then
        begin
            increment := -1;
            position := position + 1
        end
        else increment := 1
    end; { of with }

    getstring(commandline, string, legalstring);
    if legalstring then
    begin
        commandline.position := commandline.position + 1;
        endparse(commandline, noerror);
        checkempty(sentinel, noerror);
        if noerror then
        begin
            { search the file for the string }
            pline := currentline;
            locate(string, pline, count, increment, sentinel, commandline, noerror);
            if noerror then
            begin
                { move to line containing the string }
                currentline := pline;
                if verify then printpackedline(currentline)
            end { of if }
        end { of if }
    end
    else errormessage(noerror, 'invalid parameter.                              ')
end; { of procedure find }

{ header

  process the header command

  entry conditions:
        commandline              : line containing header command

  exit conditions:
        noerror                  : updated appropriately
}
```

```
procedure header(var commandline : linedef; var noerror : boolean);
var
    index             : integer;        { character position index }
    legalnumber       : boolean;        { true if legal number scanned }
    width             : integer;        { width to print header }
begin
    getnumber(commandline,width,legalnumber);
    if not legalnumber then width := 72;
    endparse(commandline,noerror);
    if (width > 0) and noerror then
    begin
        for index := 1 to width do write(index mod 10:1);
        writeln
    end { of if }
end; { of procedure header }

{ insert

  process the insert command

  entry conditions:
            commandline          : line containing the insert command
            currentline          : pointer to currentline
            sentinel             : pointer to sentinel node

  exit conditions:
            currentline          : points to last line inserted
            isitacommand         : true if last line entered contains a command
            noerror              : updated appropriately
}

procedure insert(var commandline : linedef; var currentline : lineptr;
    sentinel : lineptr; var isitacommand : boolean; var noerror : boolean);
var
    pline             : lineptr;        { pointer to data line being inserted }
    scratchline       : linedef;        { scratch line buffer }
begin
    endparse(commandline,noerror);
    if noerror then
    begin
        while not isitacommand do
        begin
            readdataline(scratchline,commandline,isitacommand);
            with scratchline do
            begin
                if not isitacommand then
                begin
                    { insert data line into file }
                    new(pline);
                    packline(scratchline,pline);
                    insertline(currentline,pline);
                    currentline := currentline ↑.nextline
                end { of if }
```

```
                end { of with }
            end { of while }
        end { of if }
end; { of procedure insert }
```

{ *next*

  *process the next command*

  *entry conditions:*

| | |
|---|---|
|       *commandline* | *: line containing the next command* |
|       *currentline* | *: pointer to the currentline* |
|       *sentinel* | *: pointer to the sentinel node* |
|       *verify* | *: verify flag* |

  *exit conditions:*

| | |
|---|---|
|       *currentline* | *: updated to point to new currentline* |
|       *noerror* | *: updated appropriately* |

}

```
procedure next(var commandline : linedef; var currentline : lineptr;
    sentinel : lineptr; verify : boolean; var noerror : boolean);
var
    count          : integer;        { number of lines to move }
    legalnumber    : boolean;        { true if legal number scanned }
begin
    getnumber(commandline,count,legalnumber);
    if not legalnumber then count := 1;
    endparse(commandline,noerror);
    checkempty(sentinel,noerror);

    if noerror then
    begin
        movelinepointer(currentline,count,sentinel,noerror);
        if verify then printpackedline(currentline)
    end { of if }
end; { of procedure next }
```

{ *print*

  *process print command*

  *entry conditions:*

| | |
|---|---|
|       *commandline* | *: line containing print command* |
|       *currentline* | *: pointer to current line* |
|       *sentinel* | *: pointer to sentinel node* |
|       *verify* | *: verify flag* |

  *exit conditions:*

| | |
|---|---|
|       *currentline* | *: pointer updated to point to last line printed* |
|       *noerror* | *: updated appropriately* |

}

```
procedure print(var commandline : linedef; var currentline : lineptr;
    sentinel : lineptr; verify : boolean; var noerror : boolean);
var
    count          : integer;          { number of lines to print }
    increment      : integer;          { direction to move through file }
begin
    getparameter(commandline,sentinel,count,noerror);
    endparse(commandline,noerror);
    checkempty(sentinel,noerror);

    if noerror then
    begin
        if count < 0 then increment := −1 else increment := 1;
        { print current line }
        printpackedline(currentline);
        count := count − increment;

        while (count <> 0) and noerror do
        begin
            { move pointer to the next line }
            movelinepointer(currentline,increment,sentinel,noerror);
            count := count − increment;
            if noerror then printpackedline(currentline)
        end { of while }
    end { of if }
end; { of procedure print }

{ replace

  process the replace command

  entry conditions:
          commandline           : line containing the replace command
          currentline           : pointer to current line
          sentinel              : pointer to sentinel node

  exit conditions:
          currentline           : set to point to last line entered
          noerror               : updated appropriately
}

procedure replace(var commandline : linedef; var currentline : lineptr;
    sentinel : lineptr; var isitacommand : boolean; var noerror : boolean);
var
    firstline      : boolean;          { true if first line being replaced }
    scratchline    : linedef;          { scratch line buffer }
begin
    endparse(commandline,noerror);
    checkempty(sentinel, noerror);
    firstline := true;
```

```
        while (not isitacommand) and noerror do
        begin
            readdataline(scratchline,commandline,isitacommand);
            if not isitacommand then
            begin
                { replace line in file with data line }
                if firstline then firstline := false
                else movelinepointer(currentline,1,sentinel,noerror);
                freetext(currentline);
                packline(scratchline,currentline);
                if currentline = sentinel ↑ .previousline then
                        errormessage(noerror, 'end of file.                    ')
            end { of if }
        end { of while }
end; { of procedure replace }

{ stop

  process the stop command

  entry conditions:
            commandline             : line containing stop command
            sentinel                : pointer to sentinel node

  exit conditions:
            noerror                 : updated appropriately
            edit file updated and message written to output
}

procedure stop(var commandline : linedef; sentinel : lineptr;
        var f : text; var running : boolean; var noerror : boolean);
var
    currentline      : lineptr;          { temporary current line pointer }
    index            : integer;          { character position index }
    scratchline      : linedef;          { scratch line buffer }
begin
    endparse(commandline,noerror);
    if noerror then
    begin
        rewrite(f);
        currentline := sentinel;
        checkempty(sentinel,noerror);
        if noerror then
        begin
            repeat
                    currentline := currentline ↑ .nextline;
                    unpackline(scratchline,currentline);
                    for index := 1 to scratchline.length do
                        write (f,scratchline.chars[index]);
                    writeln(f)
            until currentline = sentinel ↑ .previousline
        end; { of if }
```

```
              running := false;
              errormessage(noerror, 'editor complete.                              ')
          end { of if }
      end; { of procedure stop }
```

{ top

  process the top command

   entry conditions:
        commandline             : line containing the top command
        sentinel                 : pointer to sentinel node
        verify                   : verify flag

   exit conditions:
        currentline              : set to point at first line in file
        noerror                : updated appropriately
}

```
      procedure top(var commandline : linedef; var currentline : lineptr;
          sentinel : lineptr; verify : boolean; var noerror : boolean);
      begin
          endparse(commandline,noerror);
          checkempty(sentinel,noerror);
          if noerror then
          begin
              currentline := sentinel ↑ .nextline;
              if verify then printpackedline(currentline)
          end { of if }
      end; { of procedure top }
```

{ verifyflag

  process the verify command

   entry conditions:
        commandline             : line containing the verify command
        sentinel                 : pointer to sentinel node

   exit conditions:
        noerror                : updated appropriately
        verify                  : updated appropriately
}

```
procedure verifyflag(var commandline : linedef; sentinel : lineptr;
    var verify : boolean; var noerror : boolean);
var
    command                    : commanddef;   { temporary command buffer }
    legalcommand               : boolean;      { true if legal command }
begin
    skipblanks(commandline);
    getcommand(commandline,command,legalcommand,noerror);
    if (commandline.position > commandline.length) or (not noerror) then
    begin
        { toggle verification if no parameter }
        endparse(commandline,noerror);
        if noerror then verify : = not verify
    end
    else
    begin
        { process on or off parameter }
        endparse(commandline,noerror);
        if noerror then
        begin
            if command.chars = 'on            ' then verify : = on
            else
            if command.chars = 'off           ' then verify : = off
            else errormessage(noerror, 'invalid parameter.                  ')
        end { of if }
    end { of if }
end; { of procedure verify }

begin                                          { of main·program }
    writeln;
    writeln('editor version 1.0');
    noerror : = true;

    { read file to be edited }
    readfile(edfile,currentline,sentinel);
    checkempty(sentinel,noerror);
    currentline : = sentinel↑.nextline;

    { set various default parameters }
    isitacommand : = false;
    running : = true;
    verify : = on;
```

```
{ initialize command table }
with tablecommands do
begin
    shortcommands[1]        := 'd';
    shortcommands[2]        := 'i';
    shortcommands[3]        := 'p';
    shortcommands[4]        := 'c';
    shortcommands[5]        := 'r';
    shortcommands[6]        := 'f';
    shortcommands[7]        := 's';
    shortcommands[8]        := 'v';
    shortcommands[9]        := 'b';
    shortcommands[10]       := 'n';
    shortcommands[11]       := 't';
    shortcommands[12]       := 'a';
    shortcommands[13]       := 'h';
    shortcommands[14]       := '=';
    shortcommands[15]       := ' ';
    shortcommands[16]       := ' ';
    shortcommands[17]       := ' ';
    shortcommands[18]       := ' ';
    longcommands[1]         := 'delete ';
    longcommands[2]         := 'insert ';
    longcommands[3]         := 'print ';
    longcommands[4]         := 'change ';
    longcommands[5]         := 'replace';
    longcommands[6]         := 'find ';
    longcommands[7]         := 'stop ';
    longcommands[8]         := 'verify ';
    longcommands[9]         := 'bottom ';
    longcommands[10]        := 'next ';
    longcommands[11]        := 'top ';
    longcommands[12]        := 'append ';
    longcommands[13]        := 'header ';
    longcommands[14]        := '        ';
    longcommands[15]        := '        ';
    longcommands[16]        := 'abort ';
    longcommands[17]        := '        ';
end; { of with }
```

```
{ main loop of the editor }
while running do
begin
    noerror : = true;
    if isitacommand then isitacommand : = false
    else readcommand('#',commandline);
    processprefix(commandline,currentline,sentinel,noerror);
    if noerror then
    begin
        getcommand(commandline,command,legalcommand,noerror);
        if noerror then
        begin
            commandordinal(command,ordinal,tablecommands,noerror);
            if noerror then
            case ordinal of
                1 : delete( commandline,currentline,sentinel,verify,
                        noerror);
                2 : insert( commandline,currentline,sentinel,
                        isitacommand,noerror);
                3 : print( commandline,currentline,sentinel,verify,noerror);
                4 : change( commandline,currentline,sentinel,verify,
                        noerror);
                5 : replace(commandline,currentline,sentinel,
                        isitacommand,noerror);
                6 : find( commandline,currentline,sentinel,verify,noerror);
                7 : stop( commandline,sentinel,edfile,running,noerror);
                8 : verifyflag(commandline,sentinel,verify,noerror);
                9 : bottom( commandline,currentline,sentinel,verify,
                        noerror);
                10 : next( commandline,currentline,sentinel,verify,noerror);
                11 : top( commandline,currentline,sentinel,verify,noerror);
                12 : append( commandline,currentline,sentinel,verify,
                        noerror);
                13 : header( commandline,noerror);
                14 : equal( commandline,currentline);
                16 : abort( commandline,noerror,running)
            end { of case }
        end { of if noerror }
    end { of if noerror }
end { of while }
end. { of the entire editor }
```

*Style Review 9-1* _____

> ### Data Structure Design
>
> The first course in programming concentrates almost exclusively on the com-
> mands of some particular programming language (e.g., assignment, looping,
> conditional). The data structuring alternatives are usually dealt with in much
> less detail, frequently being limited to scalar variables and arrays. This tends
> to instill the idea that to develop a successful program we concentrate solely
> on the selection of an algorithm and its encoding in our particular language.
>
> We hope that this section of the text has helped you to realize that the
> selection of the appropriate data representation has as much impact on the
> characteristics of the final program as the choice of algorithm.
>
> There are many possible data representation alternatives. In Chapters 7
> to 9 we introduced arrays, strings, records, stacks, queues, sets, singly and
> doubly linked lists, circular lists, trees, binary trees, and threaded trees. There
> are other alternatives that we have not covered. For every problem the right
> selection of a data representation can simplify the task enormously and greatly
> increase efficiency.

## EXERCISES FOR CHAPTER 9

1. Files can be composed of records as in:

   **type**
      *studentrecords*                                =
         **record**
            *lastname*                    : **packed array** [1..10] **of** *char;*
            *firstname*                 : **packed array** [1..10] **of** *char;*
            *age*                         : 0..150
         **end**;
   **var**
      *x*                                   : **file of** *studentrecords;*

   Explain how to create a file x with 25 students. How do you determine the last
   name of the tenth student on the file?

2. Given two files of integers sorted into increasing order, write a complete Pascal
   program that merges the two files into a third file in sorted order.

3. Write a Pascal procedure that counts the number of components in a file.

4. What is on file f after the following segment is executed?

```
var
    f           : file of integer;
    i           : 1..10;
begin
    for i := 1 to 10 do
    begin
        if odd(i) then rewrite(f);
        write(f,i)
    end
    .
    .
    .
```

5. Write a Pascal program that "reverses" the components of a file. That is, make the first component the last one on the new file, make the second component the second to the last one on the new file, and so on. Do not test your program on any large file!

The following questions pertain to the text editor.

6. In procedure insertline, why are the parameters currentline and newline not passed by reference?

7. The last statement of procedure readline is readln(f). Could we substitute get(f)?

8. Rewrite procedure removetrailingblanks to start from the end of the line instead of the beginning.

9. Rewrite procedure change to change multiple occurrences of the old string to the new string on the current line. The syntax of the command might look like this.

   C/old string/new string/number of occurrences

10. Rewrite procedure find to locate the nth occurrence of the string in the edit file. A possible syntax might be:

    F/string/n

11. Add an overlay command to the editor. It should perform the following. The current line is printed, and then a line of input is accepted from the terminal (the overlay line). The overlay line will be positioned directly beneath the line printed

out. Each character of the overlay line (except blanks) will replace the corresponding character in the current line. For example:

| Current line | = | lool | at thit | lind |
|---|---|---|---|---|
| Overlay line | = | k | s | e |
| Result | = | look | at this | line |

12. Add a command that can *copy* all lines from the current line to a line containing some string *after* some target line.

13. The text does not describe the purpose of the header command. Look at the code for procedure header and determine what it does.

# Bibliography for Part IV

Findlay, W., D. A. Watt. *Pascal—An Introduction to Methodical Programming*. Potomac, Md.: Computer Science Press, 1978.

Harrison, M. *Data Structures and Programming,* Glenview, Ill: Scott, Foresman, 1973.

Horowitz, E., and S. Sahni. *Fundamentals of Data Structures*. Potomac, Md.: Computer Science Press, 1976.

Jensen, K., and N. Wirth. *Pascal User Manual and Report*. Second Edition. New York: Springer-Verlag, 1974.

Knuth, D. E. *The Art of Computer Programming Volume 1: Fundamental Algorithms*. Reading, Mass.: Addison-Wesley, 1968.

Lewis, T. G., and M. Z. Smith. *Applying Data Structures,* Houghton Mifflin, 1976.

Wirth, N. *Algorithms + Data Structures = Programs*. Englewood Cliffs, N.J.: Prentice-Hall, 1976.

# Part V

# PROGRAM IMPLEMENTATION CONCERNS

# Chapter 10

# DEBUGGING AND TESTING

## 10.1  PROGRAM DEBUGGING

Debugging, the process of locating and correcting errors in a program, can be painfully slow unless proper guidelines are followed from the outset of the coding phase. Studies of the time required to implement large programs indicate that *over half* a programmer's time is spent finding and correcting errors. Although we do not claim that following our guidelines will result in error-free programs, we do believe it will significantly decrease the amount of time spent debugging.

Sometimes bugs are so deeply ingrained in a program that frustrated programmers despair of ever finding and correcting them. Instead, they incorporate methods to avoid these bugs into the documentation of the program. In extreme cases programmers have decided to rewrite entire programs from scratch because a bug could not be effectively or economically corrected. These costly mistakes can and should be avoided.

There are three primary reasons that debugging is so time consuming. First, many programmers honestly believe that they can write error-free code that will work properly the first time and therefore do not prepare for their occurrence during the early stages of coding. This is a myth, albeit a seductive one. Even with the most careful programming, coding errors are inevitable. The best way to find these bugs is to have appropriate tools such as a good interactive symbolic debugger. In lieu of that, the best alternative is to put *write statements* in strategic places throughout the code as you develop the program. When the program is executed, you will be able to follow exactly what is happening and determine why things go awry.

Second, debugging is considerably easier if the program is clearly written. We have consistently emphasized the importance of style and clarity of expression. Programs with complex, intricate, and ''jumpy'' logic are difficult not only to read but also to debug. (Figure 4.5 should have proved that.) For example, see how long it takes you to locate and fix the bug(s) in the following program fragment. [The purpose of the fragment was to find a key in a list of integers and to write out the index (location) of the first occurrence of the key in the list.]

```
label
      10,20;
const
      listsize     = 20;
var
      i             : 1. .listsize;
      key           : integer;
      list          : array [1. .listsize] of integer;
      temp          : 1. .listsize;
begin
      { initialize all elements of the array list }
          .
          .
          .
      key := 5; i := 1;
 10:if list[i] = key then begin
            temp := i; goto 20
      end else begin
      i := i + 1; if i = listsize then goto 20; goto 10 end;
 20:writeln(' the key ',key,' was found in the ',
                'list at location ',temp);
          .
          .
          .
```

There are two bugs in this program segment: (1) it will not work if the key does not appear in the list—in which case temp is undefined, and (2) it will not find the key if it is the last item in the array list. Program segments like this are difficult to debug because of their cumbersome style and structure.

When programs contain hundreds or thousands of lines of code, it becomes extremely difficult, if not impossible, to follow the flow of control. Using too many **goto** statements creates more logical paths than the mind can follow. It is much easier to locate bugs when control structures and systematic indentation delineate the control paths (see Figure 4.5). For example, consider the previous program segment coded instead with a **while** statement.

```
{ declarations and initializations }

found := false;
key := 5;
i := 1;
while (i <> listsize) and (not found) do
    if list[i] = key then found := true
    else i := i + 1;
if found then
    writeln(' the key ',key,' was found in the ',
        ' list at location ',i);
```

Now it is easier to find the remaining bug because the flow of control in the program is made obvious by the **while** statement. The structured coding conventions presented in Chapter 4 represent one of the best debugging techniques available—the *avoidance* of bugs as much as possible.

The third main reason that debugging is so problematic is that it has not been taught systematically, as have other aspects of programming. Debugging is often taught in a haphazard manner, if at all. Students rarely receive formal instruction about what to do when their programs fail. This can have especially frustrating consequences when there are no error messages or other clues about what went wrong.

Our objective is to classify the approaches to debugging so that you will have a logical, organized method for dealing with any errors you encounter. The following sections treat syntax errors, run-time errors, and logic errors.

Before we cover these topics, we will mention one approach to debugging that is habit forming and *wrong:* trying something mindlessly because you do not know what else to do. We are all guilty of fixing the first bug we spot and resubmitting the program. Rushing through the program and trying quick fixes makes the program needlessly confusing and seldom corrects the original bug(s).

Bugs can be tracked down in a logical manner. It makes no sense to change or rerun a program until you have spotted a likely cause of the error. But remember, if you have no idea where the bug might be, placing write statements in proper places in your program can greatly simplify the hunt for the elusive bug.

One final comment: ignore any suggestion to rerun a program without change because the problem may have been a ''machine error.'' Such errors are rare. It might be comforting to blame the machine, but it is almost always a waste of time.

### 10.1.1  Syntax Errors

*Syntax errors* are very common and relatively easy to correct. A syntax error is any violation of the grammatical rules of the language. For example, in Pascal there are two possible ways to write an **if** statement. Their syntax is:

**if** *boolean expression* **then** *statement*

or

**if** *boolean expression* **then** *statement* **else** *statement*

The expression placed between the **if** and the **then** must evaluate to either true or false (a type boolean result). The statement may be any legal Pascal statement (including another **if** statement). If two or more statements appear after the **then** or the **else,** they must be surrounded by a **begin/end** pair. With this in mind, locate the syntax errors in these **if** statements.

```
var
    a           : boolean;
    b           : boolean;
    c           : real;
    d           : real;
begin
    .
    .
    if a then writeln(' a is true');
    else writeln(' a is false) { misplaced semicolon }
    .
    .
    if a and b then c := 10; d := 5 else c := 7
        { 2 statements placed where only 1 is allowed }
    .
    .
    if c + d then writeln(' c + d = ',c + d)
        { real expression used instead of boolean expression }
    .
    .
end.
```

Syntax errors are easier to correct than other errors because they often result in a clear error message from the compiler when it translates the program into machine language. The compiler usually produces messages indicating where and what the mistakes may be.

Figure 10.1 is a listing of a Pascal program containing numerous syntax errors. It shows how errors are printed in one particular Pascal system. You may be using a Pascal compiler that prints error messages in a different manner, but the information should be comparable.

```
1    program find(input,output);
2    const
3         listsize                  = 20;
4    var
5         found                     : boolean;
6         i                         : integer;
7         key                       : integer;
8         list                      : array [1..listsize] of integer;
9    begin
10        for i := 1 , listsize do list[i] := i;
***                      ↑ 6       ↑ 55
11        found := false;
***              ↑ 104
12        key := 5;
13        i := 1;
14        while (i <= listsize) and (not found) do
***                                      ↑ 104 ↑ 135
15            if list[i] = key then found := true; else i := i + 1;
***                              ↑ 104               ↑ 6
16        if found then
***                  ↑ 104
17            writeln(' the key ',key,' was found in the ',
18                        'list at location ',i)
19    end;
***        ↑ 6
***   incomplete program
```

*compiler error message(s):*

| | | |
|---|---|---|
| 6 | : | *unexpected symbol.* |
| 55 | : | *'to' or 'downto' expected.* |
| 104 | : | *identifier not declared.* |
| 135 | : | *type of operand must be boolean.* |

*Figure 10.1.* Sample Pascal program with syntax errors.

Figure 10.1 highlights most of the important points to consider when locating and correcting syntax errors.

1.  Some of the error messages provided by Pascal are clear. For example, error message 55 appearing after line 10 says, " 'to' or 'downto' expected." Recalling the syntax of the **for** statement:

    **for** v := initial-value **to** (or **downto**) final-value **do** statement it should be clear that we have neither a **to** nor a **downto** in line 10. Error message 104 (''identifier not declared'') occurs every time the variable found appears in the program. The problem is that the variable found has been misspelled in the variable declaration (we typed a zero instead of the letter ''oh'').

2.  Unfortunately, many error messages produced by the Pascal compiler are not as easy to understand. This is because the error has clouded our intentions and prevented the compiler from determining what type of statement we wanted. Some messages tell us nothing more than that an error exists. For example, on line 15, error message 6 ('unexpected symbol') is pointing at the reserved word **else,** which is actually legal, but not in this context. To determine what is wrong, we should first compare the statement in question with the syntax of a statement of that type (an **if** statement). Usually this will lead us directly to the error. Here the problem is that we put a semicolon before the **else.** (Recall that semicolons in Pascal are only used to separate statements from each other.)

    Sometimes error indicators point to statements with no errors. If this occurs, you should check the statement immediately preceding the one supposedly in error. Many syntax errors occur by incorrectly terminating a statement (omitting a required semicolon).

    If this still does not identify the syntax error, check all other statements related in any way to the one in error. For example, check all the pertinent **var** declarations. Also check for matching **begin/end** pairs, **repeat/until** pairs, or **then** and **else** clauses of **if** statements.

    Line 19 also contains the cryptic phrase 'incomplete program.' What is the syntax error on this line?

3.  There is usually not a one-to-one correspondence between error messages from the compiler and corrections to be made. Often a single error will generate multiple error messages. For example, line 10 has two error messages, but we can correct both at once by changing the comma to the reserved word **to.** Our error in declaring the variable found resulted in four error messages, one each time the variable found was used in the program.

    Be aware that many error messages will have already been taken care of by earlier corrections of other syntax errors.

4.  Frequently, the compiler overlooks a syntax error because of other syntax errors on the same line. Therefore it may require more than one additional run to eliminate all of them.

## 10.1.2  Run-Time Errors

Even when a program is syntactically correct, there is still the possibility of a *run-time error,* which causes abnormal program behavior during execution.

In Chapter 3 (Section 3.2.2) we discussed *defensive programming*—anticipating trouble spots and guarding against them. For example:

**if** *(count* <> *0)* **and** *((sum* **div** *count)* > *minimum)* **then** . . .

is syntactically correct. However, if count has the value zero, the program will be terminated during execution because division by zero "does not compute." In the **if** statement, both alternatives are evaluated. To avoid the run-time error, use:

**if** *count* <> *0* **then**
        **if** *(sum* **div** *count)* > *minimum* **then** . . .

Other common run-time errors include:

1.  Using a variable before assigning a value to it.
2.  Assigning values outside the prescribed bounds of the declared type of a subrange variable.
3.  Using an index outside the prescribed bounds of the array's lower and upper bound (array indices out of bounds).
4.  Failing to insure that the value of a case expression corresponds to one of the case labels.
5.  Passing illegal parameters to standard functions [e.g., $x < 0$ for sqrt(x) or ln(x)].
6.  Converting a real number, x, to an integer when round(abs(x)) > maxint. (Refer to Chapter 3 for other run-time errors.)

Run-time errors produce an error message and a *postmortem dump,* which help programmers determine what happened. Although the format of this dump may vary from one system to another, it usually includes:

1.  A message indicating what caused the run-time error, and the statement and program unit in which the error appeared.
2.  The names and current values of all (simple) variables in the main program and any procedures and/or functions activated from the main program.

The identification of run-time errors is greatly facilitated by the postmortem dump. As an example, consider the program and the dump it produced shown in Figure 10.2.

By studying the information in the postmortem dump, we can locate the source of the error. It was caused by attempting to store the value 0 into array element [1,1] where the declaration specifies only values in the range 1 to 10.

```
1    program post(input, output);
2    const
3         lowerbound                    = 1;
4         upperbound                    = 10;
5    type
6         bounds                        = lowerbound. .upperbound;
7         arraytype                     = array [bounds, bounds] of bounds;
8    var
9         i                             : bounds;
10        j                             : bounds;
11        matrix                        : arraytype;
12
13   procedure init(var arrayname : arraytype);
14   var
15        i                             : bounds;
16        j                             : bounds;
17   begin
18        for i := lowerbound to upperbound do
19             for j := lowerbound to upperbound do
20                  arrayname[i,j] := i − 1
21   end; { of procedure init }
22
23   begin
24        init(matrix);
25        for i := lowerbound to upperbound do
26        begin
27             for j := lowerbound to upperbound do
28                  write(matrix[i,j]:5);
29             writeln
30        end
31   end.
```

*program terminated at line 20 in procedure init.*
*value out of range.*

--- init ---

i =                1                                                    j =                1

*init was called from line 24 in program post.*

--- post ---

i =           undef                                                    j =           undef

**Figure 10.2.** Sample Pascal program with a run-time error.

A good interactive symbolic debugger would provide all this information and more. For example, the symbolic debugger on the UCSD (University of California—San Diego) Pascal system displays the following information on the screen when this program is executed.

*value range error*
*invalid value was 0, outside range of [1. .10]*
*in source line: arrayname[i,j] := i − 1*
*variables in module init:*

    *i := 1*                   *j := 1*

Thus users have the essential information for diagnosing the problem. After displaying this information, the symbolic debugger prompts users with the line:

*dump:*   *up*   *top*   *down*   *edit*   *file*   *quit*

Users can select one of the listed commands by pressing the first letter of the command. The up, top, and down commands are used to traverse the chain of procedure and function calls active at the time of the error. These commands permit users to trace the execution of a program back to the main program, inspecting variables that are local to each procedure or function.

The edit command activates the editor with the cursor positioned at the line in which the error occurred.

The preceding discussion explained how Pascal helps you to recover from abnormal run-time errors. However, it is wrong to expect either Pascal or the system to help you recover. You should *prevent* run-time errors by using the foolproof programming techniques introduced in Chapter 3. A run-time error is usually a sign of a poorly written program.

By preventing run-time errors, we can recover in the program itself and continue processing instead of stopping irrevocably. Looking back at the discussion on "graceful degradation" in Section 3.2.4, we see that we always want to try to continue on with useful processing. Complete termination of the program is always the *last* option we want to consider. Even if the program does nothing more than write an appropriate error message before stopping, it will usually be easier to determine what went wrong because we will be interpreting our own error message, not some cryptic message printed by the system. This is especially true when the program is used by someone other than the original author.

*Style Review 10.1* _____

***The Programming Environment***

When we think of the programming support provided by a computer system, we too often think only in terms of what languages are available. Today, however, programmers are thinking more in terms of the overall *program-*

*ming environment*—the entire collection of supporting software that assists in the development of correct and efficient computer programs. The language is only part of this environment. Tools such as the interactive debugger mentioned in Section 10.1.2 are also a significant part of the overall programming package.

An idealized programming environment would consist of a well-integrated collection of the following programs.

1. Language processors.
2. Text editors and formatters for assisting in program and data preparation.
3. Filing systems for long-term storage and retrieval of programs and data.
4. Debugging tools, such as interactive debuggers, snapshot dumps, and traces (Chapter 10).
5. Monitoring tools for measuring execution speed and profiling program execution flow (Chapter 11).
6. Special subprogram libraries.
7. On-line documentation and assistance (Chapter 12).

You should try to learn as much as possible about the programming environment of your local computer facility.

### 10.1.3   Logic Errors

Another kind of error you will almost certainly encounter is the *logic error*. This is an incorrect translation of either the problem statement or the algorithm. These errors can be extremely difficult to locate.

A good example of a logic error is the following translation of the formula for computing the two roots of the quadratic equation in the form:

$$ax^2 + bx + c = 0$$

The roots are:

$$\text{roots} = \frac{-b \pm \sqrt{b^2 - 4ac}}{2a}$$

```
discriminant := sqr(b) − 4.0 * a * c;
if (discriminant >= 0.0) and (a <> 0.0) then
begin
    r1 := −b + sqrt(discriminant) / (2.0 * a);
    r2 := −b − sqrt(discriminant) / (2.0 * a)
end
```

The syntax is correct, but the logic is flawed. The order of evaluation of the operators does not properly translate the quadratic formula.

Another example of a logic error is the following Pascal segment that sums the numbers from 1 to 100, inclusive.

```
sum := 0;
i := 1;
repeat
    sum := sum + i;
    i := i + 1
until i >= 100
```

Do you see the error? (What happens when i = 100?)

Both examples demonstrate why logic errors can be difficult to correct. Usually there is no clue as to what went wrong, except that the program produced an incorrect result. Hand-simulating a program (playing computer and executing the program step by step) may help locate problems, but this can be a very time-consuming process. Unfortunately, hand-simulation often does not uncover the problem either because the program is too complex or the mistake is too subtle. It is very frustrating to spend hours poring over a listing without finding any good reasons for the incorrect results.

The best and simplest way to uncover problems in a program is to use write statements, preferably when the program is first being written. This technique is called *program tracing*. A convenient method for tracing the flow of control in a program is to include write statements as the first and last statements of each procedure. It will also help to write out the values of any procedure parameters on entry and exit. For example:

```
procedure exchange(var p1 : integer; var p2 : integer);
    .
    .
    .
begin
    if debug then writeln(' exchange entered with p1 =', p1,' p2 =',p2);
    .
    .
    .
    if debug then writeln(' exchange exited with p1 =', p1,' p2 =',p2)
end; { of procedure exchange }
```

In this case we assume there is a global boolean flag called debug that is set to true whenever we want the computer to perform the hand-simulation automatically. It is set to false when we temporarily do not need this information. This programming technique is a good habit to develop because constantly entering and removing these debugging commands can itself introduce errors into the program. Also, we are never really sure when to remove these commands because we are never sure when new errors may appear. Therefore leaving them in the program permanently and having them turned on and off by this simple expedient:

*debug* : = *true*

or

*debug* : = *false*

can be very useful. This type of built-in debugging aid is frequently called a *debugging instrument*.

However, simply printing out values on procedure entry and exit may not be enough. As discussed in Chapters 3 and 4, most well-written Pascal programs and procedures are composed of single-entry, single-exit program segments. There are no jumps into or out of the middle of a segment. Segments can be nested, but each one should follow the single-entry, single-exit restriction.

When you are searching for an error, it is helpful to narrow its location to a specific segment if possible. This can be accomplished by initially bracketing a few of the outer program segments with write statements. If this does not isolate the error(s), it is best to include more write statements to surround the inner segments as well. All important variables referenced in a segment should be written out before and after the segment.

For example:

```
if debug then writeln(. . .);
        begin
            [②
     ①  [③
            [④
        end;
if debug then writeln(. . .)
```

If all values are correct going into block 1, but there are errors upon exiting block 1, we have definitely localized the error to block 1. This is because of the single-entry, single-exit characteristic of the block. There is no way to jump into (**goto**) a point in the middle of the block; we must enter through the writeln at the top. If localizing the error to block 1 does not pinpoint the error, we can simply repeat the process by bracketing some of the inner blocks (2, 3, or 4) with writeln commands to see if they are correct. We work our way down, narrowing the possible location of the error to smaller and smaller program segments until we have limited it to such a small area that we can easily spot it by visual check or hand-simulation.

Two other guidelines should be followed when inserting write statements in a program.

1.  Immediately echo-print all input data. Your program may be producing incorrect results not because it is wrong, but because you are inputting bad data. Although some of the debug output statements may eventually be removed or turned off when the program is complete, the echo-print statements should remain. They will associate a particular result with the data set that produced it.

2.  Avoid placing output statements in loops that are executed many times. You may be inundated with so much output that meaningful values are buried or overlooked. If a write statement is necessary in a loop you may consider printing a value every nth iteration.

```
for i := 1 to 5000 do
begin
    if ((i mod 50) = 0) and debug then writeln(. . . .);
    .
    .
    .
end { of for }
```

As we mentioned, most Pascal compilers have one built-in debugging aid—the *postmortem dump*—that activates whenever a program terminates abnormally. An example of the output produced by a postmortem dump was shown in Section 10.1.2. Some versions of Pascal include a halt procedure that terminates the program and produces a postmortem dump. Some Pascal compilers allow users to call the postmortem dump facility as a procedure, in which case the program is not terminated. This type of dump is usually called a *snapshot dump* because it gives you a picture of your program at one point in time and then keeps going.

Another useful debugging tool is the *automatic trace* facility available on some systems. In effect, the system automatically inserts write statements after every Pascal statement, producing a complete trace of the program. Since this may result in excessive output, the system gives users a method of toggling into and out of trace mode with statements such as:

```
trace(true)
```

and

```
trace(false)
```

Indiscriminate use of this trace facility can waste much computer time and paper. You must be careful to turn the trace facility off at points where you are ''sure'' no bugs exist.

*Style Review 10.2*

> **Classification of Errors**
>
> 1.  *Syntax Error*. Any violation of the grammatical rules of the language. This is usually the easiest kind of error to find and correct.
> 2.  *Run-Time Error*. Causes abnormal program behavior during execution (e.g., using a variable before assigning a value to it). The use of defensive programming techniques discussed in Section 3.2.2 can greatly reduce the possibility of this type of error. Interactive symbolic debuggers and postmortem dumps can help locate any run-time errors that do occur.
> 3.  *Logic Error*. An incorrect translation of either the problem statement or the algorithm. Using "debugging" write statements and tracing facilities from the outset will help you locate this type of error.

## 10.2  PROGRAM VERIFICATION

Some computer scientists have argued that computer programming could and should become more like mathematics. If mathematicians can "prove" that theorems are correct, why shouldn't programmers be able to prove that programs are correct? There is currently a great controversy over whether or not large programs can, in fact, be proved correct (or verified). What follows is an oversimplified explanation of program verification.

The goal of program verification is to determine if a program will actually perform according to its specifications. This verification process is performed not by a person, but by a machine. The input to a program verifier is other programs. Of course, simply inputting a program is not enough, because the program verifier will have no idea what the program is supposed to do. A description of the program *specifications* is also required.

The verifier attempts to "prove" that the program meets its specifications by partitioning it into smaller units and proving that they each meet their intended purpose. At the heart of a program verifier is an automatic theorem prover—a program (possibly aided by a person) that can prove fairly complicated theorems or lemmas. Some of these theorem provers are so advanced that they can prove many of the theorems from elementary calculus, linear algebra, and number theory.

Program verification is complicated, but recent developments are simplifying it somewhat. New programming languages devised with program verification in mind include special features and some formalized semantics that make the verification process easier. Theorists believe that correctness should be an integral part of a program's design. This constructive approach to verification is very promising, especially for programs in which computations can be carried out in parallel (as opposed to the more usual case of sequential execution).

The following example outlines one technique used in program verification. Consider this Pascal assignment statement:

    a := b / (d + c)

in which all variables are type real. We are implicitly assuming that the sum of d and c does not equal 0; otherwise, the statement cannot be performed. We also expect that after the statement is executed, the variable a will contain the value of b / (d + c). We are implicitly assuming a *precondition,* which expresses what we know to be true *before* the statement is executed, and a *postcondition,* which expresses what we know to be true *after* the statement is executed. The pre- and postconditions can be written as comments before and after a statement.

    { *pre-condition* }
        *statement*
    { *post-condition* }

Suppose this assignment statement were placed in a program where we knew that the variables on the right side of the assignment operator (:=) were strictly positive. Then our pre- and postconditions would look like this.

    { (b > 0) and (c > 0) and (d > 0) }
    a := b / (c + d)
    { (b > 0) and (c > 0) and (d > 0) and (a = b / (d + c)) }

We are making one further assumption about the assignment statement: that there are no out-of-range values.

Next we consider a simple conditional statement that determines the absolute value of a variable x, abs(x).

    **if** x >= 0
    **then** y := x
    **else** y := −x

We assume nothing about the value in the variable x (except that it is in range) before the **if** statement is performed. We indicate this by using the condition { true } which implies no knowledge of the state of the program. From the semantics of the **if** statement, we can easily derive the preconditions for the **then** and **else** clauses.

```
{ true }
if x >= 0 then
begin
    { x >= 0 }
    y := x
end
else
begin
    { not (x >= 0) }
    y := -x
end
```

(Notice that we have added **begin-end** pairs for clarity only.) The postconditions are derived as they were for assignment statements.

```
{ true }
if x >= 0 then
begin
    { x >= 0 }
    y := x
    { (x >= 0) and (y = x) }
end
else
begin
    { not (x >= 0) }
    y := -x
    { (not (x >= 0)) and (y = -x) }
end
```

The postcondition for the **if** statement is obtained by noting that only one branch of the **if** is ever executed. Therefore one of the two postconditions must be true. The final postcondition becomes:

$\{ ((x >= 0) \text{ and } (y = x)) \text{ or } ((not (x >= 0)) \text{ and } (y = -x)) \}$

which is the same as:

$y = abs(x)$

It is possible to formulate general rules for inferring the postcondition from the precondition for each different statement in Pascal. The sample statements used in the discussion of verification have all been trivial. Verifying all but the shortest programs by hand can be tedious and time consuming. This is why automating the verification process is so important. We hope such programs will be generally available soon but, in the meantime, you can make your programs easier to read by inserting comments that state the important pre- and postconditions of statements.

For a more formal and detailed discussion of formal program verification, refer to *A Discipline of Programming* by Dijkstra listed in the Bibliography for Part III.

## 10.3 PROGRAM TESTING

Exhaustive testing of a program is another means of showing that it is "correct." Unfortunately, testing can only reveal the presence of errors, never their absence. To be effective, testing must be done systematically. One approach is to devise sets of test data that exercise every control path in the program. The assumption is that if each section of a program is executed and there is an error in the program, the final output will be wrong. (There are cases in which this assumption is wrong; for example, we could have a case of offsetting errors.) Usually, though, exercising all parts of a program will uncover most errors.

The most important aspect of program testing, therefore, is the careful selection of data sets that thoroughly test the program. Too often *quantity* of test data is accepted in place of *quality*. The mere fact that a program works on 50 test runs means nothing if those 50 sets were not chosen carefully. In fact, if all 50 test cases exercised the same logical part of the program, all we would know is that the specific part works. We could not say anything at all about the remainder of the code. Our objective must be to select values that cause the execution of all flow paths through the program. These independent flow paths are most easily identified when we have written our program using the structured coding conventions discussed in Chapter 4. (This is another important argument for writing well-structured code.)

We can only test all the flow paths in a program unit if it is reasonably small. Otherwise, the number of unique paths will be so large that exhaustive testing is totally unrealistic.

Every **if, while,** and **for** statement in a program creates two distinct paths:

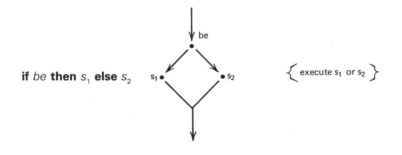

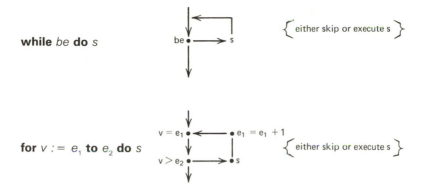

**while** *be* **do** *s*      { either skip or execute s }

**for** $v := e_1$ **to** $e_2$ **do** *s*      { either skip or execute s }

and a **case** creates n unique paths where n is the number of alternatives within the **case** statement. As programs get larger, the number of paths grows exponentially. This is another reason for keeping modules short (Section 6.2.3) and testing each module as it is developed.

Our first group of test data should attempt to test every path in the program. For example, the simple table look-up program in Figure 10.1 should be tested with at least:

1. One data set where key is found.
2. One data set where key is not found.
3. One data set where listsize = 0.

Looking at the generalized pattern matching procedure in Figure 3.5, we would want to be sure to test the following cases.

1. No match between the pattern and the text.
2. A partial match between the pattern and the text (i.e., a match in the first k characters where k < length).
3. A match that does not use the special "anymatch" character.
4. A match that does use the "anymatch" character.

However, these tests alone are insufficient.

Aside from the exhaustive testing of the valid test cases, there are three *special* cases of which you should be particularly aware. These must be carefully considered during the testing process.

1. *The Boundary Cases.* These are the data sets that fall at the very extremes of the legal range of data or fall exactly at the crossover point from one case to another. One of the most common errors in programming is the *off-by-one* error in which we do not correctly handle boundary or termination conditions. Whenever your program includes statements of the form:

> **while** *i* <= *limit* **do** . . .
> **if** *(x > 0.0)* **or** *(x < 10.0)* **then** . . .

it is wise to test your program with boundary values such as:

> i = limit
> x = 0.0
> x = 10.0

In the pattern matching program, a typical boundary case would be the pattern at the very beginning or end of the text.

> pattern = abc      text = abcdef. . . or . . .xxxxabc

Another boundary condition might be a pattern whose length is exactly equal to the length of the text.

2.  *The Null Case.* Make sure that your program works correctly for the degenerate case of ''nothing to do.'' (This point was discussed in Section 3.2.) Where appropriate, check your program with test cases such as an empty file, an empty table, an array of length 0, and a pointer value of **nil**. We should definitely test our pattern matching procedure with a pattern of length 0 and a text of length 0.

3.  *The Illegal Cases.* To test the robustness of your program (Section 3.2) and demonstrate that it does indeed display the characteristics of ''graceful degradation,'' you should be sure to include illegal data that violate the specifications of the problem. For all illegal cases, no matter how pathological, your program should do something meaningful instead of terminating abnormally. Therefore, as one final test of the pattern matching program, we would include:

    (a)  A pattern of illegal size (length outside the range 1 to max).
    (b)  A text of illegal size.
    (c)  A starting point for the search that does not lie within the text.

The importance of testing programs in short, compact units becomes clear when we look back at our review of the test cases applied to patternmatch, a short procedure of 25 to 30 lines. Adequate testing of this short module required 12 different test cases. As modules grow in size to 100, 200, or more lines, the number of cases needed to test them grows even more rapidly, and soon it is no longer feasible to apply the systematic procedures presented here. Untested program statements are sure to have undetected bugs. The result is an incorrect program. Without testing, the bug(s) will not be detected until a user supplies a data set that activates the bug and causes the program to ''bomb out.''

It cannot be stressed too early that program testing should be done exhaustively on small program modules.

The top-down design method introduced earlier tests each new module separately, as it is developed. The testing of each module will proceed as we have described. However, the fact that each new module works properly does not guarantee that the overall program will work correctly. Errors may exist in the *interface* between two modules, and this will not be apparent until they are combined. (See the railroad track diagram in Section 5.2.3.) Therefore additional testing of the overall program is essential for determining whether the correct individual modules fit together properly into a correct whole.

When we have completed our testing, what can we say? As mentioned at the beginning of this section, we cannot say the program is absolutely correct. We can say that, for a wide range of carefully selected test cases, the program operates satisfactorily. From this statement we then extrapolate that the program will operate satisfactorily on *all* data sets. If we have chosen the test data very carefully and have systematically carried out the design and testing phases, for the most part, this extrapolaton will be valid and future modifications caused by undetected bugs should be minimal.

*Style Review 10.2* _____

***Program Testing***

Test each individual module of a program.
Test all possible flow paths through the program module.
Test the boundary cases of a module.
Test the null cases.
Test the illegal cases to insure program robustness.
Try incorporating the pre- and postconditions of important statements as comments in a module.

## 10.4    CONCLUSION

Until we start breeding perfect programmers, bugs will be a part of life. In this book we have been advocating stylistic and structural guidelines for the development of effective programs. Adherence to these guidelines will greatly reduce the number of bugs in a program but will not eliminate them altogether. This chapter has presented systematic ways of searching for and exterminating bugs.

## EXERCISES FOR CHAPTER 10

1.  Assume there is a bug in the generalized pattern matching procedure in Figure 3.5. Where would you place debugging statements to locate the problem?

2.  What type of test data would you use to demonstrate that the pattern matching procedure in Figure 3.5 works as documented?

3.  Repeat Exercises 1 and 2 with the sort procedure from Figure 2.5.

4.  Find the bug(s) in the following fragment to scan a number.

```
if input ↑  =  '−' then
begin
    sign := −1;
    get(input)
end
else
if input ↑  =  '+' then
begin
    sign := 1;
    get(input)
end;
get(input)
while not eoln do
begin
    number := 10 * number + ord(input ↑ ) − ord('o');
    get(input)
end;
```

5.  The following procedure transposes the contents of an n-x-n array. Is it correct? If not, explain how to go about debugging it systematically. (See Chapter 7, Exercise 11 for the definition of a matrix transpose.)

```
procedure transpose (var a : arraydef; n : nbounds);
var
    i : nbounds;
    j : nbounds;
begin
    for i := 1 to n do
        for j := 1 to n do a[j,i] := a[i,j]
end; { of procedure transpose }
```

6.  Write a simple **if** statement equivalent to z := min(x,y). Develop all the pre- and postconditions for your statement.

*Chapter 11* _____

# PROGRAM EFFICIENCY

## 11.1  INTRODUCTION

*Program efficiency* is a measure of the amount of resources a program requires to produce correct results. Programmers and end-users are always interested in measurements that evaluate and rank programs. If two programs produce identical results, end-users will usually choose the one that requires less processor time or memory space because they think it is more efficient.

However, there has been a significant change in recent years concerning how programmers and users feel about efficiency. This change in attitude has occurred in three main areas.

1.  What are the important resources we should try to optimize?
2.  Where are the important efficiency gains to be made?
3.  How important is efficiency in the first place?

A few years ago, the only resources we measured and optimized were *machine resources:* processor time, memory cells, and mass storage space. Program efficiency was measured by running standardized test cases (called *benchmarks*) and analyzing how long they took to execute and how much memory they required.

Today, however, efficiency is defined in much broader terms that encompass human as well as machine resources and divide machine efficiency into two categories: the inherent efficiency of the method itself, and the efficiency of its use in a specific computer program. We are now dealing with a wider scope of measurement than simply "how many milliseconds did it take to run?"

The four major components of efficiency are:

1.  *User Efficiency.* The amount of time and effort users will spend to learn how

to use the program, how to prepare the data, and how to interpret and use the output.

2.  *Maintenance Efficiency*. The amount of time and effort maintenance programmers will spend reading a program and its accompanying technical documentation in order to understand it well enough to make any necessary modifications.

3.  *Algorithmic Complexity*. The inherent efficiency of the method itself, regardless of which machine we run it on or how we code it. For example, a sequential search through an N-element table will take, on the average, N/2 searches to find what it is looking for (if it is always there), and this quantity cannot be reduced by programming cleverness or newer and faster computers. A method that creates a two-dimensional n-x-n array to store a table will require at least $n^2$ cells, even if we do not use most of these spaces.

4.  *Coding Efficiency*. This is the traditional efficiency measure. Here we are concerned with how much processor time and memory space a computer program requires to produce correct answers.

Section 11.2 will discuss points 1 and 2—*the human aspects of efficiency*. Sections 11.3 and 11.4 will discuss points 3 and 4—the *machine aspects of efficiency*.


## 11.2   THE HUMAN ASPECTS OF EFFICIENCY

In the area of efficiency, there has been a significant shift in focus from the machine to the programmer. The clever programmer who could reduce the execution time of a loop by 345 microseconds or eliminate 40 memory locations from a data structure is no longer always considered an invaluable resource. In fact, that programmer's operations may not be cost effective. They may actually reduce overall efficiency because of the programming time spent on achieving these reductions and the possible increase in programmer time necessary to make future modifications.

The reason for this change in attitude is simple: money! Ten or twenty years ago, the most expensive aspect of programming was computer costs. Consequently, we tended to "optimize for the machine." Today, however, the greatest costs involved in program development are not those gray boxes in the corner but the two-legged creatures who use them. Figure 11.1 compares the trends in starting salary for programmers and average medium-scale machine costs over the last 15 years.

At today's prices, a medium-size commercial installation might pay $100,000 to $200,000 for hardware with a 5-year life expectancy. If that same installation has a data-processing staff of five (one DP director, three programmers, and one technical/clerical person) the salaries over that same 5-year period will total about $500,000—more than double the hardware costs. Most studies indicate that "people costs" in a typical computer center are 50 to 75% of overall costs. As another example, consider

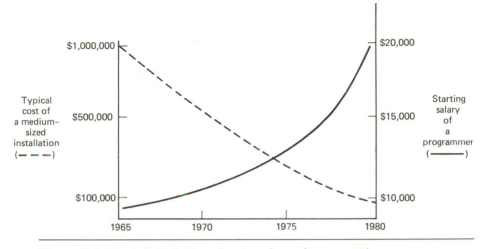

***Figure 11.1.*** Trends in hardware and personnel costs in programming.

the case of the programmer earning $10 per hour and the large computer that rents for $12 per minute (both typical values). If the programmer works to save 500 microseconds per loop execution in a loop executed 1000 times, execution time will be reduced by 1/2 second, resulting in a processor cost savings of 10¢. Even if the program is run weekly, the annual savings will be about $5—approximately 1/2 hour of programmer time. If the programmer spent days or even hours on that optimization, the result would be a net loss. Excessive programmer time spent on small gains in machine time is usually a losing proposition.

The best way to reduce costs is not to make minor reductions in machine time but to make major reductions in people time. Programmers cost more money than hardware, and any discussion of efficiency that considers only the machine and its resources is missing this fundamental point.

Computer programs should be written with these goals in mind.

1. To be correct and reliable.
2. To be easy to use for its intended end-user population.
3. To be easy to understand and easy to change.

Point 1 is crucial and too often forgotten. People want programs that work properly under all conditions and keep working properly for a long time. Correctness is essential; efficiency is not. As Van Tassel's law observes: If the program need not be correct it can be made as efficient as desired![1]

Point 2 refers to *end-user efficiency*. After correctness, the most important char-

[1]D. Van Tassel, *Program Design, Style, Efficiency, and Testing*, Englewood Cliffs, N.J., Prentice-Hall, 1978.

acteristic of a program is ease of use. An end-user is not going to be impressed by a program with poor documentation, confusing input/output formats, and inadequate results—regardless of how fast it runs. The extra time spent by that user in figuring out how to use the program and its results is certainly more important and valuable than small reductions in execution time.

Key aspects of end-user efficiency discussed in this text are:

a. Program robustness (Section 3.2).
b. Program generality (Section 3.4).
c. Portability (Section 3.5).
d. Input/output behavior (Section 3.6).
e. User documentation (Section 12.2).

Another significant human cost is Point 3—the programmer time needed to maintain and modify an existing program (*maintenance efficiency*). A complex, logically contorted program that executes rapidly but is difficult to modify will almost certainly cost more over the life of the program. The salaries of the programmers spending time trying to understand and change the program will probably far exceed the savings in machine costs. (Remember the time you spent debugging the small program fragment in Figure 4.5.)

The key points in achieving maintenance efficiency are:

a. A clear, readable programming style (Chapters 2 and 3).
b. Adherence to structured coding conventions (Chapter 4).
c. A well-designed, functionally modular solution (Chapters 5 and 6).
d. A thoroughly tested and verified program with built-in debugging and testing aids (Chapter 10).
e. Good technical documentation (Section 12.3).

In summary the Schneider-Bruell corollary to Van Tassel's law is: Cater to people, not machines. You can worry about the machine later!

However, we would be remiss if we did not acknowledge some important *exceptions* to our corollary. There are special situations in which we must write the fastest, smallest, most efficient program we can, regardless of its effect on usability or maintenance. Three major examples stand out.

1. *Real-time programs* must produce answers in a specified time period or else the results are worthless. Most often these programs are used to control ongoing physical processes (e.g., assembly lines, check-out counters, nuclear power plants). In these programs, time is critical. For example, a program automatically controlling the flight path of a space vehicle would continuously monitor altitude, position, pitch, yaw, and other values and then, if necessary, compute how to adjust engine thrust to correct these errors. If the

program could not compute the necessary actions in time, the vehicle could go into an irreversible roll or spin. In this case we would have to sacrifice any program characteristic, except correctness, to reduce running time.

2. Programs that do not fit into the memory space available on your computer. All computers have some fixed, finite memory capacity into which each program must fit. If your computer has N words of memory but your program requires more, it probably will not run.[2] The only alternative, short of buying more memory or a newer computer, is to "squeeze" the program down and make it shorter—again, without too much regard for its effect on the end-user or the maintenance programmer.

3. Programs that are run very often are likely candidates for extensive machine-level optimization. However, "very often" in computer terms does not mean weekly or even daily but, more likely, tens, hundreds, or even thousands of times a day. Programs with these rates of usage are most likely "systems programs" and are used by the computer itself and not by the end-user. Examples might include a Pascal compiler that translates hundreds of programs each day, or a computer's accounting system that may log in and bill thousands of interactive users daily. With this type of program, even small reductions in running time, when multiplied by the number of times the program is used, can result in very significant savings.

These cases, however, are "special" and, in general, we would state the following. Do not worry initially about the machine-level efficiency of the program you are writing. Choose a reasonably good method that solves the problem and write a clear, readable, and correct program that both the user and the maintenance programmer will find easy and efficient to use.

## 11.3   ALGORITHMIC EFFICIENCY

When we start discussing the topic of efficiency of a computer program, we must approach it from two different points—the inherent efficiency of the algorithm itself and the efficiency of the program to implement that algorithm. As we will see, the former is significantly more important than the latter.

The most obvious way to measure the efficiency of an algorithm is to run it for a specific set of data and measure how much processor time or memory space is needed to produce a correct solution. However, this would produce a measure of efficiency for only one very special case and will probably be inadequate to predict how the algorithm will perform with a different set of data. An algorithm for finding

---

[2]Some computers will still execute programs that are larger than the available memory capacity. They use sophisticated memory management techniques called "overlays" or "virtual memory." However, we are not considering the existence of such techniques when making this point.

a name in a telephone book by searching sequentially from A to Z would work well for a book with 50 or 100 entries but would be totally unacceptable for use with the New York City directory.

We need a way to formulate generalized guidelines that predict that, for any arbitrary data set, one particular method will probably be better than another. Specifically, we would like to associate a value n (the *size* of problem) with either t (the processing *time* needed to get the solution) or s (the total memory *space* required by the solution). The value, n, is a measure of the size of the problem we are attempting to solve. For example, if we are searching or sorting a list, the size n would most likely be the number of items in the list. If we were inverting an r-x-r matrix, the size of the problem would be r—the dimensions of the matrix.

The relationships between n, t, and s can sometimes be given in terms of explicit formulas.

$$t = f(n)$$
$$s = g(n)$$

However, such formulas are rarely used. First, they are difficult to obtain because they often rely on machine-dependent parameters that we may not know. Second, we do not care to use $f(n)$ and $g(n)$ to compute exact timings of specific data cases. We simply want a general guideline for comparing and selecting algorithms for *arbitrary* data sets.

We can get this type of information by using what is called *O-notation*.

$$t \approx O(f(n)) \ \{ \text{ this is read ``t is on the order of } f(n)\text{'' } \}$$
$$s \approx O(g(n))$$

where the function $f(n)$ is the *time complexity* or *order* of the algorithm and $g(n)$ is the *space complexity*.

Formally, O-notation states that there exist constants M and $n*$ such that if $t \approx O(f(n))$, then $t <= Mf(n)$ for all $n > n*$. This formidable definition is not as difficult as it looks. It simply states that the computing time (or space) requirement of the algorithm grows no faster than a constant times the function $f(n)$. That is, as n increases (i.e., the problem gets bigger), the time (or space) needed to solve that problem increases at a rate that is of the order of the function $f(n)$. If the order of the algorithm were, for example, $O(n^3)$, as the size of the problem doubled, the time to solve the new problem could increase about eightfold ($2^3$). The *order* of an algorithm approximates the amount of resources necessary to solve a problem as the size of the problem gets larger. For reasonably large problems we will want to select algorithms of the *lowest order* possible. We will not know exactly how much time or space is required, but we will know that as the size of the problem increases there will be a point ($n*$)

at which our method will always take less time (or space) than a method that is of a higher order. And, as the problem gets bigger, the gains become even more significant. Figure 11.2 shows this quite vividly.

For problems of size n < n∗, the choice of algorithm is not too critical. However, as n becomes much larger than n∗, the $O(n)$ algorithm is always superior to the $O(n^2)$ and $O(n^3)$ algorithms and becomes better and better as n increases.

This is the general guideline for which we have been looking, and this is why O-notation is the fundamental technique for describing the efficiency properties of algorithms.

Let us look at some specific examples. The simple sequential search procedure of Figure 11.3 looks at every entry in an n-item list to locate a specific key. We can easily see that both the time and space complexity of this method are $O(n)$. In the "worst case" we will never have to look at more than n items before we either find the key or determine that it is not there. If the list were to double in size, the average time required to search it would obviously double, and so would the amount of space needed to store it.

However, if the list is already sorted, we can do much better than $O(n)$ using the *binary search* technique.

The binary search algorithm works by comparing the key against the item located in the middle position of the list. If it matches, we have found our item. If not, we discard the half of the list that cannot contain the desired key and repeat the process. We will eventually find what we are looking for or discard the entire list. The binary search program is shown in Figure 11.4.

With each comparison we halve the size of the list under consideration. Therefore the list will be ½, ¼, ⅛, . . . its original size. In the worst case the process will continue until the list is empty. We can rephrase this by saying that the greatest

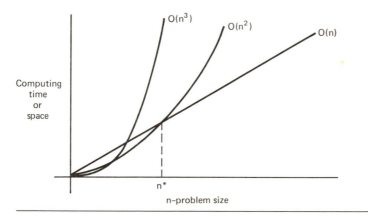

**Figure 11.2.** Comparison of three complexity measures.

```
{ procedure to sequentially search a list for a specified key

entry conditions:
    list                         : n element array of integers
    n                            : size of the list, n > = 1
    key                          : integer value we are searching for

exit conditions:
    position                     : 0 if the key was not found; 1 − n to denote
                                   the location of the first occurrence of the key
}
procedure sequential(list : arraytype; n : integer; key : integer;
    var position : integer);
var
    found                        : integer; { to control loop termination }
    i                            : integer; { array subscript }
begin
    found : = false;
    i : = 1;

    while (not found) and (i < = n) do
    begin
        if list[i] = key then found : = true
        else i : = i + 1
    end { of while }
    if found then position : = i else position : = 0
end; { of procedure sequential }
```

*Figure 11.3.* Sequential search procedure.

number of comparisons required by the binary search method will be the value k, where k is the first interger, such that:

$2^k > = n$ (the size of the list)

Another way to write this is:

$k > = \log_2 n$

and we say that the time complexity, or order, of the binary search method is $0(\log_2 n)$. This is a lower-order algorithm than the O(n) sequential search because its time complexity grows more slowly as n gets larger (see Figure 11.5).

The larger a list (i.e., the greater the value of n), the more time is saved by using the binary search. For example, look at the table on the bottom of page 449.

{ *this procedure searches an n element list for a specified key. the entry and exit
conditions are identical to those listed for the procedure sequential shown in
Figure 11.3*
}

**procedure** *binary(list : arraytype; n : integer; key : integer;*
    **var** *position : integer);*
**var**
       *bottom*   *: integer;*           { *subscript into list* }
       *found*    *: boolean;*         { *to control loop termination* }
       *middle*   *: integer;*           { *subscript into list* }
       *top*       *: integer;*           { *subscript into list* }
**begin**
    *bottom := 1;*
    *top := n;*
    *found := false;*
    **while (not** *found*) **and** *(top <= bottom)* **do**
    **begin**
        *middle := (top + bottom)* **div** *2;*
        **if** *list[middle] = key* **then** *found := true*
        **else**
        **if** *list[middle] > key* **then** *bottom := middle − 1*
        **else** *top := middle + 1*
    **end**; { *of while* }

    **if** *found* **then** *position := middle* **else** *position := 0*
**end**; { *of procedure binary* }

**Figure 11.4.** Binary search procedure.

We see that for small values of n the gains are not significant but, for large lists, the improvements can be monumental. A list with 800,000 items will require (in the worst case) 800,000 comparisons using sequential look-up, while the binary search will never need more than 20—an improvement factor of 40,000!

As a second example, we will analyze the selection sort program in Figure 2.4, which sorts a list of n items. This method looks through all n items to find the largest

|  | *Number of Comparisons* | |
| :---: | :---: | :---: |
| *n* | *O(n)* | *O(log$_2$n)* |
| 8 | 8 | 3 |
| 800 | 800 | 10 |
| 800,000 | 800,000 | 20 |

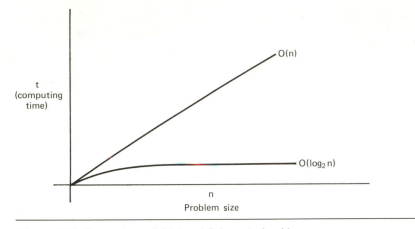

**Figure 11.5.** Comparison of O(n) and O(log₂ n) algorithms.

and places it in location 1. It then looks through the n − 1 remaining items to find the second largest, and so forth. The total number of comparisons needed is:

$$n + (n - 1) + (n - 2) + \ldots + 2 = \frac{n(n + 1)}{2} - 1$$

$$= \frac{1}{2} n^2 + \frac{1}{2} n - 1$$

Since, for large values of n, $n^2 \gg n$, we can disregard the terms $\frac{1}{2} n$ and 1 and say that the complexity, or order, of the selection sort is $O(n^2)$. This is a very typical complexity displayed by many sorting methods.

Another approach to sorting is called the *merge sort*. Instead of sorting the entire list, we break the original list into a number of smaller sublists, sort each one separately, and merge them back together into one sorted master list.

Assume that we break the n-element list into m sublists and sort each one using the $O(n^2)$ selection sort algorithm just discussed. Each of the sublists will have n/m items, so the complexity of the sort phase will be $O(n^2/m^2)$. Since we have m of these smaller piles, the total time to sort them all will be $O(n^2/m^2 * m) = O(n^2/m)$. After sorting, we must merge all m sublists. This will involve looking at the top item in each of the m piles to find the largest (see Figure 4.6). This requires m comparisons. This process must be repeated n times so the approximate time for the merge phase is $O(mn)$. The overall time for the merge sort is thus $O(n^2/m + mn)$. Now if we choose the number of sublists m to be:

$$m = \sqrt{n}$$

the complexity of the merge sort becomes $O(n^2/\sqrt{n} + n\sqrt{n}) = O(n^{3/2})$ and our time complexity is $O(n^{3/2})$—a lower order than the previous $O(n^2)$ approach. The coding of this algorithm will be an exercise at the end of this chapter. However, we can still do better than $O(n^{3/2})$. Theoretically, the fastest possible method for sorting a list will be of the order $O(n \log_2 n)$.[3] There are a number of methods that will run this fast, including the heap sort,[4] and the Quicksort[5] method of C.A.R. Hoare. $O(n \log_2 n)$ is a lower order than $O(n^{3/2})$, and the differences between these methods can become extremely significant.

| $n$ | Selection Sort $O(n^2)$ | Merge Sort $O(n^{3/2})$ | Quicksort $O(n \log_2 n)$ |
|---|---|---|---|
| 10 | 100 | 32 | 40 |
| 1,000 | 1,000,000 | 32,000 | 10,000 |
| 100,000 | $10^{10}$ | 3,200,000 | 1,700,000 |

For a list of 100,000 items the number of comparisons required for sorting the list could differ by as much as four orders of magnitude, depending on the algorithm used. (The Quicksort algorithm is left as an assignment at the end of the chapter.)

So far in our discussion most of the complexities have been of the form $O(n)$, $O(n^{3/2})$, or $O(n^2)$. Algorithms with these types of time complexities are called *polynomial time algorithms* because their time and space complexity functions are polynomial functions of relatively small degrees. The computational demands of these algorithms are usually reasonable, even for large problems, because their time and space needs do not grow "unreasonably" fast. However, not all algorithms are of this type. A second fundamental class contains the *exponential time algorithms*, also called *NP-algorithms* (for *non-polynomial*). For these problems, no polynomial time algorithm has been discovered. The typical complexity displayed by the class of NP algorithms is $O(2^n)$, $O(n^2 2^n)$, or $O(n^n)$. The time and space demands of these algorithms grow extremely fast and consume vast amounts of resources, even for small problems. In most cases it is not feasible to attempt to solve any realistically sized problem using an exponential algorithm, no matter how clever the programmer or how fast the computer. This class of algorithms is not just of academic interest. They occur frequently in several areas of computer science, applied mathematics, and operations research. An important new area of research has developed specifically to study this category of "computationally intractable" problems. (For an excellent discussion of this subject, see Reference 6 in the Part V bibliography.)

Our purpose in this discussion has not been to teach you how to analyze algorithms but to make you aware of this fundamental principle regarding computer efficiency:

[3]E. Horowitz and S. Sahni, "Fundamentals of Data Structures," *Computer Science Press, 1976,* Section 7.4.

[4]Ibid., Section 7.6

[5]Ibid., Section 7.3

*The single most important consideration in developing an efficient computer program is choosing an efficient algorithm.*

Selecting a better technique often results in order-of-magnitude improvements in running time or memory space requirements. This contrasts with the relatively smaller gains to be made by worrying about how to code that algorithm. When efficiency is critical and the problem size is large, finding the lowest-order algorithm should always be our first concern.

As a final (and dramatic) example, we can point to the terminal room simulator program developed in Chapter 6. That program used discrete event simulation, a technique that may appear at first glance to be clever and ingenious. In this case, however, it is a grossly inefficient way to achieve the desired result. The answers shown in Figure 6.14 took about 40 seconds to generate on a large and expensive computer system. There is a totally different technique called *analytic queueing theory* which, in this case, would have produced almost identical results 100 to 1000 times more rapidly! This is another illustration of the critical importance of selecting the best algorithm.

The three examples presented here all analyzed the *time complexity* of an algorithm. However, as stated at the beginning of the chapter, memory space is also an important resource, and we can develop and analyze *space complexity* measures in a virtually identical fashion. For example, both the sequential and the binary search methods require a table large enough to hold all n items and are therefore O(n) in terms of space complexity.

Memory space and computer time are usually inversely related; we can frequently reduce space requirements by increasing processing time or conversely, reduce processing time by making more memory available. This is called the *space-time tradeoff*.

As an example, we will develop one more searching method that is far superior to the sequential and binary search. However, this superiority is achieved through an increase in memory. The method, called *hashing*, is based on the idea that the key that we are looking for can be uniquely transformed, by a hashing function, into a value that immediately locates the key in the table. If we were using social security numbers as our key, we could have a table that was 999,999,999 elements long, and the social security number could be a subscript directly into the table.

**var** *table:* **array** [0..999999999] **of** *peoplerecords;*

To find the individual with social security number 123-45-6789, we would merely reference:

*table* [123456789]

This hashing function is trivial; it is simply the identity function, which leaves the key unchanged. This function correctly maps each social security number into a unique table entry.

The time complexity of such a method is O(1), since we will always go directly to the correct location. However, this reduction in search time has been achieved by requiring enough memory space to hold *all* the possible things we can look up—a potentially enormous number! If our key were a decimal number with a maximum of d digits, our space needs would be $O(10^d)$. And we would need this many memory cells, even if (as with the nine-digit social security number) most of the cells are unused. This is a perfect example of the space-time tradeoff.

In reality, the hashing methods used in most computer programs try to find a reasonable compromise between available space and acceptable running time. We usually allocate less space than the number of unique keys, but enough space so that the table is not too full (say, 50 to 75% full). However, we can now no longer guarantee that each key will take us to a unique location in the table (e.g., 123-45-6789 and 987-65-4321 may now take us to the same array location because there are less than 999,999,999 spaces). This is called a *collision*, and a typical way to handle it is to say that if the original location we went to has already been used by another key, we will search sequentially through the table for the next available space and use that. Figure 11.6 shows the procedures enter and find which search a table using the modified hashing technique just described.

Analysis of both space and time demands have shown that hashing is, under almost all conditions, the best search technique available and is the most common algorithm used in computer programs.[7] The following table shows the average number of comparisons needed to find a value in an n-element table using the three methods just described.

| | Average Number of Comparisons | | |
|---|---|---|---|
| *n* | *Sequential* | *Binary* | *Hashing (50% full)* |
| 5,000 | 2,500 | 12 | 1.5 (table size = 10,000) |
| 50,000 | 25,000 | 15 | 1.5 (table size = 100,000) |
| 500,000 | 250,000 | 19 | 1.5 (table size = 1,000,000) |

As before, we see that the gains to be made through the proper selection of an algorithm are enormous—usually orders of magnitude. And this gain will exist regardless of how we code these algorithms. The simplest, most straightforward hashing program coded by a first-semester freshman will run many times faster than a sequential search method coded by a team of crack programmers, even if it does not contain a single wasted microsecond.

[7]The actual time complexity of our hashing method is $O(1 + (p/2) * (1/(1-p)))$ where $0.0 <= p < 1.0$ and p, the *density*, is a measure of how full the table is. The analysis is complex and not shown here.

{ *this function takes a key > 0, randomizes it using addition, multiplication, and division and produces a value in the range 1..len* }

**function** *hash(key : integer; len : integer) : integer;*
**var**
    *temp                     : integer;*
**begin**
    *temp :=* ((837 * *key*) **div** 19) + 4091;
    *hash :=* (*temp* **mod** *len*) + 1
**end**; { *of function hash* }

{ *procedure to enter a new key in the table using a hashing scheme. assume all the keys are positive so a 0 value means the space is empty. the function hash returns a value in the range 1..len. the variable done is set to true if the value was entered and is set to false if the table was full* }

**procedure** *enter(***var** *table : arraytype; len : integer;*
    *key : integer;* **var** *done : boolean);*
**var**
    *count                    : integer;* { *to index through the table* }
    *location                : integer;* { *subscript into table* }
**begin**
    *done := true;*
    *location := hash(key,len);*
    **if** *table[location]* = 0 **then** *table[location] := key*
    **else**
    **begin**
    { *the location we hashed to was full. now we will look sequentially for an open space and stop when we find it or have looked at the whole table* }
        *count := 0;*
      **repeat**
          *count := count + 1;*
          *location := ((location + 1)* **mod** *len) + 1*
          { *the last assignment causes us to search the table in a circular fashion. after table[len] comes table[1]* }
      **until** *(table[location]* = 0) **or** *(count* = *len);*

      **if** *table[location]* = 0 **then** *table[location] := key*
      **else** *done := false*
    **end** { *of else clause* }
**end**; { *of procedure enter* }

{ *this procedure searches table of size len looking for a specific key, using the*
*method of hashing. it returns a 0 if the key was not found or a value in the*
*range 1..len to indicate its location in the table* }
**procedure** *find(table : arraytype; len : integer;*
      *key : integer;* **var** *position : integer);*
**var**
       *count*                *: integer;*         { *used to loop through the*
                                                                   *table* }
**begin**
      *position := hash(key,len);*
      **if** *table[position]* <>*key* **then**
      **begin**
         *count := 0;*
         **repeat**
             *count := count + 1;*
             *position := ((position + 1)* **mod** *len) + 1*
         **until** *(table[position] = key)* **or** *(count = len);*

         **if** *count = len* **then** *position := 0;*
      **end** { *of if* }
**end**; { *of procedure find* }

---

*Figure 11.6.* Hashing procedures enter and find.

## 11.4 CODING EFFICIENCY

We arrive finally at the topic of coding efficiency and hope that the discussion leading up to this section has helped to put it in its proper perspective. Our concern for writing very fast, "tight" code should come *last*, not first (except for the special cases discussed earlier). We should always try to select the most efficient method available and then develop a straightforward, clear, readable (and correct!) implementation of that method. Then, if we can improve the code to make it run faster without impairing the clarity, *and* if we are willing to spend the time, *and* if we (or someone else) is willing to pay the salaries involved, *and* if it will result in a net savings, we might consider using the techniques we will be discussing in this section. Remember, however, the typical gains to be made will be nothing like the improvements we discussed earlier. If we are lucky, the reductions in space (or time) to be gained by code manipulations will be on the order of 10, 20, or 30% (unless the program was horrendously inefficient to begin with).

    As an example of code optimization, consider the bubble sort algorithm in Figure 6.3 (ignoring the side effect in the first **for** loop). We can improve the implementation of that algorithm by noting that, after pass i, the last i items in the list are in their proper place and do not need to be checked. The algorithm could be recoded as shown.

```
j := n; { top of the already sorted portion of the list }
repeat
    sorted := true;
    j := j - 1;
    for i := 1 to j do
        if (list[i] > list[i + 1]) then
        begin
            sorted := false;
            { exchange the items that are out of place }
            temp := list[i];
            list[i] := list[i + 1];
            list[i + 1] := temp
        end { of if statement }
until sorted
```

The preceding changes can improve the running time of the bubble sort from 0 to 50%, depending on the initial contents of the list.

However, before you begin to modify a program, you should be aware of a fundamental characteristic of computer programs called the *80/20 rule*, which states that 80% of the execution time of a program will be spent in only 20% of the program modules. In all programs there will be some modules that are executed only a few times while others will be executed thousands of times. For example, referring to the simulation program in Chapter 6, the procedure initialize will be executed once per run. The procedure arrivalevent will be executed once for each new student who enters the terminal room. For the data in Figure 6.14 this is about 375 arrivals per simulation run. It makes sense to concentrate our limited efforts on only those frequently used routines where the payoff is greatest. We can get the needed information through a programming technique called *profiling* the code. This is simply a count of how many times for a typical data set each module of a program was entered. After completing the program this information is printed out in an *execution profile* (Figure 11.7).

| Unit | Times Executed |
|------|----------------|
| A | * |
| B | ** |
| C | ************************** |
| D | *** |
| E | **************** |
| Main | * |

*Figure 11.7.* Execution profile of a program.

In this situation we should obviously attempt to optimize only the statements within modules C and E and leave the others as is.

The execution profile of a program is sometimes available directly as a service of the computer system (much like the trace or debug features described in Chapter 10).

**program** *main*

> .
> .
> .

**begin**
> *profile(true); { turn counter on to measure module execution }*

However, even if this service is not automatic, we can still get this important information ourselves by *instrumenting* our program. We define the following global data structures:

**type**
> *modulenames = (name1,name2,...,nameN);*

**var**
> *tally:* **array** *[modulesnames]* **of** *integer;*

where name1, name2, ..., nameN are related to the names of all the subprograms used in our program. (Using the exact same name is illegal.) Now, as the first line of each subprogram used, we simply write:

**procedure** *modulename1*

> .
> .
> .

**begin**
> **if** *measuring* **then**
> > *tally[name1]* := *tally[name1] + 1*

(We assume that at the beginning of the program or in an initialize routine we will clear all entries in tally to 0.) Now we can turn on our profiling instrument by the inclusion of one line at the beginning of our program.

> *measuring* := *true*

We can turn off our instrument when it is not needed by rewriting that initialization so that it reads:

> *measuring* := *false*

However, we should leave these profiling statements in our program in case they are needed again in the future.

*Style Review 11.1* _____

> ### Program Instruments
>
> In Chapters 10 and 11 we have seen three different but related uses of program instruments.
>
>     **if** *debugging* **then** . . .
>     **if** *testing* **then** . . .
>     **if** *measuring* **then** . . .
>
> When included in programs, these instruments can be extremely helpful tools to maintenance programmers during any future modification operations.

Just as we measure the usage of modules within a program, we should also analyze the frequency of execution of statements *within* a module. As before, we will see that there is enormous variation in the number of times different statements are executed. For example, the following code fragment multiplies two matrices: a, which is $p \times m$, and b, which is $m \times n$. We produce a result matrix c, which is $p \times n$, as well as a p-element vector, row, that contains the sum of the elements in each row of the new matrix c.

| line | | times executed |
|------|------|------|
| 1 | **for** *i* := *1* **to** *p* **do** | |
| 2 | **begin** | |
| 3 |    *rowtotal* := *0;* | $p$ |
| 4 |    **for** *j* := *1* **to** *n* **do** | |
| 5 |    **begin** | |
| 6 |      *sum* := *0;* | $pn$ |
| 7 |      **for** *k* := *1* **to** *m* **do** | |
| 8 |        *sum* := *sum* + *(a[i,k]* * *b[k,j]);* | $pnm$ |
| 9 |      *c[i,j]* := *sum;* | $pn$ |
| 10 |      *rowtotal* := *rowtotal* + *c[i,j]* | $pn$ |
| 11 |    **end**; { of loop on j } | |
| 12 |    *row[i]* := *rowtotal* | $p$ |
| 13 |   **end** { of loop on i } | |

For values of $p = m = n = 25$, the assignment statements on lines 3 and 12 will be executed 25 times, the assignment on lines 6, 9, and 10 will be done 625 times, and the assignment on line 8 will be done 15,625 times. Again, it is obvious where we should apply our efforts. Always try to concentrate on the most frequently executed portions of a program unit—usually the innermost loops in a nested set and the condition clause (**then** or **else**) with the greatest chance of being entered.

A partial list of suggestions to help increase coding efficiency and reduce program running time is presented next.

1.  Avoid the unnecessary computation of repeated subexpressions. Sequences such as:

    *a:* = *a* + *sqrt(sqr(sin(x))* − *sqr(cos(x)))*;
    *b:* = *b* + *sqrt(sqr(sin(x))* − *sqr(cos(x)))*;
    *c:* = *c* + *sqrt(sqr(sin(x))* − *sqr(cos(x)))*

    can be rewritten as:

    *diff :* = *sqrt(sqr(sin(x))* − *sqr(cos(x)))*;
    *a :* = *a* + *diff;*
    *b :* = *b* + *diff;*
    *c :* = *c* + *diff*

2.  Avoid unnecessary use of the real data type. Statements such as:

    *counter :* = *0.0; { we will count from 0 to 50 by ones }*

    .

    .

    .

    *counter :* = *counter* + *1.0;*

    .

    .

    .

    **if** *counter* > *50.0* **then** . . .

    will run at about a factor of 3 to 10 times slower than their integer counterparts.
3.  Reduce the *strength* of an operation. This means replace one operation with another that does the same thing but faster. On most machine + and − are executed faster than ∗, which in turn is faster than /, which is faster than a procedure call. For example, on the PDP-11/45 computer manufactured by the Digital Equipment Corporation, the following are the instruction times for various arithmetic operations ($\mu = 10^{-6}$ seconds).

    | | |
    |---|---|
    | *add, sub* | *.85 μsec* |
    | *floating point add* | *2.40 μsec* |

| | |
|---|---|
| *mul* | *3.84 μsec* |
| *div* | *7.44 μsec* |
| *procedure call and return* | *8.50 μsec* |

Thus, in most cases, we would gain by rewriting the following assignments as shown.

*Slower*                                    *Faster*

*a := b * 2*                                *a := b + b*
*a := b / 2.0*                              *a := b * 0.5*
*a := succ(b)*                              *a := b + 1*
*a := sqr(b)*                               *a := b * b*

4.  Reduce the number of repetitive subscript evaluations. Subscript evaluations are very time consuming, and duplications should be eliminated. Instead of:

*sum := a[i,j,k − 1] + a[i,j,k + 1];*
*diff := a[i,j,k − 1] − a[i,j,k + 1]*

we could reduce the number of subscripts evaluated from four to two by writing:

*left  := a[i,j,k − 1];*
*right:= a[i,j,k + 1];*
*sum:= left + right;*
*diff := left − right*

5.  Remove constant operations from inside loops. The computation of $\sum_{i=1}^{N} ax_i$ could be written as:

*sum := 0;*
**for** *i := 1* **to** *n* **do**
    *sum := sum + a * x[i]*

but we have unnecessarily included an extra multiplication within the loop. It would have been better to write the loop as:

*sum := 0;*
**for** *i := 1* **to** *n* **do**
    *sum := sum + x[i];*
*sum := a * sum*

This has been a partial listing of coding techniques to improve the execution-time efficiency of programs. There are more, but many will depend on an analysis of timing characteristics of a particular computer to determine whether they will actually result in a net savings. However, be aware that the gains to be made are usually not enormous. If the program fragments in the fifth suggestion were run on a PDP-11/45, by removing one fixed-point multiply from inside the loop, we have saved 3.84 microseconds per cycle. If n = 10,000 (a reasonably large value), we would have 9999 fewer multiplications and would save:

$$9999 \times (3.84 \times 10^{-6}) \text{ seconds} = 0.038 \text{ seconds}$$

Not very impressive!

There is one final comment to be made about code optimization of the type just described; it is frequently done for you by the compiler itself. An *optimizing compiler* is a language translator that attempts to produce fast, efficient machine language code. It usually accomplishes this at the expense of increased compile time and numerous passes over the source program. This type of compiler will attempt to do what we have just discussed—reduce the strength of operations, remove constants from loops, and the like. Most students do not have access to such a compiler because their needs are quite different—fast compilations and helpful error messages. However, optimizing compilers are used frequently in production programming environments, especially for the final translation of routines that will be used often. Thus, when available, they further strengthen our statement that programmers should not spend much of their valuable time ''diddling code'' to achieve minor gains in space or time. Work on selecting a good method and writing a readable and reliable program. If coding efficiency is still of concern, translate your program into machine language using the best optimizing compiler available.

*Style Review 11.2* _____

### Diddling Code

We cannot stress too emphatically the point made at the end of this chapter: The futility of ''diddling around'' with a working program trying to shorten the code by a few statements or trying to get it to execute a few seconds quicker. Unless the program was highly inefficient to start with or the change is obvious, these efforts are rarely cost effective. They can actually *increase* costs over the life of the program because of the increased programmer time needed to implement future changes.

This does not imply that working programs should never be reviewed and, possibly, improved. It simply means that any changes to increase performance should address *gross* inefficiencies that can produce a real cost savings, and the change should not reduce the program's clarity, readability, and good structure.

## EXERCISES FOR CHAPTER 11

1. The most common computing times for algorithms are:

   $O(1)$
   $O(\log_2 n)$
   $O(n)$
   $O(n \log_2 n)$
   $O(n^2)$
   $O(n^3)$
   $O(2^n)$

   Graph these time complexities in the range $1 <= n <= 128$ and compare their rates of growth. (Assume the coefficients in all cases are 1 and that all lower-order terms are disregarded.) What is the difference between $O(n^2)$ and $O(2^n)$ for $n = 4, 8, 64$? What is the point n∗ where the above time complexities are ordered in this sequence? That is, for any $n > n∗$, $O(1) < O(\log_2 n) < O(n) < O(n \log_2 n) < O(n^2) < O(n^3) < O(2^n)$.

2. In reality, time complexities have coefficients other than 1 and have lower-order terms that are not considered. For example, the actual time or space complexity of an $O(n^3)$ algorithm might be:

   $(1/2)n^3 + (1/4)n^2 + 1$

   However, this does not change the fact that if f(n) is of a lower order than g(n), there will always be a point n∗ such that, for all $n > n∗$, $O(f(n)) < O(g(n))$. Find the point n∗ where the first complexity function is always less than the second.

   (a) $(1/2)n^2 + (1/2)n - 1$    $(1/8)n^3$
   (b) $\log_2 n$    $10{,}000n$
   (c) $3n^3 + 50$    $(1/50)2^{n/2}$
   (d) $5 \times 10^6$    $(1/10{,}000)n$

3. The Quicksort algorithm of C.A.R. Hoare operates as follows. We must locate a *dividing point* in the list such that all values on one side of the dividing point are less than the element at the dividing point and all values on the other side are greater than it. This gives us two lists, which can then be sorted independently to produce a single sorted list (since all elements in the one list are larger than the elements in the other).

   To find the dividing point, we use two pointers, left (1) and right (r) initially pointing to the leftmost and rightmost elements. We move r in the leftward direction, comparing it to the element pointed to by 1. Whenever we find a right value smaller than the current left value, we *interchange* the values pointed to by l and r. We now start moving l to the right, looking for a left value larger than the current right value. If we find it, we interchange again. We repeat this process, always interchanging and moving in the opposite direction. The element at which l and r meet is the dividing point.

We now have two separate, but possibly unsorted, lists and can repeat the same operations independently on each sublist. We are finished whenever each sublist is sorted; that is, when it has 0 or 1 elements.

Write a Pascal procedure *quicksort* to implement the Quicksort algorithm just described. Run it on lists of 50, 100, and 200 elements and compare the sorting time of this method to the time needed by the selection sort of Figure 2.4.

4. Write a hashing procedure that uses the technique called *chaining* to handle the problem of collisions. In chaining each entry of the hash table is considered as the head of a chain of values that hashed to that location. For example, if a, b, and c hashed to location 2, d hashed to location 3, and e and f hashed to location 4, a 5-element hash table would look like this.

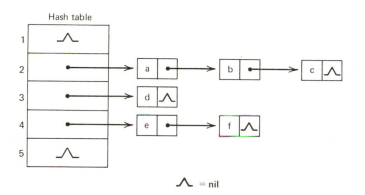

Write the routines enter and find for creating and using a hash table in this form.

5. If our hash table size from Exercise 4 is n elements and the table currently holds k values, what is the average number of comparisons needed to find an item using the chaining method? How does this compare with the average time needed by the procedures shown in Figure 11.6? (The complexity is given in footnote 7.) How much space is needed to store k elements in both methods? (Assume each value and each pointer occupy one memory cell.) What does this analysis say about the space/time trade-off?

6. The functions pack and unpack are standard Pascal functions that can be useful in reducing the total space needed to store large arrays. If the following declarations have been made:

    A : **array** [a..b] **of** T;
    B : **packed array** [x..y] **of** T;

    and if $(b-a) >= (y-x)$, the call:

    pack(A,a,B)

    means pack elements A[a] through A[y − x + a] into B[x] through B[y]. The statement:

    unpack(B,A,a)

    means unpack elements B[x] through B[y] into A[a] through A[y − x + a].

    Write the procedures pack and unpack as described. If a single element of a packed array can hold k elements of type T while an element of an unpacked array holds 1, what are the space gains to be made by performing:

    pack(A,a,B)

    if A and B are as first defined?

7. Write the merge sort algorithm discussed and analyzed in Section 11.3.

8. Profile the lexical scanner program given at the end of Chapter 7. For a typical data set, how many times is each program unit executed? For which routines should we make the greatest effort to optimize the program efficiency?

9. (a) What is the *time* complexity of the procedure concatenate shown in Section 8.2.4?
    (b). What is the *space* complexity of the algorithms for representing a sparse matrix as a doubly linked list as discussed in Section 8.5?

10. (a) A polynomial:

    $$P = a_nx^n + a_{n-1}x^{n-1} + \ldots + a_1x + a_0$$

    can be evaluated in many different ways. The straightforward way is:

    $$P = a_n * \underbrace{x * x * \ldots * x}_{n\text{ times}} + a_{n-1} * \underbrace{x * \ldots * x}_{n-1\text{ times}} + \ldots + a_0$$

    Write a procedure to evaluate a polynomial with coefficients a[0]...a[n] at the point x. Determine how many multiplications, additions, and assignments are required as a function of the degree, n, of the polynomial.

(b) An alternative way to evaluate P is to *factor* the polynomial in the following manner (called *Horner's rule*).

$$P = ( ...((a_nx + a_{n-1})x + a_{n-2})x + ... + a_1x) + a_0$$

Write a procedure to evaluate a polynomial with coefficients a[0]..a[n] at point x using Horner's rule. How many multiplications, additions, and assignments are required as a function of the degree, n?

(c) Using the PDP-11 timing statistics in Section 11.4 (assume an assignment is $0.85\mu sec$ and that polynomials use integer add and multiply), approximately how much time is saved in evaluating a polynomial of degree n = 5 by Horner's rule? If n = 25?

11. In Exercise 8 in Chapter 4, you were asked to determine if, for a given set of nodes and links, there is a *path* between any two arbitrary nodes, i and j. You were given a matrix M[i,j] that describes the physical connections between nodes:

M[i,j] = 1 if there is a link from node i to node j
M[i,j] = 0 if there is no link from node i to node j

For example, if our connections are:

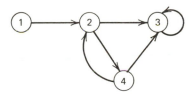

M would look like:

| 0 | 1 | 0 | 0 |
|---|---|---|---|
| 0 | 0 | 1 | 1 |
| 0 | 0 | 1 | 0 |
| 0 | 1 | 1 | 0 |

There is a path from node 1 to node 3 ($1 \rightarrow 2 \rightarrow 3$), but no path from 4 to 1 or from 3 to 2. If you have not done so already, develop a Pascal procedure that, given M and some specific i and j, determines if there is a path from node i to node j. Determine the time complexity of your algorithm as a function of the number of nodes.

12. Profile the simulator program at the end of Chapter 6. Identify how many times each module is executed. Identify the modules that should be optimized first. Which modules should we not bother with at all?

13. Try to think of techniques for improving coding efficiency in addition to the five suggested in Section 11.4. Using timing data for the computer available to you, determine exactly what savings there will be in computing time (or memory space) for typical cases.

14. If we wanted to store the complete contents of a CRT screen with 80 columns and 24 rows, a straightforward approach would be to use the following structure.

    **var** *screen* : **array** [*1..24,1..80*] **of** *char;*

    This will require saving 1920 characters. Design a data structure that saves the contents of a screen using less memory space. (*Hint*. What if most of the spaces are blanks? Review the sparse matrix techniques discussed in Section 8.5.) Using your data structure, write a Pascal procedure to display, on an 80 × 24 screen, the information that was saved. The characters should be placed in their proper position on the screen.

15. A matrix of the type M from Exercise 11 is also called a *directed graph*. An interesting problem with directed graphs is to find out if one can *tour* the graph. (Formally, this is called finding a Hamiltonian cycle.) Touring means visiting every node exactly once and ending back at the original starting point. For example:

Tour: 1 → 3 → 4 → 2 → 1                    No tour possible

Write a procedure that takes an n-x-n directed graph and a starting point i and determines if there is a tour of the graph beginning at node i. The best-known algorithm (Reference 6, Part V, pp. 348-350) has an exponential time complexity of $O(n^2 2^n)$. This value grows unbelievably fast, so the problem is really "unsolvable" for large graphs. Try out your procedure with n = 4, 8, 16, and 32 and see what happens to computing time.

# Chapter 12

# DOCUMENTATION AND PROGRAM MAINTENANCE

## 12.1 PROGRAM DOCUMENTATION

Just because this is the last chapter of the book, one should not develop the mistaken impression that documentation is performed only after all other programming steps have been completed. Documentation is developed *continually* during program design and implementation. Each step in the development process contributes textual material that will eventually become part of the overall program documentation package. Figure 12.1 lists some programming steps (from Figure 1.1) and the corresponding documentation usually developed during that phase.

Documentation is usually presented as the final phase of program development *only* because at the completion of our project we take all existing documentation of the type listed in Figure 12.1 and produce a more finished and "elegant" product. This includes rewriting, condensing, editing, typesetting, and binding—whatever it takes to produce usable and readable finished copy.

Documentation materials fall into two distinct classes—*user documentation* and *technical documentation*. The next two sections will discuss each of these classes separately.

## 12.2 USER DOCUMENTATION

### 12.2.1 What It Contains

User documentation is intended solely for *end-users*. These people may or may not be familiar with computers, with Pascal, or even with the program they are trying to run. They simply want to provide some data to the program and get back the correct results. The inner workings of the program are immaterial to their needs. Therefore

| Program Development Phase | Typical Documentation |
|---|---|
| 1. Problem definition | 1. Written specification statements<br>User needs statements<br>User manuals |
| 2. Outlining and structuring the solution | 2. Macro-level (system) flow charts<br>Development tree diagrams<br>High-level program and data structure descriptions<br>Cost estimates for software development |
| 3. Selecting solution methods | 3. Time and space complexity analysis<br>Micro-level flow charts<br>Pseudocode |
| 4. Coding | 4. Listings<br>Program comments<br>Lower-level procedure and data structure descriptions |
| 5. Debugging and testing | 5. Sample outputs<br>Error logs<br>Formal verification documents<br>Listings of all changes and modifications<br>Timings and benchmarks |

**Figure 12.1.** When documentation is developed.

user documentation concentrates almost exclusively on the *input/output characteristics* of the program and presents a general overview of what it does. Lower-level implementation details about specific procedures, parameters, or data structures are usually inappropriate.

The major piece of user documentation is a separately bound book, manual, or write-up called the *user's guide* or *user's manual*. It includes, in nontechnical language, all the information a user needs to employ the program properly. The exact format and contents are a matter of personal writing style, but certain information should always be included.

1.  The program *name*.
2.  A brief, nontechnical *description* of what the program does. This should include an explanation of the results that are produced and, if appropriate, the algorithm or method used. Important assumptions made in solving the problem should also be described.
3.  The *input data* required by the program. This should include what values are required, what limits are placed on the values, where and how they should

be entered (the input format), the order in which they must be entered, and from what input device they may be entered. This is especially critical with a very generalized program (see Section 3.4) in which users may have to provide a great deal of data to customize the program to exact personal needs. Also, if default input values are assumed, they should be specified here.

4. The *normal output* of the program when presented with valid data. This should probably include an actual output listing from the program along with explanatory notes for each value describing what it is, how it is to be interpreted, and in what units it is being presented.

5. The *exception reports*. These are the error messages, warnings, or other abnormal output the program produces when it encounters invalid or suspicious data. For each error message the manual should explain what causes that error and how users can repair it. Between points 4 and 5, every message, normal or abnormal, that could be produced by the program should be listed and explained.

6. *Program limitations*. These are bounds or constraints inherent in the program itself that cannot be exceeded by users. Examples might include limits on accuracy (because of the physical characteristics of the computer), or maximum amounts of input (because of fixed **array** declarations).

7. The *command sequence* needed to execute the program on a specific computer system. This usually will be in the form of a listing of a complete input deck ready for submission, or a complete terminal session.

8. The name, address, and telephone number of the *author* of the program and the person currently responsible for program maintenance and assistance.

The manual should provide everything users need to prepare the necessary data, run the program, and interpret the results.

Section 12.2.2 is an example of the user documentation for the simulation case study developed in Section 6.3. The write-up, which assumes that the program is currently stored in a program library called lib, describes how to use the program on the CDC Cyber/74 at the University of Minnesota Computer Center. Naturally, this location-dependent information will be different for user's guides at different installations.

*Style Review 12.1* _____

### Programmers and Documentation

Most programs are poorly documented because most programmers hate to write. Coding, debugging, and testing are considered challenging and, if not outright fun, at least interesting. But writing is viewed as drudgery, to be done as quickly as possible or avoided altogether. This is why many programming teams now include a clerical staff to develop and maintain documentation (Figure 5.5). However, when programmers are not backed up by clerical staff, they must realize that they are responsible for clear, thorough documentation. This is as important as any other aspect of programming and must be allocated adequate time and effort.

## 12.2.2   Example of User Documentation
## (For the Simulation Program in Section 6.3)

1. *Program name*. SIMULATOR
2. *Description*. This program is a discrete event simulation model of a computer terminal room. The model assumes that people enter the terminal room and wait in a queue to use one of "nterms" identical computer terminals. When a terminal becomes available, the next person in line is assigned to it. After completing the terminal session, the person may either leave the room or wait in a consultant's queue for assistance. After being helped, the person goes back to the end of the terminal queue. The overall organization of the terminal room is summarized by the following diagram.

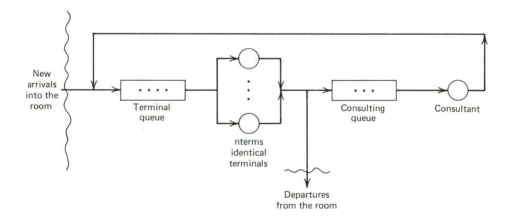

The simulation model runs for a simulated time period specified by the user. Upon completion of the simulation, the program outputs a number of statistics related to the behavior of the system. Users may then study the output and decide whether to rerun the simulation with different parameters or to terminate the program.

3. *Assumptions made by the program.* The simulation program makes certain fixed assumptions about four key model parameters.

   (a) Arrival rates into the terminal room.

   (b) Length of a terminal session.

   (c) Length of a consultation session.

   (d) Percentage of students who utilize the consultant after a terminal session.

   The assumptions are that these four parameters will follow the statistical distributions in Figure 6.11 of *Advanced Programming and Problem Solving with Pascal* (John Wiley & Sons). If these assumptions are unacceptable, the program code must be modified. Contact the person responsible for program maintenance for exact instructions on changing the program (see Section 8).

4. *Input.* The two input parameters are:

   (a) *nterms.* The number of terminals in the terminal room —(integer) nterms $>= 1$.

   (b) *maxtime.* The simulaton time in minutes—(real) maxtime $> 0.0$.
   (*Caution.* Large values of maxtime can produce excessive machine costs. Using values of maxtime $> 5000.0$ should be carefully considered in terms of available funds.)

   After getting the results of a single simulation run, the program will ask the following question. (A ''?'' is a prompt by the computer, requesting input.)

   Do you wish to run the simulaton program again?
   Type yes or no
   ?

   Users simply type ''yes'' or ''no'' followed by a carriage return. If a ''yes'' is entered, users will be allowed to enter new values for nterms and maxtime and rerun the model.

5. *Output.* In addition to echo printing the input data, the program produces seven results at the conclusion of each simulation run.

   (a) The total number of students who entered the room during the simulation. This counts *all* students regardless of whether they exited the terminal room by the end of the simulation run.

   (b) The total number of students who exited the terminal room during the simulation run.

   (c) The average overall time per student (in minutes) spent in the terminal room including waiting time, terminal sessions, and consultation. This average includes only those students who exited the room during the simulation. It does *not* include those still within the room.

   (d) The average waiting time per student (in minutes) in the terminal queue. This is the waiting time *per terminal session.* A student who uses the terminal five times will encounter this delay *each time*.

(e)  The average waiting time per student (in minutes) in the consulting queue. Again, this is the average for *each* consulting session.

(f)  The percentage of time (0 to 100%) the ''nterms'' terminals are busy. This is a percentage averaged over *all* terminals. That is, if there are two terminals, one of which was used 50% of the time and another that was used 25% of the time, the value printed by the program would be 37.5%.

(g)  The percentage of time (0–100) that the consultant was busy.

6.  *Error conditions*. The following error messages are produced by the system.

(a)  ∗∗∗ You made an error in the input. Please try again. ∗∗∗
*Cause*. The value of either nterms or maxtime was entered improperly. They must be entered as shown.

nterms : integer value, nterms > 0
   must be the *first* value entered
   must be followed by at least one blank

maxtime : real value, maxtime > 0.0
   must be the *second* value entered on the same or the following line

(b)  ∗∗∗ Fatal Error—We encountered an empty calendar. Dump followed by a system halt. ∗∗∗
*Cause*. This is a major program malfunction that should never occur. If it does, there is nothing users can do personally to recover. Contact the individual responsible for program maintenance for advice on how to proceed (see Section 8).

7.  *Command sequence*. The program simulator is currently stored in a translated version on the public file called lib. To run the program, users must type the following commands. (The symbol ''c'' denotes any alphanumeric character; ''d'' is any digit 0 to 9. Symbols in *italics* are messages from the computer.)

*U of Minnesota MERITSS System*
*User Number:* ddddddd (enter your account number)
*Password:* ccccccc (enter your 1–7 character password;
                                    however, it will *not* be printed)

*User logged in*
*Recover/System:* get,simulator/un = lib.
*Ready*
run,simulator.

Welcome to the simulator program

    .

    .

    .

If there are any problems executing the program on the UCC computer facilities, users should contact the University Computer Center, User Services HELPline at (612) 555-1212. They are open 8 A.M. to 4 P.M., Monday to Saturday

8. *Program maintenance.* For more information about this program or how to modify the code, contact:

Dr. Steven C. Bruell
Dr. G. Michael Schneider
c/o Department of Computer Science
University of Minnesota
Minneapolis, Minnesota 55455
(612) 555-1212
　　　　Effective Date: November 27, 1980
　　　　Revision No: 1.1

*Style Review 12.2*

### When to Write the User's Manual

Most people would assume that a user's guide of the type shown in Section 12.2.2 would be written *after* the program has been coded and tested. Actually, it is frequently written *before* the coding is even begun. The user's guide is a specification document that describes exactly what the program will provide to users. If we write this before we code, users will know exactly what we plan to do. If it is not what is wanted, we can easily change the specifications. Once the program has been written, these changes are more time consuming and expensive.

For example, if users found that the simulator as coded in Section 6.3 was not exactly what was needed, it would require costly and time-consuming software modification. However, if we wrote the user's guide in Section 12.2.2 *before* we coded that program, we could let users know exactly what we were going to do before we invested time in coding and testing. A change at this point would only affect a specification document, not a working program. Writing the user's manual during program design is common practice among programmers and systems analysts.

## 12.3 TECHNICAL DOCUMENTATION

Technical documentation is the material the *maintenance programmer* needs to change, correct, or understand the program. This material is not intended for end-users but for technical specialists. The *structure* of the program (not simply its input/output characteristics) is of primary importance to these people.

Technical documentation must describe two distinct aspects of a program—*low-level coding details* and *high-level program structure*.

The low-level detail is totally contained in the *program listing*. The listing is actually one of the most important pieces of documentation available to programmers. This is why the rules on programming style in Chapters 2 and 3 are so important. They result in a simple, readable program listing that greatly facilitates maintenance. In addition, the modularity rules presented in Chapters 4 and 5 and Section 6.2 also enhance the clarity of the listing by dividing the program into separate, logically coherent pieces that can be treated as single units. Writing programs with the following points in mind will greatly facilitate extracting middle- and low-level details from the program listing.

1. Good descriptive commentary.

2. Clear mnemonic names.

3. Helpful indentation.

4. Clear, understandable control structures.

However, except for relatively short programs, a listing alone may be insufficient. For large programs with thousands of lines of code, it takes too long to study a listing of dozens, or even hundreds of pages. What we need is separate technical documentation that gives an overview of the *high-level* or *global* structure of the program. By using this documentation, usually referred to as a *program reference manual* (PRM) or a *program logic manual* (PLM), we will know where to look in the listing for additional information. The PRM is both a "technical summary" and a program "table of contents," since it contains both high-level descriptive material and numerous pointers to lower-level detail.

Again, the contents of a PRM is a matter of personal taste and corporate policy. However, the following types of information are almost always included.

1. *Program Name and Purpose*. This will be similar to what we included in the user's guide, points 1 and 2. It is simply a few paragraphs describing what the program does.

2. *The Historical Development of the Program and Its Current Status and Location*. This includes the names and addresses of the original authors and anyone who had responsibility for program maintenance. It includes the completion date of the original work as well as dates and descriptions of listings showing the changes, the testing of these changes, and the date of final acceptance.

3. *The Overall Program Structure*. This will probably be given in terms of a

development tree, similar to Figure 6.8 or the list of modules contained in the text editor of Secton 9.3.4. This shows how the individual modules are related to each other.

4. *Description of Each Module.* Every module contained in the program development tree of part 3 should be individually described in terms of its high-level structure. This should include:
   (a) Module name and type.
   (b) Calling sequence.
   (c) Purpose.
   (d) Entry and exit conditions.
   (e) Where it is called from.
   (f) What other routines it activates.
   (g) What data items are modified by this module.
   (h) Reference to the location of this module in the listing (i.e., line number).
   Figure 12.2 shows how the module putinqueue from the simulator program of Section 6.3 might typically be documented in a PRM.

5. *Description of Key Data Structures.* All important data structures should be individually described. This should include:
   (a) Data structure name.
   (b) Type.
   (c) Purpose.
   (d) Which modules access or modify this structure.
   (e) Reference to where in the listing this data structure is originally declared and initialized.
   Figure 12.3 shows the PRM description of the data structure called calendar, also from the simulator.

6. *Built-in Maintenance Aids.* If the program includes any debugging and testing aids or efficiency instruments for timing or profiling the code (as described in Chapters 10 and 11), the PRM should include a guide to their use.

The *User's Guide,* the *Program Reference Manual,* and the most current *program listing* are the major items in the set of program documents.

## 12.4 PROGRAM MAINTENANCE

*Program maintenance* is the process by which programs are corrected and updated. Unlike student programs, which have a short life (usually one successful run), most large production programs are used for a long time; 1 to 5 years is not unusual and 10 to 15 years is not unheard of. Events that cause changes to the original program include:

1. *Newly Discovered Bugs.* Errors that were not detected during the original testing phase may be discovered during actual program operation.

1. *Name.* putinqueue

2. *Calling Sequence.* **procedure** *putinqueue(p : studentptr;* **var** *q : queue);*

   where p is a pointer to a student record and
       q is the head of a queue of student records.

3. *Purpose.* This procedure will add the student record pointed at by p to the *end* of the queue pointed at by the head node of q. q may be either the terminal queue or the consultant queue

4. *Entry Conditions*

   p : points to student being placed in line
   q : points to the queue of 0 or more students

   *Exit Conditions*

   p : the next field of the record pointed to by p is set to **nil**
   q : has had student p placed at the end of the line

5. *Where called from*

   | | | |
   |---|---|---|
   | arrivalevent | line 12 | (to add a newly arriving student to a terminal queue) |
   | termevent | line 19 | (to add a student to a consulting queue) |
   | consultevent | line 9 | (to add an existing student to a terminal queue) |

6. *Other Procedures Called*

7. Major Data Items Accessed(A) or Modified(M)

   (A) p -- pointer to student record
   (M) p ↑ .next -- next field of student record is set to **nil**
   (M) termqueue, consultqueue

8. *Locaton in Listing.* lines mmmm − nnnn

9. *Error Conditions that May Occur*

---

**Figure 12.2.** Technical description of module putinqueue.

2. *Specification Changes.* The original problem changes because of new laws, user needs, or changing consumer demand. The program must be modified to reflect this change.

3. *Specification Expansion.* The original program must be expanded to provide additional information that had not been anticipated.

1. *Data Structure Name.* calendar

2. *Type.* A pointer to the head of a linked list of event records.

3. *Purpose.* Calendar is used to hold the list of events that must be processed by the simulation program. The events are stored on the calendar sorted by increasing time in the eventtime field. The calendar should never by empty.

4. *Modules that Access or Modify Calendar.*

   getnextevent: removes the first item on the calendar and resets the head pointer accordingly.

   schedule: adds a new event to the calendar. It is added in the proper location so that calendar is sorted by increasing time in the eventtime field.

5. *Declaration*
   Module main, line 40

   *Initialization*
   Module initialize, lines 6-7

**Figure 12.3.** Technical documentation of calendar data structure.

4.   *New Equipment.* The program must be rewritten to take advantage of a newer computer or compiler.

The costs involved with program maintenance are usually grossly underestimated by most programmers. The October 1972 issue of the *EDP Analyzer* reports that, in a typical programming environment, over 50% of the time is dedicated to maintaining existing programs. *Less than 50%* of the time is devoted to designing and coding new programs! A more recent article (''Looking at Software Maintenance,'' *Datamation,* November 1976) states that: ''Most programmers with ten years experience have spent at least 60% of their time in maintenance.'' Other statistics related to program development indicate that over the life of a large, complex program, maintenance costs will exceed the original development and coding costs, sometimes by as much as 5 to 1.

It is only quite recently that programming project managers have begun to appreciate the magnitude of maintenance costs. This had led to a demand for techniques that reduce maintenance costs, speed up changes, and minimize the risk of introducing errors. Most of the programming techniques, styles, methods, and approaches we have presented in this text have been designed for precisely that purpose. In reviewing these techniques, we can now appreciate why they are so critical.

1.   *Clarity and Readability (Chapter 2).* The program listing is the primary piece of technical documentation. The clarity of that listing significantly affects the time it takes to implement a program change.

2.   *Portability and Generality (Chapter 3).* Software tends to last longer than

hardware. Therefore a program written with portability in mind will create much less havoc when the inevitable new computer arrives. Likewise, when user needs change, the truly general programs will require little or no programmer time to adapt to the new specifications.

3. *Structured Code (Chapter 4)*. The most common maintenance operation is tracking down bugs that went unnoticed during the testing phase. A program implemented as a series of single entry-single exit blocks will facilitate debugging. Likewise, a robust program will not "bomb out" without a useful error message. It will terminate gracefully with helpful information and will greatly facilitate locating and correcting any bugs.

4. *Modularity (Chapters 5 and 6)*. When a change can be localized to a single module, it makes maintenance much easier. We do not need to worry about the effects of that change "rippling" throughout other modules in the system.

5. *Debugging and Testing (Chapter 10)*. A formalized debugging and testing scheme reduces the likelihood of undetected bugs in the finished program, thus reducing the maintenance required on that program.

6. *Documentation (Chapter 12)*. The technical documentation of a program is necessary because many other programmers will be working with and maintaining that program over its life span.

*Style Review 12.3* _____

### *Programmers and Maintenance*

Our earlier comments about programmers and documentation also apply to maintenance. Programmers *hate* maintenance. Most programmers consider it demeaning and uncreative, certainly the least desirable programming assignment. After all, who wants to spend time trying to find and fix someone else's "dumb" mistakes?

However, as we have mentioned throughout this text, the importance of many of the programming guidelines becomes apparent only during the maintenance phase. Nothing demonstrates the importance of good programming habits like maintaining a poorly written, undocumented, bug-filled program! It is usually a frustrating but enlightening experience.

This leads us to a suggestion for all programming managers and supervisors. *All* new programmers should be assigned to a maintenance responsibility for their first 3 to 6 months on the job. This experience will serve them well when they have personal responsibility for the design and implementation of *new* software.

The needs of programmers during the maintenance phase is a common thread running throughout our discussion and motivating much of what we have presented. The stylistic guidelines and implementation methods we have stressed make the most sense when you remember these key points.

1. Programs change often.
2. Programs tend to be used for long periods of time, sometimes 5, 10, even 15 years.
3. The most significant program cost can be maintenance.
4. A program will probably be maintained by 5 to 10 different programmers over its lifetime, and most of them will be initially unfamiliar with its contents.

## 12.5 CONCLUSION

Computer programming is a complex operation with a number of important phases, each with its own techniques and guidelines. Computer programmers must be familiar not only with the syntax of a particular programming language, but also with project management techniques, algorithmic analysis, advanced data structures, human factors engineering, testing and verification methods, and technical writing. The difference between coding and programming is like the difference between grammar and creative writing. A knowledge of the former will allow you to write correctly, but a knowledge of both will allow you to write elegantly. If you have mastered the techniques and guidelines in this text, you are on your way to becoming a *programmer* in the fullest sense of the word.

## EXERCISES FOR CHAPTER 12

1. Write a complete user's manual for the "Idiot's Delight" program in Chapter 7. When writing the manual, you should assume that users:

   (a) Do not know Pascal.
   (b) Will not have access to the program listing.
   (c) Are naive about computers but do know how to log on and use a terminal.

   Your user's manual should contain everything users need to use the program at your installation correctly.

2. Write a user's manual for the text editor program discussed at the end of Chapter 9, with the assumptions listed in Exercise 1. This is a much more substantial undertaking, and it would probably be appropriate to develop this documentation in a small group (two to four people).

3.  Write a complete set of technical documentation for the "wcwreverse" program in Section 7.4. First, study the listing (comments, style, structure) and state whether you feel it is acceptable as written. If not, suggest changes that would make the listing a more useful technical document. Next, write the PRM document for "wcwreverse." Follow the general guidelines in Section 12.3, but feel free to add other information that you feel is critical.

4.  Exchange your technical documentation from Exercise 3 with someone else in the class and, using that documentation, make the following modification to "wcw-reverse." Change the program so it will accept an *arbitrary* number of characters, c, between the strings w and w'. That is, the program will recognize strings of the form:

$$wccc. \; . \; . \; . \; . \; .cccw'$$

$$\underbrace{\qquad\qquad\qquad}$$

1 or more

5.  Develop and write the technical documentation for a program that others in the class may not have seen or studied recently (a homework assignment, a program from another Pascal textbook, or a program from one of the earlier chapters). Now exchange programs and documentation and have them perform some modification to that program. Keep track of how helpful the PRM and the listing were in making the modification. Make suggestions to the original author on how and where the documentation could have been more helpful, complete, and detailed.

Perform the following modification operations.

6.  The users of procedure trapezoidrule (Figure 2.4) are worried about integrating over a function that is undefined somewhere within the interval.

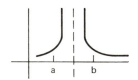

They want you to modify the program so that if the value of $f(x)$ at some point $x_0$ exceeds a preset maximum:

$$|f(x_0)| > \text{max}$$

where $|\;|$ means absolute value, then the program will print out the following message:

*** Warning, integral not defined over this interval ***

7.  The users of the code for the merge operation (Figure 4.6) are worried that if the original lists alist and blist are not properly sorted, the merged result (clist) is incorrect. Modify the fragment so that if alist or blist is not properly sorted, the following error message is printed.

    ** Warning -- list xxx not properly sorted into ascending order. The value out of sequence is xxx. The merge operation is terminated **

8.  The users of the grade analysis program "statistic" (Figure 2.6) are unhappy with one aspect of the output: the histogram is too long. If, for example, the scores were 0 to 200, the histogram would be 201 lines long. (About four pages!) Rewrite the program so users can provide a value for a quantity called "range," which specifies the *width* of the intervals to be used when printing the histogram. As a result, the histogram will specify the number of values that fall in this range. If, for example, our scores were 0 to 200 and r was 45, the histogram would look like this.

    | | |
    |---|---|
    | 180–200 | ** |
    | 135–179 | ******** |
    | 90–134 | ******* |
    | 45–89 | **** |
    | 0–44 | * |

    The range, r, need not be an even multiple of the possible scores. The last interval will simply include all remaining scores (as in the range 180 to 200). Naturally r must be greater than 0 and less than the total possible values. If it is not, produce an appropriate error message.

# Bibliography for Part V

Aho, A., J. Hopcroft, and J. Ullman. *The Design and Analysis of Computer Algorithms*. Reading, Mass.: Addison-Wesley, 1974.

Brown, A. R., and W. A. Sampson. *Program Debugging*. New York: American Elsevier Publishing Co., 1973.

Conway, R., and D. Gries. *An Introduction to Programming*. Englewood Cliffs, N.J.: Winthrop, 1975. (See Chapter V, "Program Testing," Chapter VII, "Performance Evaluation," and Chapter VIII, "Confirmaton of Correctness.")

Goodman, S. E., and S. T. Hedetniemi. *Introduction to the Design and Analysis of Algorithms*. New York: McGraw-Hill, 1977.

Hetzel, W. C., ed., *Program Test Methods*. Englewood Cliffs, N.J.: Prentice-Hall, 1973.

Horowitz, E., and S. Sahni. *Fundamentals of Computer Algorithms*. Potomac, Md.: Computer Science Press, 1978.

Knuth, D. E. *The Art of Computer Programming*. Reading, Mass.: Addison-Wesley, Vol. I, *Fundamental Algorithms* (1968), Vol. II, *Seminumerical Algorithms* (1969), and Vol. III, *Sorting and Searching* (1973).

Poole, P. C. "Debugging and Testing," in *Advanced Course in Software Engineering*. New York: Springer-Verlag, 1973.

Rusten, R., ed., *Debugging Techniques in Large Systems*. Englewood Cliffs, N.J.: Prentice-Hall, 1971.

Van Tassel, D. *Program Style, Design, Efficiency, Debugging, and Testing*. Second Edition, Englewood Cliffs, N.J.: Prentice-Hall, 1978.

Yeh, R. T. *Current Trends in Programming Methodology, Vol. II: Program Testing and Validation*. Englewood Cliffs, N.J.: Prentice-Hall, 1977.

The following journals contain articles related to the material covered in this section:

Software: Practice and Experience

*IEEE Transactions on Software Engineering*

The monthly newsletters of the following special interest groups (SIG) of the ACM: SIGSOFT (software engineering), SIGDOC (documentation), and SIGMETRICS (performance evaluation).

APPENDIX

# SYNTAX OF THE
# Pascal LANGUAGE*

Represents Pascal reserved words or syntacic entities that are not defined further (e.g., a letter or a digit)

Represents a Pascal operator

Represents a syntactic entity that is defined by another flow diagram.

Program

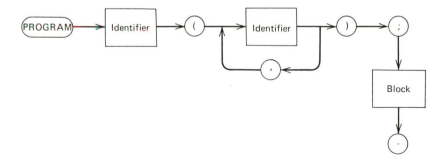

*This appendix is taken from Appendix D, p. 116–118 of Jensen and Wirth, *PASCAL Users Manual and Report,* Springer-Verlag, 1974, with their permission.

**484**

Block

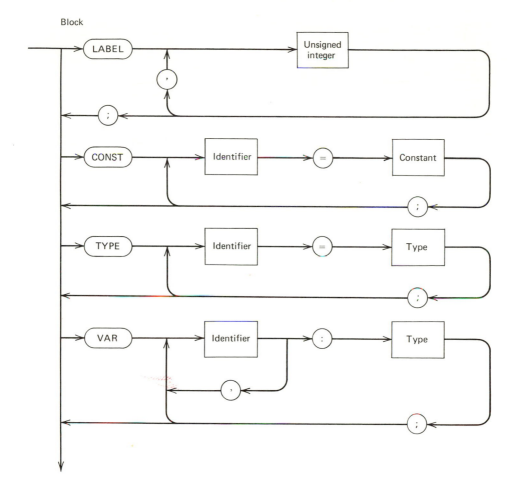

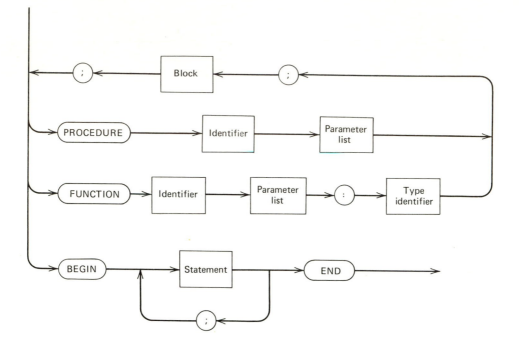

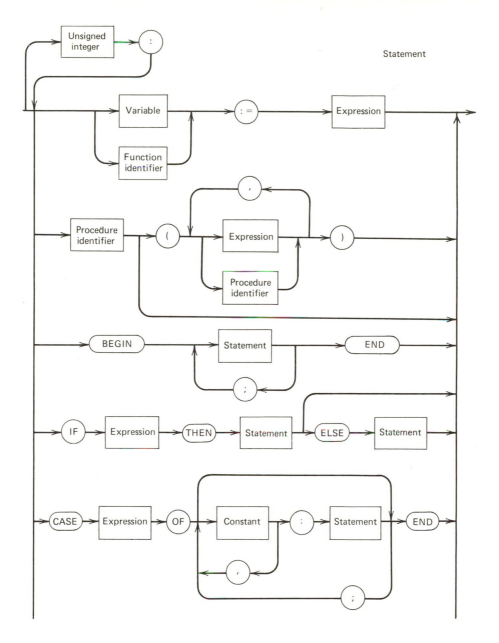

Statement

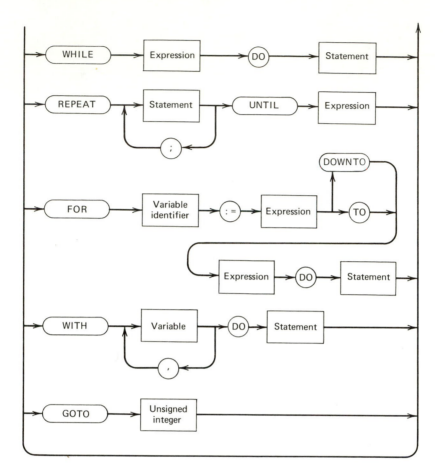

Type

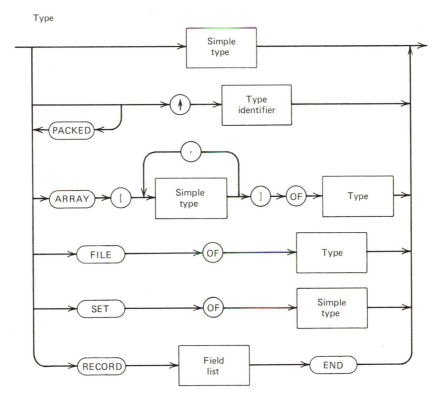

Simple type

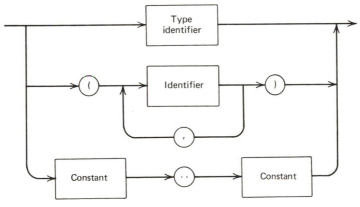

Parameter list

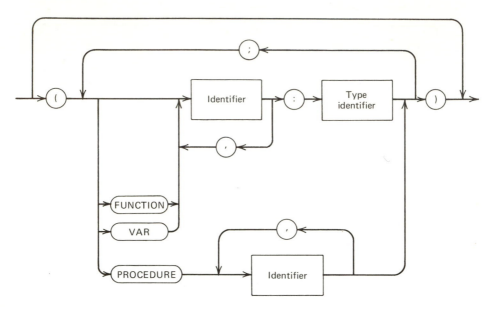

Field list

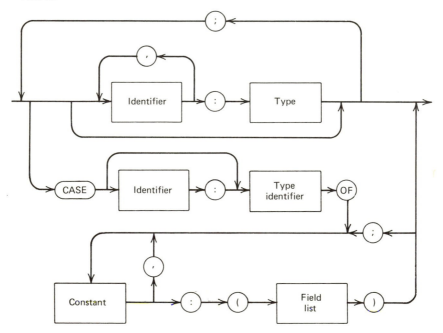

Expression

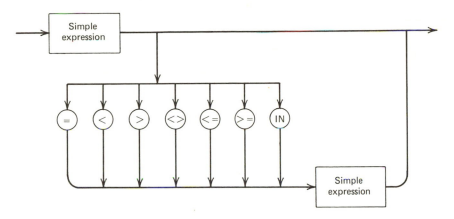

Simple expression

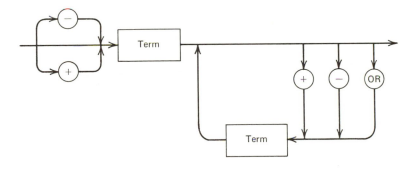

Term

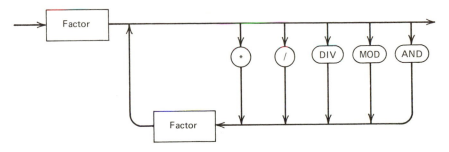

Factor

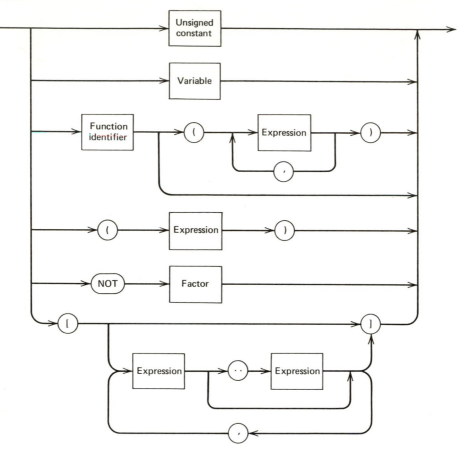

Variable

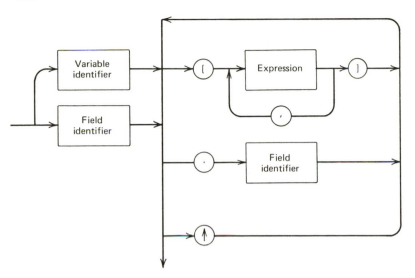

Unsigned constant

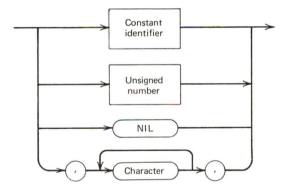

Constant

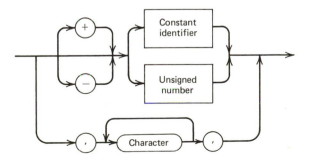

Identifier

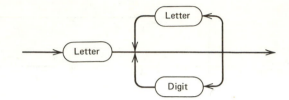

Unsigned integer

Unsigned number

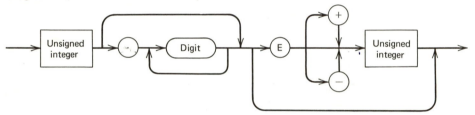

# STANDARDIZED
# Pascal IDENTIFIERS

## B.1 RESERVED WORDS

| | | | |
|---|---|---|---|
| and | end | nil | set |
| array | file | not | then |
| begin | for | of | to |
| case | function | or | type |
| const | goto | packed | until |
| div | if | procedure | var |
| do | in | program | while |
| downto | label | record | with |
| else | mod | repeat | |

## B.2 STANDARD IDENTIFIERS

*Constants*

| | | |
|---|---|---|
| *false* | *true* | *maxint* |

*Types*

| | | | | |
|---|---|---|---|---|
| *integer* | *boolean* | *real* | *char* | *text* |

*Files*

| | |
|---|---|
| *input* | *output* |

*Functions*

| Name | Parameter | Types Result | Description |
|---|---|---|---|
| abs(x) | integer or real | Same as parameter | Absolute value |
| arctan(x) | integer or real | real | Inverse tangent |

| chr(x) | integer | char | Character whose ordinal number is x |
|---|---|---|---|
| cos(x) | integer or real | real | Cosine |
| eof(f) | file | boolean | End-of-file indicator |
| eoln(f) | file | boolean | End-of-line indicator |
| exp(x) | real or integer | real | $e^x$ |
| ln(x) | real or integer | real | Natural logarithm |
| odd(x) | integer | boolean | True if x is odd False otherwise |
| ord(x) | User-defined scalar, char, boolean | integer | Ordinal number of x in the scalar data type of which x is a member |
| pred(x) | Scalar, but not real | Same as parameter | Predecessor of x |
| round(x) | real | integer | x rounded |
| sin(x) | real or integer | real | Sine |
| sqr(x) | real or integer | Same as parameter | Square of x |
| sqrt(x) | real or integer | real | Square root |
| succ(x) | Scalar, but not real | Same as parameter | Successor of x |
| trunc(x) | real | integer | x truncated |

### Procedures

| Name (parameters) | Description |
|---|---|
| dispose(p) | Returns the dynamic variable referenced by pointer p to the available space list |
| get(f) | Advances file f to the next component and places the value of the component in f ↑ |
| new(p) | Allocates a new variable that is accessed through pointer p |
| pack(a,i,z) | Takes the elements beginning at subscript position i of array a and copies them into packed array z beginning at the first subscript position |
| page(f) | Causes the printer to skip to the top of a new page before printing the next line of text file f |
| read(. . .) readln(. . .) | Reads information from text files. |
| reset(f) | Positions file f at its beginning for reading |
| rewrite(f) | Empties file f and allows it to be written into |

| | |
|---|---|
| unpack(z,a,i) | Takes the elements beginning at the first subscript position of packed array z and copies them into array a beginning at subscript position i |
| write(. . .) ⎫<br>writeln(. . .) ⎭ | Writes information to text files. |

## B.3   SUMMARY OF OPERATORS

*Types*

| Operator | Description | Operand(s) | Result |
|---|---|---|---|
| := | Assignment | Any, except file | — |
| + | Addition | integer or real | integer or real |
| | Set union | Any set type | Same as operand |
| − | Subtraction | integer or real | integer or real |
| | Set difference | Any set type | Same as operand |
| * | Multiplication | integer or real | integer or real |
| | Set intersection | Any set type | Same as operand |
| **div** | Integer division | integer | integer |
| / | Real division | integer or real | real |
| **mod** | Modulus | integer | integer |
| **not** | Logical negation | boolean | boolean |
| **or** | Disjunction | boolean | boolean |
| **and** | Conjunction | boolean | boolean |
| <= | Implication | boolean | boolean |
| | Set inclusion | Any set type | boolean |
| | Less than or equal | Any scalar type | boolean |
| = | Equivalence | boolean | boolean |
| | Equality | Scalar, set, or pointer | boolean |
| <> | exclusive **or** | boolean | boolean |
| | Inequality | Scalar, set, or pointer | boolean |
| >= | Set inclusion | Any set type | boolean |
| | Greater than or equal | Any scalar type | boolean |
| < | Less than | Any scalar type | boolean |
| > | Greater than | Any scalar type | boolean |
| **in** | Set membership | Left operand: scalar<br>Right operand: set with base type the type of the left operand | boolean |

# CHARACTER SETS

The charts in this appendix depict the ordering for several commonly used character sets. Numbers are base 10 and only printable characters are shown.

Many other character sets and collating sequences not included here are in current use.

## C.1 CDC SCIENTIFIC, WITH 64 CHARACTERS

| Left<br>Digit | Right<br>Digit | 0 | 1 | 2 | 3 | 4 | 5 | 6 | 7 | 8 | 9 |
|---|---|---|---|---|---|---|---|---|---|---|---|
| 0 | | : | A | B | C | D | E | F | G | H | I |
| 1 | | J | K | L | M | N | O | P | Q | R | S |
| 2 | | T | U | V | W | X | Y | Z | 0 | 1 | 2 |
| 3 | | 3 | 4 | 5 | 6 | 7 | 8 | 9 | + | − | * |
| 4 | | / | ( | ) | $ | = | | , | . | ≡ | [ |
| 5 | | ] | % | ≠ | ⌈ | ∨ | ∧ | ↑ | ↓ | < | > |
| 6 | | ≤ | ≥ | ¬ | ; | | | | | | |

## C.2   ASCII (AMERICAN STANDARD CODE FOR INFORMATION INTERCHANGE)

| Left Digit(s) | Right Digit | 0 | 1 | 2 | 3 | 4 | 5 | 6 | 7 | 8 | 9 |
|---|---|---|---|---|---|---|---|---|---|---|---|
| 3 | | | | | ! | " | # | $ | % | & | ' |
| 4 | | ( | ) | * | + | , | — | . | / | 0 | 1 |
| 5 | | 2 | 3 | 4 | 5 | 6 | 7 | 8 | 9 | : | ; |
| 6 | | < | = | > | ? | @ | A | B | C | D | E |
| 7 | | F | G | H | I | J | K | L | M | N | O |
| 8 | | P | Q | R | S | T | U | V | W | X | Y |
| 9 | | Z | [ | 1/8 | ] | ∧ | — | ` | a | b | c |
| 10 | | d | e | f | g | h | i | j | k | l | m |
| 11 | | n | o | p | q | r | s | t | u | v | w |
| 12 | | x | y | z | { | \| | } | ~ | | | |

Codes 00 to 31 and 127 (decimal) represent special control characters that are not printable.

## C.3   EBCDIC (EXTENDED BINARY CODED DECIMAL INTERCHANGE CODE)

| Left (Digit(s)) | Right Digit | 0 | 1 | 2 | 3 | 4 | 5 | 6 | 7 | 8 | 9 |
|---|---|---|---|---|---|---|---|---|---|---|---|
| 6 | | | | | | | | | | | |
| 7 | | | | | | ¢ | . | < | ( | + | \| |
| 8 | | & | | | | | | | | | |
| 9 | | ! | $ | * | ) | ; | ¬ | — | / | | |
| 10 | | | | | | | | ^ | , | % | — |
| 11 | | > | ? | | | | | | | | |
| 12 | | | | : | # | @ | ' | = | " | | a |
| 13 | | b | c | d | e | f | g | h | i | | |
| 14 | | | | | | | j | k | l | m | n |
| 15 | | o | p | q | r | | | | | | |
| 16 | | | | s | t | u | v | w | x | y | z |
| 17 | | | | | | | | \ | { | } | |
| 18 | | [ | ] | | | | | | | | |
| 19 | | | | | A | B | C | D | E | F | G |
| 20 | | H | I | | | | | | | | J |
| 21 | | K | L | M | N | O | P | Q | R | | |
| 22 | | | | | | | | S | T | U | V |
| 23 | | W | X | Y | Z | | | | | | |
| 24 | | 0 | 1 | 2 | 3 | 4 | 5 | 6 | 7 | 8 | 9 |

Codes 00 to 63 and 250 to 255 represent nonprintable control characters.

# INDEX OF USEFUL PROGRAMS AND PROGRAM FRAGMENTS

# SUBJECT INDEX

**503**